recent advances in phytochemistry

volume 28

Genetic Engineering of Plant Secondary Metabolism

RECENT ADVANCES IN PHYTOCHEMISTRY

Proceedings of the Phytochemical Society of North America
General Editor: Helen A. Stafford, *Reed College, Portland, Oregon*

Recent Volumes in the Series:

A Continuation Order Plan is available for this series. A continuation order will bring delivery of each new
volume immediately upon publication. Volumes are billed only upon actual shipment. For further
information please contact the publisher.

recent advances in phytochemistry

volume 28

Genetic Engineering of Plant Secondary Metabolism

Edited by

Brian E. Ellis
University of British Columbia
Vancouver, British Columbia, Canada

Gary W. Kuroki
Applied Biosystems
Foster City, California

and

Helen A. Stafford
Reed College
Portland, Oregon

PLENUM PRESS • NEW YORK AND LONDON

Library of Congress Cataloging-in-Publication Data

Genetic engineering of plant secondary metabolism / edited by Brian E.
  Ellis, Gary W. Kuroki, and Helen A. Stafford.
       p.   cm. -- (Recent advances in phytochemistry ; v. 28)
    "Proceedings of the Thirty-Third Annual Meeting of the
  Phytochemical Society of North America on Genetic engineering of
  plant secondary metabolism, held June 27-July 1, 1993, in Pacific
  Grove, California"--T.p. verso.
    Includes bibliographical references and index.
    ISBN 0-306-44804-1
    1. Plant genetic engineering--Congresses. 2. Plants--Metabolism-
  -Congresses. 3. Metabolism, Secondary--Congresses.   I. Ellis,
  Brian E.   II. Kuroki, Gary W.   III. Stafford, Helen A., 1922-   .
  IV. Phytochemical Society of North America. Meeting (33rd : 1993 :
  Pacific Grove, Calif.) V. Series.
  QK861.R38 vol. 28
  [QK981.5]
  581.19'2 s--dc20
  [581.1'5]   94-12197
                                                                    CIP

Proceedings of the Thirty-Third Annual Meeting of the
Phytochemical Society of North America on
Genetic Engineering of Plant Secondary Metabolism,
held June 27-July 1, 1993, in Pacific Grove, California

ISBN 0-306-44804-1

©1994 Plenum Press, New York
A Division of Plenum Publishing Corporation
233 Spring Street, New York, N.Y. 10013

PREFACE

In this volume of Recent Advances in Phytochmistry you will find a record of the pioneering attempts of plant biochemists and molecular biologists to modify the patterns of secondary metabolism in plants, as presented at the 33rd annual meeting of the Phytochemical Society of North America, in Asilomar, California, on June 27 - July 1, 1993. The studies described here represent a marriage of the newest of technologies with one of the oldest human activities, exploitation of plant chemistry. They also represent the beginning of a new era of phytochemical research, an era that will undoubtedly begin to provide answers to some of the long-standing questions that have absorbed plant biochemists for the past century.

There is, for instance, a common deflating experience to which every worker in the area of plant secondary metabolism can probably relate. After hearing about the latest research findings regarding some aspect of remarkable compound "X", someone in the audience finally directs the inevitable question at the hapless speaker. "Tell me, is anything known as to the biological role of compound "X" in the plant?" The answer, in most cases, must be "essentially nothing"! This is a frustrating scenario for both the speaker and the audience, since the very fact that a complex biosynthetic pathway remains encoded in a plant genome points to an associated selective advantage. The problem is that establishing the nature and scale of that advantage is a very complex task. It is generally accepted that plants rely heavily on their chemical virtuosity to improve their chances of survival in a hostile environment. That environment, however, is very dynamic. Over the reproductive lifetime of any given plant, its various tissues are likely to face a largely unpredictable array of environmental challenges. The plant's chemistry must therefore be correspondingly flexible and responsive, with diffferent metabolic products playing key roles at different points in time and space. The result is a delicate minuet of challenge and response. For the plant biochemist, this pattern makes establishment of a rigorous correlation between the influence of any given compound and the outcome of a relevant plant-environment interaction a truly daunting task.

One relatively straightforward test of the potential positive contribution of a given secondary metabolite to plant survival would be comparison of the performance of isogenic lines that differ only in their ability to synthesize the compound in question. Unfortunately, few stable genetically defined systems of this sort have been identified in the plant kingdom. With the advent of efficient genetic engineering technology for plant systems, however, it will no longer be necessary to seek out those rare, naturally occurring, null genotypes. It is now possible in principle (albeit, not yet simple in practice) to select a suitable point in any biosynthetic pathway of interest, to isolate a cDNA encoding the enzyme in question and to introduce into the plant a sense or anti-sense version of that gene that is capable of suppressing the expression of a related trait. In this fashion, truly isogenic genotypes can be created that allow much more penetrating tests to be made of the biological roles of plant secondary metabolism. Examples of the research thrust in the areas of polyketides, flavonoids, glucosinolates, terpenoids, lignin and alkaloids will be found in the present volume.

Another aspect of plant secondary metabolites that has fascinated plant biochemists is their use for human purposes, whether as food, pigments, pharmaceuticals, poisons or entertainment. The quantity of the active principle is of prime importance for the value of such plant material, and increasing these levels is an on-going goal. Again, genetic engineering offers the potential to extend the normal range of variability for such traits, or to generate novel accumulation patterns of the desired metabolites. Studies in this volume describing efforts to modify flower pigmentation, monoterpene composition and lignin accumulation represent the first generation of what is certain to become a major research area during the coming decades.

Finally, it is worth comtemplating the opportunities for creating totally novel variants on existing plant chemical patterns. This potential is emphasized by the dramatic progress that has recently been made in understanding, and then manipulating, the details of polyketide antibiotic biogenesis in the streptomycetes. The power that such a combination of protein engineering and plant molecular biology will eventually unleash in the arena of plant secondary metabolism is certain to revolutionize phytochemistry as we enter the twenty-first century. It is to these revolutionaries that this volume is dedicated, particularly those eager graduate students and postdoctoral fellows who have always enlivened the annual symposia of the Phytochemical Society of North America. To use a metaphor appropriate to the venue of the 1993 symposium, they are riding the crest of a wave that will give them the trip of their lives.

February 1994   Brian E. Ellis
                                                          Gary W. Kuroki
                                                          Helen A. Stafford

CONTENTS

Chapter One

PROGRESS IN THE GENETIC ENGINEERING OF THE PYRIDINE AND TROPANE ALKALOID BIOSYNTHETIC PATHWAYS OF SOLANACEOUS PLANTS

Richard J. Robins,*[1] Nicholas J. Walton,[1] Adrian J. Parr,[1]
E. Lindsay H. Aird,[1] Michael J. C. Rhodes[1], and John D. Hamill[2]

[1]Department of Genetics and Microbiology, Institute of Food
Research, Norwich Research Park, Colney, Norwich NR4 7UA, UK

[2]Department of Genetics and Developmental Biology, Monash
University, Clayton, Melbourne, Victoria 3168, Australia

* Author for correspondence. FAX: ++-44-603-507723

Genetic Engineering of Plant Secondary Metabolism,
Edited by B.E. Ellis *et al.*, Plenum Press, New York, 1994

INTRODUCTION

It has long been an objective to increase or decrease the yield of plant-derived secondary products. Over the last few hundred years this has been achieved in a few examples by selective breeding within the natural germplasm pool. In the 1960s, however, the potential of using plant cell *in vitro* cultures to enhance variation was developed, and since then substantial effort has been put into the empirical manipulation of such cultures to yield high levels of metabolites.[1] This work largely entailed varying the culture conditions and examining the effects on the products of interest.[2] Less common, but much more effective, were selection procedures that targeted specifically the desired product.[3] Nevertheless, success has been very limited.

The development of methods by which genes might be isolated and re-introduced into plants or plant cultures, however, was rapidly seized upon as a potentially powerful alternative approach.[4] Instead of inducing coarse changes by environmental influences, it would now be possible to target very specifically the desired pathway. Major alterations to product accumulation might be possible by the insertion of one or two genes coding for 'key' enzymes. It was considered that all that was needed for this approach to succeed was (a) the pertinent genes of the pathway, (b) a suitable transformation technique enabling their re-introduction in a more actively expressed form and (c) a tissue culture or plant regeneration system in which to exploit the engineered organism. The naivety of such an argument has subsequently been demonstrated and, although it may not prove to be fundamentally flawed, further complicating factors - such as co-suppression[5] and the need to consider the sub-cellular localization of the activity[6] - have now been identified.

Consider first the availability of genes. Even now, relatively few genes are available from secondary product pathways in plants. Indeed, in many instances the enzymology involved is poorly described or completely unknown. A severe problem with this area is that plant secondary products constitute a very wide range of chemical types.[7] Hence, the pathways are non-generic in nature, meaning that the associated enzymes must often be obtained *de novo* for each target product. Isolating such enzymes, often present at only low levels and subject to seasonal variation, can be a daunting task. A valuable outcome of the years of research with *in vitro* cultures is that some of the fundamental biochemistry has been elucidated for a number of pathways.[8] Nevertheless, a detailed level of biochemical understanding sufficient to form the basis of a genetic manipulation strategy has only been obtained in the phenylpropanoid pathway, leading to lignin and flavonoids.[8] With the best-studied alkaloid-forming pathways - those for indole alkaloids or for isoquinoline alkaloids - many of the enzymes have been assayed, but little is known of their regulation.[8] In other alkaloid-producing pathways, such as the tropanes, only very limited understanding was available until recently.

Transformation techniques, in contrast, advanced rapidly. The principal advantage here was that generic technologies were being established.[9] Furthermore, they were applicable to the alteration of major agronomic traits, such as plant development or the introduction of herbicide resistance. The most commonly applied method is that developed from the natural infection mechanism of a soil pathogen, *Agrobacterium*. This organism has evolved a plasmid-encoded system by which genes are passed from the bacterial cell into the plant cell and therein integrated into the plant genome. By the development of modified plasmids, vectors were created into which a gene, with a suitable promoter, could be inserted and transferred to the plant.[10]

Some background was also in place to assist in the exploitation of the engineered plant material. The potential value of using transformed cultures for the production of secondary products was already recognised by the mid-1980s.[4] One particular group of alkaloids, the tropanes, had proved particularly resistant to attempts to promote biosynthesis in dispersed cell culture.[11] It was argued that, since these compounds were root-derived, they might more readily be produced *in vitro* by using organised roots in culture. Both untransformed[12] and transformed root cultures[13], the latter generated using *A. rhizogenes,* were found to accumulate high yields of tropane alkaloids. Similarly, the pyridine alkaloids in *Nicotiana* accumulated to high levels[14], although these were additionally amenable to production in aggregated liquid cultures. The advantages of using transformed roots are their independence of plant-growth regulators, their often rapid growth rates and, of course, the fact that the gene-delivery method is intrinsic to the establishment of the culture[15].

Since 1987, when Hamill and colleagues[4] outlined the value of transformed roots for secondary product formation, a wide range of products have been shown to be made by transformed roots. Three years later, the same authors were able to demonstrate that nicotine accumulation in *N. rustica* transformed roots could be increased by genetic engineering[16]. The current paper will summarize this work and discuss it in the light of more recent work aimed at engineering the production of related pyridine alkaloids in *Nicotiana*. It will then review the progress made towards engineering the more complex metabolic pathway leading to the tropane alkaloids. Finally, some consideration will be given to the requirements for future strategies for the manipulation of pathways by genetic engineering.

THE BIOSYNTHETIC ROUTE TO THE PYRIDINE AND TROPANE ALKALOIDS

Nicotiana rustica and *N. tabacum* root cultures principally contain nicotine, while root cultures of *Datura, Hyoscyamus* and related genera contain predominantly hyoscyamine and scopolamine (Fig. 1). In the genus *Duboisia* both alkaloid types occur in the same species. This is understandable since they are closely related and share a common biosynthetic origin, putrescine, for one ring system (Fig. 2). Putrescine is a primary metabolite, which may be derived from either arginine or ornithine in plants. It is involved in the formation of the

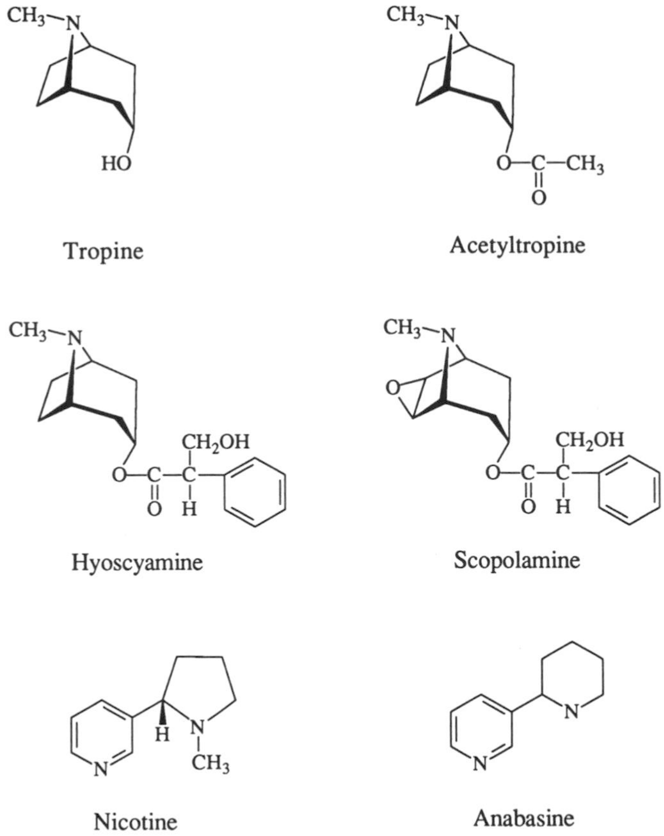

Tropine

Acetyltropine

Hyoscyamine

Scopolamine

Nicotine

Anabasine

Fig. 1. The structures of the major tropane and pyridine alkaloids accumulated by root cultures of *Datura, Hyoscyamus, Atropa* or *Nicotiana*.

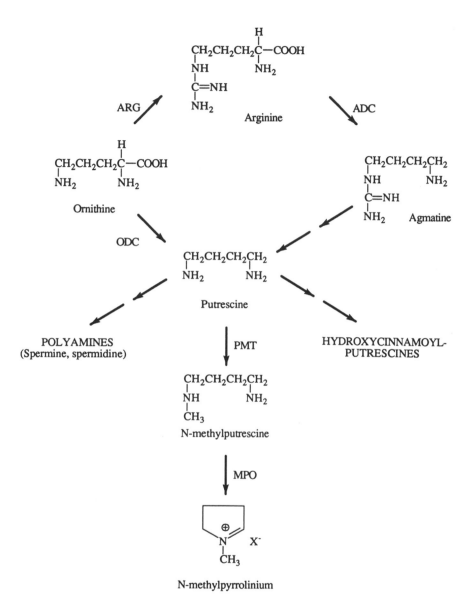

Fig. 2. The biosynthetic route from ornithine and/or arginine to N-methylpyrrolinium, which is common to tropane and pyridine alkaloid formation. ARG = arginase; ADC = arginine decarboxylase; ODC = ornithine decarboxylase; PMT = putrescine N-methyltransferase, MPO = N-methylputrescine oxidase

cell-function-regulating polyamines, spermine and spermidine,[17] and in the putative long-distance signalling compounds, the hydroxycinnamoylputrescines, in the Solanaceae.[18] Thus, the first step of the alkaloidal pathways is the utilization of putrescine for N-methylputrescine formation. N-methylpyrrolinium is then formed by the oxidative deamination of N-methylputrescine. After this stage the pathways diverge, this common intermediate providing both the pyrroline moiety of nicotine and part of the tropane ring of hyoscyamine.

In nicotine formation (Fig. 3), N-methylpyrrolinium is condensed with a putative intermediate derived by the decarboxylation of nicotinic acid. In tropine formation (Fig. 4), a C-3 unit is introduced and a second ring formed to give the bicyclic tropane structure found in tropinone. The subsequent stereospecific reduction of this ketone followed by esterification of the 3α-alcohol, tropine, with a moiety of tropic acid completes hyoscyamine biosynthesis. Tropic acid is derived from phenylalanine. Further metabolism, involving the introduction of a 6β-hydroxyl group followed by oxidation to the 6β,7β-epoxide, results in scopolamine formation.

As can be seen from Figure 4, the pathway by which scopolamine is made is more complex than that for nicotine. Not only is a C-3 unit, derived from acetate, required for tropine formation, but a second pathway converting phenylalanine to tropic acid is involved. However, the overlap of the synthetic routes to these alkaloids in their initial stages makes it convenient to test genetic engineering approaches in *Nicotiana* prior to their application in *Datura*.

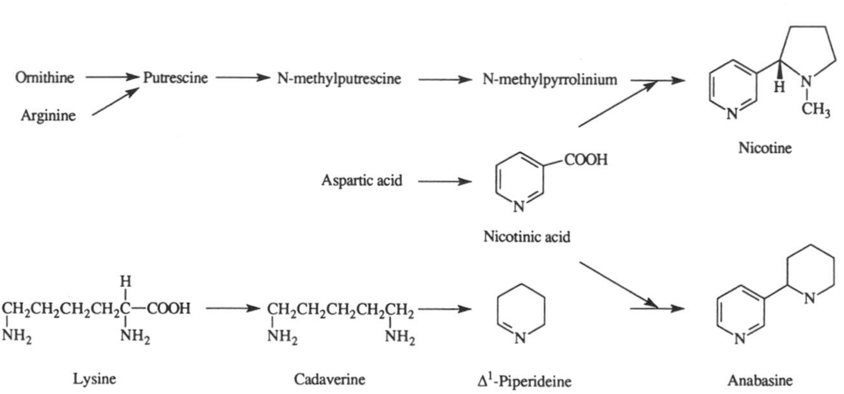

Fig. 3. The biosynthetic pathways by which nicotine or anabasine are made from ornithine/arginine or lysine respectively. Note that nicotinic acid utilization is common to both pathways.

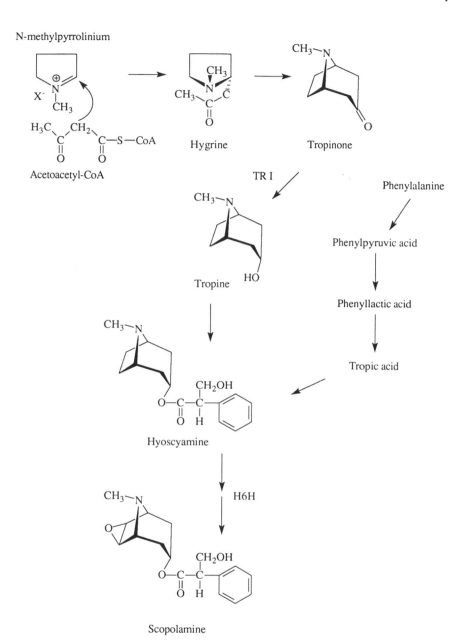

Fig. 4. The biosynthetic pathway by which hyoscyamine and scopolamine are made from N-methylpyrrolinium. TRI = tropinone reductase I; H6H = hyoscyamine 6β-hydroxylase.

THE GENETIC ENGINEERING OF THE PYRIDINE ALKALOID PATHWAY IN *NICOTIANA*

The logical planning of a genetic engineering program would seem to require the identification of those genes most responsible for, or associated with, the limitation of metabolic flux. Some indications can be obtained by (1) relating the levels of enzymes expressed in the tissue with the sizes of fluxes through the pathway, and (2) evaluating the effectiveness of exogenous precursors in altering the sizes of pools of intermediates and end products.[19]

The pathway for nicotine production was selected as having the advantage of being fairly short, involving possibly only three specific enzymes. Of these, two were described - although not purified - some time ago; namely, putrescine N-methyltransferase (PMT) and N-methylputrescine oxidase (MPO).[20] Thus, the biosynthetic route was relatively well established, although the regulation of these activities was poorly understood. Recently, the enzyme system responsible for producing nicotine from N-methylpyrrolinium and nicotinic acid has been reported, though not characterised in detail.[21] The choice was also influenced by the ease with which transformed root cultures of *Nicotiana* synthesizing and accumulating nicotine could be generated and cultured.[14] These considerations made this pathway a good model for a genetic engineering program aimed at an alkaloid-producing pathway.

Enhancing Nicotine Accumulation

Feeding the amino acids, ornithine or arginine had almost no influence on the level of nicotine in root cultures of *N. rustica*. In contrast, feeding their amine counterparts, putrescine or agmatine, respectively, led to small but demonstrable increases in alkaloid accumulation.[22] Feeding N-methylputrescine also enhanced nicotine levels, but only slightly.[23] The limited extent to which these exogenous supplements affected nicotine formation could be taken to indicate a major limitation in metabolism subsequent to N-methylpyrrolinium formation. Nevertheless, the effects of putrescine or agmatine indicated some possibility of a limitation in the supply of putrescine. If this were so, then a potential target would be the decarboxylases by which putrescine and agmatine are made.

However, further considerations were pertinent. Firstly, the enhancement obtained by feeding the amines was small and, secondly, there was evidence that arginine was the preferred origin of the putrescine incorporated into nicotine.[24] Thirdly, nicotine biosynthesis involves the incorporation of a molecule of nicotinic acid.[25] It was found that, when nicotinic acid or nicotinamide was fed, nicotine accumulation was again enhanced.[26] Thus, it was not possible clearly to identify a unique step at which alkaloid production was limited.

Furthermore, it was found that PMT, which catalyses the first step committed to the pathway, is closely regulated in relation to alkaloid production. In suspension cultures of N. tabacum, productive lines show high PMT activity while nonproductive lines have only trace levels of this enzyme.[27] When suspension cultures are converted from non-productive to production conditions, PMT shows a marked sub-culture-related induction of activity.[28] If cells are passaged into conditions ineffective in inducing nicotine accumulation, this activity is not induced. Similarly, with root cultures, nicotine production is lost if roots are grown in media containing phytohormones.[29] The visible effect of this is to cause swelling of the roots, followed by fragmentation of the structure and the formation of a suspension culture.[30] This is reversible, integral roots being obtained when cells are passaged into phytohormone-free medium. The transition to a suspension culture is accompanied by a rapid loss of PMT activity from the culture,[29] and this deficiency in both PMT activity and nicotine accumulation is maintained while cells are kept on the supplemented medium. Based on this effect, it was concluded that PMT was important in the overall regulation of the pathway and therefore constitutes another good target for genetic engineering.

Nevertheless, in view of the effect of feeding putrescine, an attempt was made to enhance nicotine formation by engineering the supply of this metabolite. In the absence of a plant-derived clone for either ODC or ADC (see Fig. 2), the odc from Saccharomyces cereviseae was obtained and inserted into an expression cassette with the powerful enhanced cauliflower mosaic virus 35S protein promoter.[16] Using binary plasmid infection,[31] the gene was introduced into N. rustica and a series of clonal root cultures generated. As control, the same infection vector was used but expressing cat, a gene conferring resistance to the antibiotic, chloramphenicol. When a range of clones expressing one or the other of these genes was examined, it was found that the level of ODC was enhanced in a significant number of odc positive constructs relative to control clones (Fig. 5A). Both populations showed a wide range of activities. Most significant, however, was the finding that the level of ODC in odc lines persisted into the late stationary phase of the culture, in contrast to control and other transformed root cultures. Other enzymes were not similarly affected. This indicated that the introduced gene was being expressed in a constitutive manner. mRNA isolated from the odc clones showed the same phenomenon, with a high

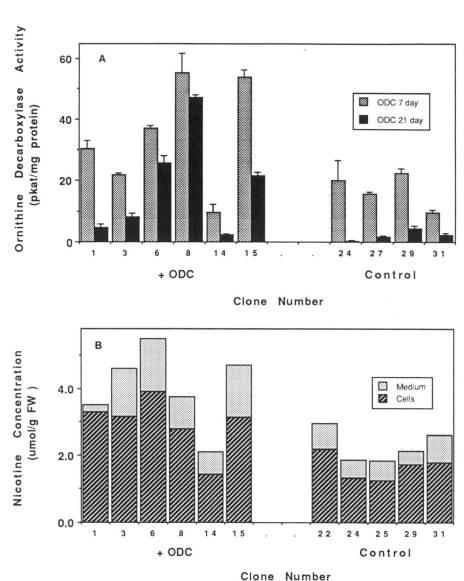

Fig. 5. The levels of (A) ornithine decarboxylase (ODC) and (B) nicotine in transgenic *N. rustica* root cultures expressing the *S. cereviseae odc* gene (+ODC) or the *cat* gene (Control). In (A), ODC activity ± s.e. is shown in clones at 7 and 21 days following sub-culture: in (B) the nicotine concentration is shown for 14-day-old roots (redrawn from ref. 16 with permission).

level present throughout the growth cycle. In contrast, no *odc* mRNA was detected in the controls.

When metabolic intermediates were examined, it was found that, at least in some clones, the enhanced ODC activity had increased the levels of putrescine and, in particular, N-methylputrescine. Furthermore, the average nicotine content of the cultures at 14-days-old rose from 2.28 ±0.22 to 4.04 ±0.48 μmol/g fresh mass (Fig. 5B).

The increments in both ODC activity and nicotine were small but significant. It was anticipated that higher increases in ODC could be obtained. However, the importance of putrescine as a regulator of plant developmental processes[32] indicates that the supply of putrescine is likely to be tightly regulated in plants, as in animals.[33] In addition, once the limitation to the supply of putrescine is at least partially alleviated, other metabolic steps, previously not contributing to flux limitation, may become limiting. One of these might be the enzyme system that condenses N-methylpyrrolinium with decarboxylated nicotinic acid. The activity of the enzyme identified and suggested to carry out this step, named nicotine synthase (NS), was detected only at a very low level,[21] quite inadequate to account for the rates of nicotine accumulation observed in cultures. Thus, the enzymology and mechanism of this important step in nicotine formation requires much further investigation. Another possibility is that the level of PMT becomes limiting. These alternatives can only be tested by further genetic engineering. As the isolation of both PMT[34] and MPO[35] has now been achieved, and a clone for *pmt* is likely to be obtained shortly, it may be possible to estimate the extent to which nicotine accumulation is controlled by the condensation step.

Altering the Alkaloid Profile

Another important objective of genetic engineering is to alter the spectrum of products being made. Of particular interest is the possibility of making plants or cultures able to accumulate as the major product a metabolite that is present in only low amounts, or is absent altogether, in the parent plant. Anabasine is present in many *Nicotiana* species as a minor alkaloid.[25] It is derived from lysine, via cadaverine, in a pathway apparently parallel to that for the biosynthesis of nicotine, except that there is no methylation step involved (Fig. 3). When root cultures of *N. rustica* [22] or *N. hesperis* [36] were fed cadaverine, anabasine accumulation was greatly enhanced relative to nicotine. In *N. rustica,* the nicotine:anabasine ratio changed from about 10:1 to 1:5.[22] Lysine feeding to these cultures had much less influence, indicating that a

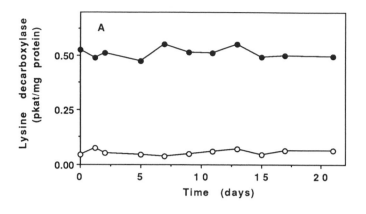

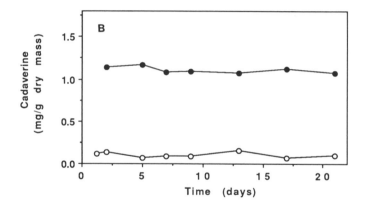

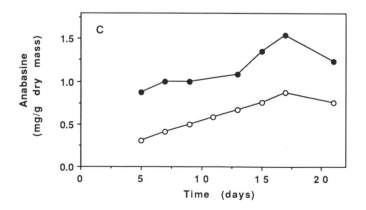

deficiency in lysine decarboxylase (LDC) might be responsible.

In order to test whether enhanced anabasine accumulation could be brought about by genetically engineering the LDC level, Berlin and colleagues isolated *ldc* from the bacterium *Hafnia alvei* [37] and inserted it into transgenic root cultures of *N. glauca* under the control of the cauliflower mosaic virus 35S promoter.[38] *N. glauca* accumulates anabasine as a major alkaloid but, even so, only low endogenous LDC activity can be detected. In two lines out of fifty four examined, the presence of mRNA from the transgene was confirmed by northern blotting. The low frequency at which enhanced expression was seen in transgeneic root cultures confirmed the previous observation with whole plants that obtaining LDC expression in transgenic material can be difficult.[6] The *ldc*-positive clones showed a roughly 6-fold enhancement of LDC activity (Fig. 6). This increased activity was correlated with a 10-fold increase in cadaverine and a 2-fold increase in anabasine accumulation. Simultaneously, nicotine accumulation was diminished; the ratio nicotine:anabasine being 75:25 in controls and about 60:40 in *ldc*-expressing transgenic roots. Thus, it was demonstrated that effects seen as a result of exogenous precursor feeding could be reproduced effectively by genetic engineering, the difference being, of course, that the latter procedure results in a permanently-altered metabolic status.

THE GENETIC ENGINEERING OF THE TROPANE ALKALOID PATHWAY IN *DATURA, HYOSCYAMUS* AND *ATROPA*

As indicated above, the pathway by which hyoscyamine and scopolamine are made (Figs. 2 and 4) presents a greater challenge to the genetic engineer than that for the biosynthesis of nicotine. Nevertheless, a similar approach was taken, initially identifying which enzyme activities might limit flux and proceeding to obtain purified protein for these. A further complication of this pathway is that it is branched and can give rise to a number of minor products (Fig. 7). Root cultures accumulate, to varying extents depending on the genus/species and culture conditions,[39] representatives of four basic types of alkaloid, viz: hygrine-derived[40] (eg. cuscohygrine); pseudotropine-derived (eg.

Fig. 6. The levels of (A) lysine decarboxylase (LDC), (B) cadaverine and (C) anabasine in control (open symbol) and transgenic (closed symbol) *N. glauca* root cultures over a growth cycle (redrawn from ref. 38, with premission).

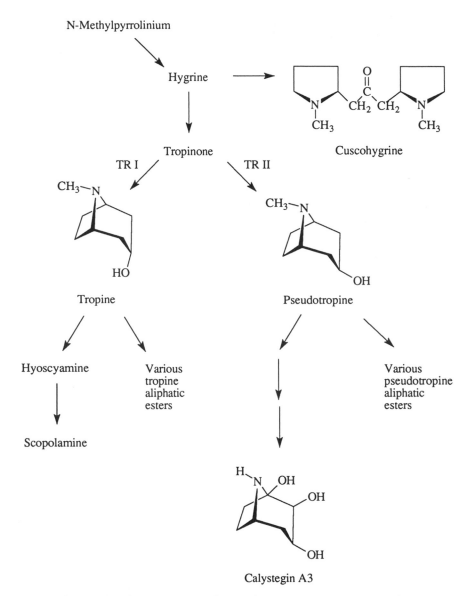

Calystegin A3

Fig. 7. The diverse groups of alkaloid products that originate from N-methylpyrrolinium in species expressing the tropane alkaloid pathway. TRI = tropinone reductase I; TRII = tropinone reductase II

calystegins,[41] aliphatic esters of pseudotropine[42]); aliphatic esters of tropine[43] (eg. acetyltropine, tigloyltropine); and aromatic esters of tropine (eg. hyoscyamine, scopolamine). Some effort has been directed towards understanding the enzymology involved at these branch points and how the relative fluxes might be influenced.[44] This knowledge will be of value in programs aimed at either increasing or decreasing the accumulation of these alternative products.

Flux Analysis

Root cultures of *Datura* and *Hyoscyamus* usually accumulate hyoscyamine as the primary product.[39] The kinetics of this accumulation are approximately growth-related, although the concentration increases by about 50% in older tissue.[45] The enzymes so far identified as involved in hyoscyamine and scopolamine formation show transient expression kinetics in both *Datura*[45] and *Hyoscyamus*[46] cultures. Thus, there is apparently some level of overall control of the amounts of activity present, indicating that altering pertinent activities could influence accumulation. In *D. stramonium* roots, the levels of activity of ODC, ADC and PMT determined *in vitro* do not exceed by a large extent the amount required to account for the level of alkaloid accumulated.[29] Thus, these might present realistic targets for genetic engineering (cf. *N. rustica*).

Potential limitations to flux were investigated by feeding ornithine, arginine, putrescine, agmatine and N-methylputrescine to root cultures.[45] None of these increased hyoscyamine levels. Instead, it was found that, to variable extents, these early-stage intermediates led to an enhanced accumulation of tropine and, to a lesser extent, of acetyltropine. This latter metabolite is normally a minor component of the alkaloid spectrum, but reached levels comparable to hyoscyamine in cultures fed 10 mM amines (although the hyoscyamine level was diminished under these conditions). If the intermediates, tropinone or tropine, were fed, hyoscyamine again failed to be elevated but tropine and acetyltropine increased up to 50-fold, exceeding the level of hyoscyamine present.[45] Thus, it appeared that in these root cultures the most important step limiting hyoscyamine accumulation was the esterification of tropine with tropic acid.

A priori, this limitation could simply reflect a deficiency in either the supply of the phenylalanine-derived unit or in its esterification. Feeding either tropic acid or phenyllactic acid failed to enhance hyoscyamine accumulation[45] and recent work has strongly suggested that free tropic acid may not be an intermediate.[47] Thus, the activation step urgently requires investigation and characterisation. Currently, however, no information is available on either the activation of tropic acid or on the ester-forming activity.

Under no conditions was more than a trace of scopolamine accumulation observed, indicating that these *D. stramonium* roots are severely limited in their capacity to carry out the hydroxylation and epoxidation steps required to form the 6β,7β-epoxide. Therefore, it could be concluded that:

1. the most substantial limitation to hyoscyamine formation is at the point of esterification;

2. the accumulation of tropine leads to the accumulation of undesirable side products;

3. the pathway has excess capacity for tropine formation, indicating that enhancing the input into N-methylputrescine could lead to a higher flux to tropine;

4. the 6β,7β-epoxide formation is restricted in cultures.

It should be emphasized that these conclusions relate to the *D. stramonium* D15/5 root culture. In plants, the limitations might be different. Indeed, in other similar root cultures, the situation is not identical. In a *Brugmansia (Datura)* hybrid culture, feeding tropinone, tropine or hyoscyamine led to enhanced scopolamine production.[43] Hence, in this culture, firstly, the epoxidation capacity was apparently present in excess, and secondly, the total alkaloid accumulation might be limited by the supply of tropine. This emphasizes the extent to which engineering requirements relate to individual cases and to the system in which exploitation will take place (see below). Nevertheless, a gene identified as important in one system, once isolated, can be applied to a number of engineering programs.

Pathway Definition

Clearly, it is desirable for the pathway to be correctly defined at the biochemical level. With hyoscyamine formation in *Datura,* one initial uncertainty was that evidence from label-feeding experiments indicated that N-methylputrescine might not be synthesised by the route established for *Nicotiana* which involes the sequential activities of ODC and PMT (Fig. 8). The incorporation of the 2-NH_2 and 5-NH_2 of ornithine into hyoscyamine appeared to be distinguishable, indicating that perhaps free putrescine, a symmetrical intermediate, was not involved.[48] Before considering engineering ODC and, perhaps, PMT, it was desirable to determine whether this alternative pathway was active, or whether N-methylputrescine was synthesised by the equivalent route to that in *Nicotiana*. The equivalence was confirmed on a biochemical basis, as neither of the necessary enzymes required for 5-N-methylornithine to be an important intermediate could be detected, whereas ODC and PMT were present in excess.[49]

Again, drawing the parallel with *Nicotiana*, it was possible that ADC rather than ODC might provide the principal source of the putrescine incorporated. This possibility was tested by growing cultures in the presence of substrate-activated 'suicide' inhibitors of ODC and ADC, namely α-difluoromethylornithine (DFMO) and α-difluoromethylarginine (DFMA) respectively.[50] These inhibitors are highly specific and are used by their respective enzymes as substrates.[51] Decarboxylation produces an unstable nucleophile which reacts with a cysteine residue at the active site, covalently binding the reaction product to the protein and irreversibly inactivating the enzyme. Growing *D. stramonium* roots in the presence of either DFMA or DFMO had little influence on fresh mass yield but DFMA, specifically, diminished hyoscyamine accumulation. This inhibitor led to depletion of the

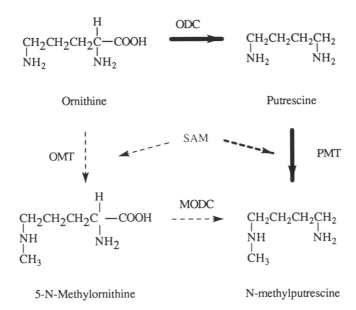

Fig. 8. Hypothetical alternative routes for N-methylputrescine formation in *Datura*. MODC = 5-N-methylornithine decarboxylase; OMT = ornithine N-methyltransferase; SAM = S-adenosylmethionine.

putrescine and hydroxycinnamoylputrescine pools as well as the N-methylputrescine, hygrine and tropine pools. Thus, it appeared that ADC activity might be more important for the tropane alkaloid pathway. However, the separation is unlikely to be absolute. Firstly, ornithine and arginine are both effectively incorporated into hyoscyamine,[49] and this is unlikely to involve the prior metabolism of ornithine to arginine. Secondly, it was found in these experiments that the levels of spermine and spermidine were maintained constant regardless of which inhibitor was present, indicating that putrescine was probably metabolised to these compounds preferentially, irrespective of the original source. However, it does indicate that ADC might be a better target for genetic engineering than ODC.

Regulation of the Pathway

In any engineering program it is especially important to identify enzymes that show a regulatory role *in vivo* and to isolate clones for these. While the concept of the 'rate-limiting enzyme" is now considered somewhat naive, the natural levels of activity of these enzymes are likely to be tightly regulated at the protein level. Thus, they can be expected to make a strong contribution to overall flux control[52] and to be sensitive to endogenous or exogenous effectors. Feed-back regulation by products down-stream in the pathway will act to damp the total flux through the system.[52] In addition, they may well be regulated at the level of gene expression. These enzymes might be expected to have some generic roles, for example by showing common properties between different or divergent pathways.

In *Nicotiana*, the enzyme PMT has already been indicated as playing a key role in affecting the overall flux into the pathway to nicotine. A similar picture is seen in *Datura*. If *D. stramonium* root cultures are treated with phytohormones, dispersion of the culture also occurs, over a period of 4 to 14 days.[53] During this period, alkaloid is rapidly lost from the system. In contrast to *N. rustica*, in which the total alkaloid is simply diluted by growth,[29] there appears to be an active degradation of tropine and hyoscyamine (Fig. 9). This difference probably reflects different turnover rates for these alkaloids. Again, there is a diminution of PMT activity within 24 h of the phytohormone treatment and, although this shows partial recovery, it is completely absent in dispersed cultures.[53] Following the removal of phytohormones after >20 transfers, alkaloid production is restored and this is accompanied by a restoration of PMT activity. As yet, it has not been possible to follow these changes at the molecular level.

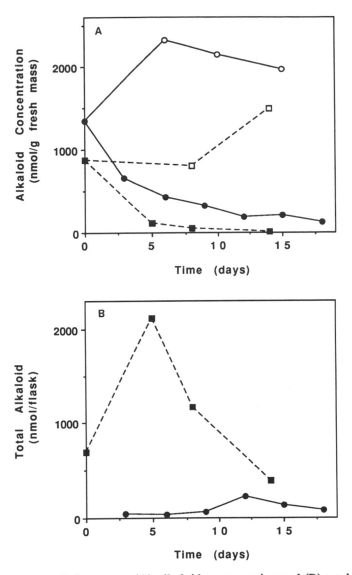

Fig. 9. The influence on (A) alkaloid concentration and (B) total alkaloid content of sub-culturing *N. rustica* (solid lines) or *D. stramonium* (dashed lines) transformed root cultures into NK medium [added naphthylacetic acid (2 mg/l) and kinetin (0.5 mg/l)] (closed symbols) or normal growth medium (open symbols).

Another line of evidence indicates a potentially important regulatory role for PMT, in this instance at the protein level. In roots treated with DFMA (and, to a lesser extent DFMO), PMT activity is diminished in a concentration-dependent manner. As DFMA-treatment causes the near-total removal of agmatine, it seemed possible that this effect might be due to an influence of agmatine on PMT levels. However, agmatine in normally grown roots was not found to stimulate PMT. But, when roots were first treated with DFMA, so as to lower the PMT level, addition of agmatine caused a substantial stimulation of the PMT activity, restoring the levels normally present in the absence of DFMA.[19] Hence, it appears that PMT may be subject to feed-forward regulation, increasing the evidence that this activity has a major role in regulating the pathway.

Enzyme Purification

The purification of several enzymes in tropane alkaloid biosynthesis is now complete or very nearly complete. Taking them in the order in which they are active in the pathway:

Putrescine N-methyltransferase. Although the basic properties of this enzyme from *Nicotiana* were described some time ago,[54] it has proved to be difficult to purify. Limited success was recently reported with enzyme extracted from *Hyoscyamus niger* roots[55] but only very recently has pure PMT been obtained.[34] The enzyme was purified using sequential polyethyleneglycol 6000 precipitation, anion-exchange chromatography, affinity chromatography using an L-ornithine-derivatised column and gel-filtration chromatography. The partially purified *H. niger* protein and the purified *D. stramonium* enzymes show similar properties with relation to substrate specificity.

N-methylputrescine oxidase. This activity is yet to be purified from a tropane producing species. It does, however, appear to be less unstable than PMT and has been isolated in pure form from *N. tabacum* transformed root cultures by sequential ammonium sulphate precipitation, anion-exchange chromatography, gel-filtration chromatography and hydrophobic-interaction chromatography on PhenylSuperose®.[35] Antiserum raised against the polypeptide shows a wide distribution of this protein within the Solanaceae and even in some non-solanaceous species, indicating that MPO might be a specific metabolic application of an oxidase generally having a wider function, or a member of a closely related family of oxidases. Interestingly, cross-reactivity is strongest in other families where a diamine oxidase is required in other alkaloid-producing pathways. MPO is, however, shown to be distinct from the well-characterized pea seedling diamine oxidase, both in immunocross-reactivity and kinetic properties.[23,35]

Tropinone reductases I and II. These activities, the tropine-forming (TR I) and pseudotropine-forming (TR II) tropinone reductases respectively, have been purified from both *D. stramonium* [56] and *H. niger.* [57] Following ammonium sulphate precipitation, the enzymes from *D. stramonium* were separated from each other by fractional elution from Butyl-Fractogel® and Fractogel-Red®; those from *H. niger* by chromatography on anion-exchange, hydroxyapatite, Phenyl-Superose® and chromatofocussing. TR I and TR II show quite different properties, while the equivalent activities from the two sources are similar.

Hyoscyamine 6β-hydroxylase. Although the last activity in the pathway, hyoscyamine 6β-hydroxylase (H6H) was the first to be obtained in purified form.[58] It was obtained from *H. niger* root cultures by a procedure involving ammonium sulphate precipitation and fractionation by Butyl-Toyopearl®, DEAE-Toyopearl® and hydroxyapatite chromatography. This preparation showed activity both as the hydroxylase and as the epoxidase (see below).

Cloning of Genes for Tropane Alkaloid Biosynthesis

The first clone for a gene specific to the tropane alkaloid pathway was that for hyoscyamine 6β-hydroxylase.[59] This clone was obtained by conventional techniques following the purification of H6H activity, as outlined above. The gene shows some similarity to other hydroxylases, including those involved in oxidative reactions in the formation of ethylene and anthocyanins.[59]

Recently, full length cDNA clones for TR I and TR II have been isolated from *D. stramonium*, using oligonucleotide probes derived from internal peptide sequence analysis and polymerase-cchain-reaction amplification.[60] A clone for TR II from *H. niger* has similarly been isolated.[61] The sequences of TR I and TR II show considerable homology.

Currently, progress is being made on obtaining full-length clones for ODC, ADC, PMT and MPO, but these are not yet available. Once this phase of gene isolation is completed, it will be possible to progress rapidly in the analysis of the effects of genetic engineering on tropane alkaloid production.

Genetic Manipulation of Scopolamine Formation

Atropa belladonna is typically a hyoscyamine-rich plant. The *h6h* gene was introduced into transgenic *A. belladonna* under the control of the cauliflower mosaic virus 35S promoter using a binary vector (pHY8) in *A. tumefaciens* LBA4404.[62] Following leaf-disc infection, calli were formed and used to regenerate plantlets. These were screened for H6H by immunoblotting. Strongly-expressing plantlets were grown on and propagated by cuttings. At flowering, these were selfed, providing a T_1 generation. All members of this generation showed the presence of the *h6h* gene by Southern hybridization. As well as the endogenous H6H, seen by cross-reactivity of antiserum in wild-type plants, all transformants examined contained two of three possible further copies of the transgene. Western blotting showed that the expression of H6H in transgenic plants was de-regulated, protein being found in all tissues, whereas it only occurred in branch roots in wild-type plants.

      When transgenic plants were examined at various developmental stages, the level of scopolamine was always much higher than in control plants at an equivalent stage. In normal plants, scopolamine is predominant in young vegetative material but hyoscyamine becomes the major alkaloid by the time of flowering. In mature transgenic *A. belladonna*, however, the alkaloid pool continued to be composed almost exclusively of scopolamine (Fig. 10).

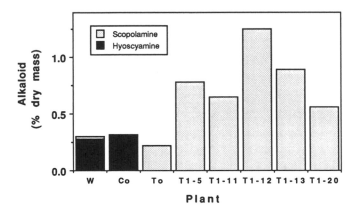

Fig. 10. Levels of scopolamine and hyoscyamine in normal *Atropa belladonna* plants (W) and transformed plants expressing the *h6h* gene (T). W = normal; Co = control transformants; To = transgenic plants, parental generation; T1-5 to T120=selfed progeny of To plants (redrawn from ref. 62, with permission).

Cultures of transformed roots of *A. belladonna*, infected with *A. rhizogenes* harbouring pHY8 have also been generated.[63] Several clones were isolated and one, T2, showed enhanced levels of H6H activity, as determined by both western blot and enzyme activity assays (Fig. 11). Surprisingly, a number of other clones failed to show higher than endogenous levels. Clone T2 contained comparable levels of hyoscyamine to control transformed roots but approximately 2-fold elevated levels of 6β-hydroxyhyoscyamine and scopolamine (Fig. 11).

Role of Gene Cloning in Defining the Pathway

The ability to clone a gene and to express it in a different host can give insights into the biochemistry of the pathway. To come full circle, the ability to express *h6h* has enabled the final uncertainty in scopolamine biosynthesis to be resolved. Previously, the isolation of H6H had enabled the intermediacy of 6β–hydroxyhyoscyamine rather than 6,7-dehydrohyoscyamine to be demonstrated unequivocally.[11] What remained unclear, however, was whether the two steps from hyoscyamine to scopolamine might be catalysed by two enzymes or one. The H6H and epoxidase activities co-purified, but failed to maintain a constant ratio.[58] However, by expressing the *h6h* clone in *N. tabacum*, Hashimoto and co-workers have confirmed that only the single polypeptide is required to carry out both reactions.[64] *N. tabacum* cannot normally form either hyoscyamine or scopolamine *de novo*, but transgenic plants expressing *h6h* were demonstrated to be able to biotransform hyoscyamine into scopolamine. The transformation was not complete, and some 6β–hydroxyhyoscyamine accumulated also. However, it was clear from this experiment that a single gene product was responsible for both the hydroxylation and epoxidation steps.

Opportunities to Enhance Minor Components

Quite recently, it has been shown that the polyhydroxynortropane alkaloids, the calystegins, are derived from pseudotropine.[41] These highly water-soluble alkaloids have structural characteristics in common with other groups of polyhydroxyalkaloids known to be glycosidase inhibitors[65] and have been found also to have this property.[66] The accumulation of calystegins in root cultures tends to be low relative to tropanes.[41] The possibility exists, however, of improving calystegin production by directing more tropinone towards pseudotropine and away from tropine. This might be achieved by engineering cultures that were both enhanced in TR II activity and down-regulated in TR I

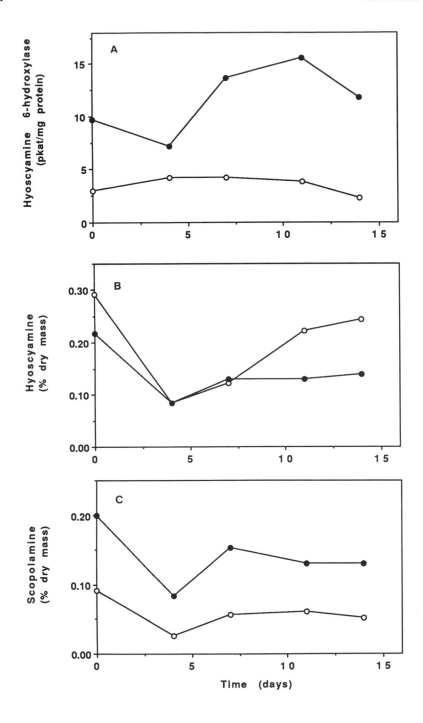

activity. Using chemical inhibitors of TR I, it has already been found that decreasing the effective TR I level leads to a limited increase in the accumulation of pseudotropine-derived products.[44] Permanent depletion of TR I might be achieved by anti-sense expression of the TR I gene.

A similar approach might increase the amount of the pseudotropine esters, which are less commonly accumulated. Increased availability of pseudotropine has been shown to stimulate the accumulation of these esters,[44] and the enzyme activity require to esterify pseudotropine is present in excess.[42] Unlike the calystegins, however, no useful function of these compounds has yet been identified, making them a less attractive target for genetic engineering.

CONCLUSION

In 1990 we wrote ' The key to success in [plant pathway engineering] lies in having sufficient prior understanding of the biochemistry and regulation of the pathway under consideration to pinpoint accurately the points in the pathway at which flux is most severely restricted".[19] To pursue this approach is a task of enormous dimensions, even for a single pathway. It could well prove much easier, and probably both quicker and more exciting, to utilize whatever genes are available that might influence the pathway and simply see what happens! Limited manipulation of a multistep secondary product pathway can be achieved by the insertion of a gene influencing a single step, as demonstrated by the examples given above. Similarly, the accumulation of serotonin (5-hydroxytryptophan) in root cultures of *Peganum harmala* is enhanced by inserting the first gene of a two-step pathway, tryptophan decarboxylase.[67] Other tryptophan-derived alkaloids in these cultures, the β-carbolines, do not, however, respond to the elevation of the activity of this one enzyme.

Fig. 11. Levels of (A) hyoscyamine 6β-hydroxylase activity, (B) hyoscyamine and (C) scopolamine in transformed roots of *Hyoscyamus niger* expressing the *h6h* gene(closed symbols) and control transformants (open symbols). (Redrawn from ref. 63, with permission).

A further problem is that, by drawing increased amounts of metabolite into the secondary pathway, the supply of the requisite primary metabolite might recently been demonstrated in plants. In transgenic *P. harmala* root cultures, the extent to which serotonin accumulates is limited by the endogenous availability of tryptophan: feeding this amino acid enhances production.[67] Undoubtedly this problem will arise again.

An additional complication, so far not seen with alkaloids but described in anthocyanin formation in *Petunia*, is that of co-suppression [5] This phenomenon, in which the insertion of extra copies of a homo-gene leads to a lower level of enzyme than in the unengineered tissue, is yet to be fully understood. It is however, a potentially serious problem if the objective of a program is to enhance the total yield, rather than to alter the composition of products.

Yet another problem is that the enzyme may need to be expressed within the engineered organism in the correct tissue or in the correct sub-cellular organelle. For example, LDC expression was not observed in transgenic *N. tabacum* plants unless the transgene was expressed in the chloroplast.[6] However, although H6H protein is normally localized in the pericycle of roots,[69] when *h6h* was expressed under a non-tissue-specific promoter, good levels of expression were found.[62,63]

To summarize, manipulation of a pathway apparently requires having the right gene(s), in the right place(s), with the right level(s) of expression and linking all this correctly to primary metabolism. None of this should be surprising. Perhaps now, however, is the time to think of alternative strategies to that outlined above for tropane alkaloid manipulation. An attractive approach might be to elucidate the intrinsic regulatory processes that determine the overall expression of a pathway *in toto*. Many pathways of secondary metabolism are not expressed 'constitutively', even in culture. It is likely that regulation at the whole pathway level occurs. In plants, many pathways are organ-specific - flower pigments and root-derived tropane alkaloids to name but two - or are temporally regulated. By isolating and altering the expression of these regulatory factors, probably transcription factors, it might be feasible to influence the whole pathway in a single operation. The development of such generic approaches seems highly desirable if the full potential of engineering of secondary product pathways in plants is to be realized.

ACKNOWLEDGEMENTS

We are grateful to our colleagues, Russell McLauchlan, Judy Furze, Ray McKee, Abigael Peerless, Chris Waspe and Peter Wilson for assistance with work contributing directly or indirectly to this chapter and for helpful discussions. Thanks are also extended to Birgit Dräger (Münster, FRG) and David Tepfer (Versailles, France) for discussions and/or permission to report unpublished work.

REFERENCES

1.   PHILLIPSON, J.D. 1990. Plants as sources of valuable products. In: Secondary Products from Plant Tissue Culture. (B.V. Charlwood, M.J.C. Rhodes, eds.), Clarendon Press, Oxford, pp. 1-21.

2.   BECKER, H., SAUERWEIN, M. 1990. Manipulating the biosynthetic capacity of plant cell cultures. In: Secondary Products from Plant Tissue Culture. (B.V. Charlwood, M.J.C. Rhodes, eds.), Clarendon Press, Oxford, pp. 43-57.

3.   BERLIN, J. 1990. Screening and selection for variant cell lines with increased levels of secondary metabolites. In: Secondary Products from Plant Tissue Culture. (B.V. Charlwood, M.J.C. Rhodes, eds.), Clarendon Press, Oxford, pp. 119-137.

4.   HAMILL, J.D., PARR, A.J., RHODES, M.J.C., ROBINS, R.J., WALTON, N.J. 1987. New routes to secondary products. Bio/Technology 5:800-804.

5.   JORGENSEN, R. 1991. Beyond antisense - How do transgenes interact with homologous plant genes? TIBTECH 9:266-267.

6.   HERMINGHAUS, S., SCHREIER, P.H., McCARTHY, J.E.G, LANDSMANN, J., BOTTERMAN, J., BERLIN, J. 1991. Expression of a bacterial lysine decarboxylase gene and transport of the protein into the chloroplasts of transgenic tobacco. Plant. Mol. Biol. 17:457-486.

7.   LUCKNER, M. 1990. Secondary Metabolism in Microorganisms, Plants, and Animals. 3rd. Edition. Springer-Verlag, Berlin, Heidelberg, 563 pp.

8.   RHODES, M.J.C., ROBINS, R.J. 1987. The use of plant cell cultures in studies of metabolism. In: Biochemistry of Plants, 13. Methodology (D.D. Davies, ed.), Academic Press,, New York, pp. 65-125.

9.   LINDSEY, K. 1991. Plant Tissue Culture Manual: Fundamentals and Applications. Kluwer Academic Publishers.

10. JONES, J.D.G., SHLUMUKOV, L., CARLAND, F., ENGLISH, J., SCOFIELD, S.R., BISHOP, G.J., HARRISON, K. 1992 Effective vectors for transformation, expression of heterologous genes, and assaying transposon excision in plants. Transgenic Research 1:285-297.

11. ROBINS, R.J., WALTON, N.J. 1993. The biosynthesis of tropane alkaloids. In: The Alkaloids (G.A. Cordell, ed.), Academic Press, Orlando, 44:115-187.

12. HASHIMOTO, T., YUKIMUNE, Y., YAMADA, Y. 1986. Tropane alkaloid production in *Hyoscyamus* root cultures. J. Plant Physiol. 124:61-75.

13. PAYNE, J., HAMILL, J.D., ROBINS, R.J., RHODES, M.J.C. 1987. Production of hyoscyamine by 'hairy root' cultures of *Datura stramonium*. Planta Medica, 53:474-478.

14. HAMILL, J.D., PARR, A.J., ROBINS, R.J., RHODES, M.J.C. 1986. Secondary product formation by cultures of *Beta vulgaris* and *Nicotiana rustica* transformed with *Agrobacterium rhizogenes*. Plant Cell. Rep. 5:111-114.

15. RHODES, M.J.C., ROBINS, R.J., HAMILL, J.D., PARR, A.J., HILTON, M.G., WALTON, N.J. 1990. Properties of transformed root cultures. In: Secondary Products from Plant Tissue Culture. (B.V. Charlwood, M.J.C. Rhodes, eds.), Clarendon Press, Oxford, pp. 201-25.

16. HAMILL, J.D., ROBINS, R.J., PARR, A.J., EVANS, D.M., FURZE, J.M., RHODES, M.J.C. 1990. Over-expressing a yeast ornithine decarboxylase gene in transgenic roots of *Nicotiana rustica* can lead to enhanced nicotine formation. Plant Mol. Biol. 15:27-38.

17. SMITH, T.A. 1990. Plant polyamines - metabolism and function. In: Polyamines and Ethylene: Biochemistry, Physiology and Interactions. (H.E. Flores, R.N. Arteca, J.C. Shannon, eds.), American Society of Plant Physiologists, Rockville, pp. 1-23.

18. MARTIN-TANGUY, J., TEPFER, D. 1990. Effects of Ri T-DNA from *Agrobacterium rhizogenes* on growth, development and polyamine metabolism in tobacco. In: Polyamines and Ethylene: Biochemistry, Physiology and Interactions. (H.E. Flores, R.N. Arteca, J.C. Shannon, eds.), American Society of Plant Physiologists, Rockville, pp. 238-250.

19. ROBINS, R.J., WALTON, N.J., HAMILL, J.D., PARR, A.J., RHODES, M.J.C. 1990. Strategies for the genetic manipulation of alkaloid-producing pathways in plants. Planta Medica 57:S27-35.

20. MIZUSAKI, S., TANABE, Y., NOGUCHI, M., TAMAKI, E. 1973. Changes in the activities of ornithine decarboxylase, putrescine N-methyltransferase and N-methylputrescine oxidase in tobacco roots in relation to nicotine biosynthesis. Plant Cell Physiol. 14:103-110.

21. FRIESEN, J.B., LEETE, E. 1990. Nicotine synthase - an enzyme from *Nicotiana* species which catalyses the formation of *(S)*-nicotine from nicotinic acid and 1-methyl-Δ'-pyrrolinium chloride. Tet. Lett. 6295-6298.

22. WALTON, N.J., ROBINS, R.J. RHODES, M.J.C. 1988. Perturbation of alkaloid production by cadaverine in hairy root cultures of *Nicotiana rustica*. Plant Science 54:125-131.

23. WALTON, N.J., MCLAUCHLAN, W.R. 1990. Diamine oxidation and alkaloid production in transformed root cultures of *Nicotiana tabacum*. Phytochemistry 29:1455-1457.

24. TIBURCIO, A.F., GALSTON, A.W. 1986. Arginine decarboxylase as the source of putrescine for tobacco alkaloids. Phytochemistry 25:107-110.

25. LEETE, E. 1983. Biosynthesis and metabolism of the tobacco alkaloids. In: Alkaloids: Chemical and Biological Perspectives, 1. (S.W. Pelletier, ed.), John Wiley, London. pp. 85-152.

26. ROBINS, R.J., HAMILL J.D., PARR, A.J., SMITH, K., WALTON, N.J., RHODES, M.J.C. 1987. Potential for use of nicotinic acid as a selective agent for isolation of high nicotine-producing lines of *Nicotiana rustica* hairy root cultures. Plant Cell Rep. 6:122-126.

27. OHTA, S., YATAZAWA, M. 1980. Metabolic key step discriminating nicotine producing tobacco callus strain from ineffective one. Biochem. Physiol. Pflanzen 175:382-385.

28. FETH, F., WAGNER R., WAGNER, K.G. 1986. Regulation in tobacco callus of the nicotine pathway, 1. The route ornithine to methylpyrroline. Planta 168:402-407.

29. RHODES, M.J.C., ROBINS, R.J., AIRD, E.L.H., PAYNE, J., PARR, A.J., WALTON, N.J. 1989. Regulation of secondary metabolism in transformed root cultures. In: Primary and Secondary Metabolism of Plant Cell Cultures II. (W.G.W. Kurz, ed.) Springer-Verlag, Berlin, Heidelberg, pp. 58-72.

30. AIRD, E.L.H., HAMILL, J.D., ROBINS, R.J., RHODES, M.J.C. 1988. Chromosome stability in transformed hairy root cultures and the properties of variant lines of *Nicotiana rustica* hairy roots. In: Manipulating Secondary Metabolism in Culture. (R. J. Robins, M.J.C. Rhodes, eds.) Cambrige University Press, Cambridge. pp. 137-144.

31. HAMILL J.D., PRESCOTT, A., MARTIN, C. 1987. Assessment of the efficiency of cotransformation of the T-DNA of disarmed binary vectors derived from *Agrobacterium tumefaciens* and the T-DNA of *A. rhizogenes*. Plant Mol. Biol. 9:573-584.

32. EVANS, P.T., MALMBERG, R. L. 1989. Do polyamines have roles in plant development? Ann. Rev. Plant. Physiol. Plant Mol. Biol. 40:235-269.

33. HEBY, O., PERSSON, L. 1990. Molecular genetics of polyamine synthesis in eukaryotic cells. TIBS 15:153-158.

34. WALTON, N.J., PEERLESS, A.C.J., ROBINS, R.J., RHODES, M.J.C, BOSWELL, H.D., ROBINS, D.J. Purification and properties of putrescine N-methyltransferase from transformed roots of *Datura stramonium* L. Planta (in press)

35. McLAUCHLAN, W.R., McKEE, R.A., EVANS, D.M. 1993. The purification and immunocharacterisation of N-methylputrescine oxidase from transformed root cultures of *Nicotiana tabacum* l. cv SC58. Planta 191:440-445.

36. WALTON, N.J., BELSHAW, N.J. 1988. The effect of cadaverine on the formation of anabasine from lysine in hairy root cultures of *Nicotiana hesperis*. Plant Cell Rep. 7:115-118.

37. FECKER L.F., BEIER, H., BERLIN, J. 1986. Cloning and characterization of a lysine decarboxylase gene from *Hafnia alvei*. Mol. Gen. Genet. 203:177-184.

38. FECKER L.F., HILLEBRANDT, S., RÜGENHAGEN, C., HERMINGHAUS, S., LANDSMANN, J., BERLIN, J. 1992. Metabolic effects of a bacterial lysine decarboxylase gene expressed in hairy root culture of *Nicotiana glauca*. Biotech. Lett. 14:1035-1040.

39. PARR, A.J., PAYNE, J., EAGLES, J. CHAPMAN, B.T., ROBINS, R.J., RHODES, M.J.C. 1990. Variation in tropane alkaloid accumulation within the Solanaceae and strategies for its exploitation. Phytochemistry 29:2545-2550.

40. PARR, A.J. 1992. Alternative metabolic fates of hygrine in transformed root cultures of *Nicandra physaloides*. Plant Cell. Rep. 11:270-273.

41. DRÄGER, B., FUNCK, C., HÖHLER, A., MRACHATZ, G., PORTSTEFFEN, A., SCHAAL, A, SCHMIDT, R. 1993. Calystegins as a new group of tropane alkaloids in Solanaceae. Plant. Cell. Tiss. Org. Cult. (in press)

42. ROBINS, R.J., BACHMANN, P., PEERLESS, A.C.J., RABOT, S. 1993. Esterification reactions in the biosynthesis of tropane alkaloids in transformed root cultures. Plant. Cell. Tiss. Org. Cult. (in press)

43. ROBINS, R.J., PARR, A.J., PAYNE, J., WALTON, N.J. RHODES, M.J.C. 1990. Factors regulating tropane alkaloid production in a transformed root culture of a *Datura candida* x *D. aurea* hybrid. Planta 181:414-422.

44. DRÄGER, B., PORTSTEFFEN, A., SCHAAL, A, McCABE, P. H., PEERLESS, A.C.J., ROBINS, R.J. 1992. Levels of tropinone-reductase activities influence the spectrum of tropane esters found in transformed root cultures of *Datura stramonium*. Planta 188:581-586.

45. ROBINS, R.J., PARR, A.J., BENT, E.G., RHODES, M.J.C. 1991. Studies on the formation of tropane alkaloids in *Datura stramonium* L. transformed root cultures, 1. The kinetics of alkaloid production and the influence of feeding intermediate metabolites. Planta 183:185-195.

46. HASHIMOTO, T., YUKIMUNE, Y., YAMADA, Y. 1989. Putrescine and putrescine N-methyltransferase in the biosynthesis of tropane alkaloids in cultured roots of *Hyoscyamus albus*, I. Biochemical studies. Planta 178:123-130.

47. ROBINS, R.J., EAGLES, J., COLQUHOUN, I., WOOLLEY, J.G., ANSARIN, M. 1992. Is free tropic acid an intermediate in hyoscyamine biosynthesis? In: Phytochemistry and Agriculture, Abstracts. (T. van Beek, ed.), p. 32.

48. LEETE, E., McDONELL, J.A. 1981. The incorporation of [1-^{13}C, ^{14}C, *methylamino* ^{15}N]-*N*-methylputrescine into nicotine and scopolamine established by means of carbon-13 nuclear magnetic resonance. J. Amer. Chem. Soc. 103:658-662.

49. WALTON, N.J., ROBINS, R.J., PEERLESS, A.C.J. 1990. Enzymes of N-methylputrescine biosynthesis in relation to hyoscyamine formation in transformed root cultures of *Datura stramonium* and *Atropa belladonna*. Planta 182:16-141.

50. ROBINS R.J., PARR, A.J., WALTON, N.J. , 1991. Studies on the formation of tropane alkaloids in *Datura stramonium* L. transformed root cultures, 2. On the relative contributions of L-arginine and L-

ornithine to the formation of the tropane ring. Planta 183:196-201.

51.   BEY, P., DANZIN, C., JUNG, J. 1987. Inhibition of basic amino acid decarboxylases involved in polyamine biosynthesis. In: Inhibition of Polyamine Metabolism (P. McCann, A.E. Pegg, A. Sjoerdsma, eds.), Academic Press, New York, pp. 1-31.

52.   KACSER, H., PORTEOUS, J.W. 1987. Control of metabolism: what do we have to measure? TIBS 12:5-14.

53.   ROBINS, R.J., BENT, E.G., RHODES, M.J.C. 1991. Studies on the formation of tropane alkaloids in *Datura stramonium* L. transformed root cultures, 3. The relationship between morphological integrity and alkaloid biosynthesis. Planta 185:385-390.

54.   MIZUSAKI, S., TANABE, Y., NOGUCHI, M., TAMAKI, E. 1971. Phytochemical studies on tobacco alkaloids XIV. The occurrence and properties of putrescine N-methyltransferase in tobacco roots. Plant Cell Physiol. 12:633-640.

55.   HIBI, N., FUJITA, T., HATANO, M., HASHIMOTO, T., YAMADA. Y. 1992. Putrescine *N*-methyltransferase in cultured roots of *Hyoscyamus albus*. Plant Physiol. 100:826-835.

56.   PORTSTEFFEN, A., DRÄGER, B., NAHRSTEDT, A. 1992. Two tropinone reducing enzymes from *Datura stramonium* transformed root cultures. Phytochemistry 31:1135-1138.

57.   HASHIMOTO, T., NAKAJIMA, K., ONGENA, G., YAMADA, Y. 1992. Two tropinone reductases with distinct stereospecificities from cultured roots of *Hyoscyamus niger*. Plant Physiol. 100:836-845.

58.   HASHIMOTO, T., YAMADA. Y. 1987. Purification and characterization of hyoscyamine 6β-hydroxylase from root cultures of *Hyoscyamus niger* L. Hydroxylase and epoxidase activities in the preparation. Eur. J. Biochem. 164:277-285.

59.   MATSUDA, J., OKABE, S., HASHIMOTO, T., YAMADA. Y. 1991. Molecular cloning of hyoscyamine 6β-hydroxylase, a 2-oxoglutarate-dependent dioxygenase, from cultured roots of *Hyoscyamus niger*. J. Biol. Chem. 266:9460-9464.

60.   NAKAJIMA, K., HASHIMOTO, T., YAMADA, Y. 1993. Two tropinone reductases with different stereospecificities are short-chain dehydrogenases evolved from a common ancestor. Proc. Natl. Acad. Sci. 90:9591-9595.

61.   NAKAJIMA, K., HASHIMOTO, T., YAMADA, Y. 1993. cDNA encoding tropinone reductase-II from *Hyoscyamus niger*. Plant Physiol. 103:000-000.

62   YUN, D.-J., HASHIMOTO, T., YAMADA, Y. 1992. Metabolic engineering of medicinal plants: Transgenic *Atropa belladonna* with improved alkaloid composition. Proc. Natl. Acad. Sci. USA 89:11799-11803.

63.   HASHIMOTO, T., YUN, D.-J., YAMADA, Y. 1993. Production of tropane alkaloids in genetically engineered root cultures. Phytochemistry 32:713-718..

64.   YUN, D.-J., HASHIMOTO, T., YAMADA, Y. 1993. Expression of hyoscyamine 6β-hydroxylase gene in transgenic tobacco. Biosci. Biotech. Biochem. 57:502-503.

65.   WINCHESTER, B., FLEET, G.W.J. 1992. Amino-sugar glycosidase inhibitors: versatile tools for glycobiologists. Glycobiology 2:199-210.

66.   MOLYNEUX, R.J., PAN, Y.T., GOLDMANN, A., TEPFER, D., A.D. 1993. Calystegins, a novel class of alkaloid glycosidase inhibitors. Arch. Biochem. Biophys. 304:81-88.

67   BERLIN, J., RÜGENHAGEN, C., DIETZE, P., FECKER, L.F., GODDIJN, O.J.M., HOGE, H.C. 1993. Increased production of serotonin by suspension and root cultures of *Peganum harmala* transformed with a tryptophan decarboxylase cDNA clone from *Catharanthus roseus*. Trans. Res. 2 (in press).

68.   STEPHANOPOULOS, G., VALLINO, J.J. 1991. Network rigidity and metabolic engineering in metabolite overproduction. Science 252:1675-1681.

69.   HASHIMOTO, T., HAYASHI, A., AMANO, Y., KOHNO, J., IWANARI, H., USUDA, S., YAMADA, Y. 1991. Hyoscyamine 6β-hydroxylase, an enzyme involved in tropane alkaloid biosynthesis, is localized in the pericycle of the root. J. Biol. Chem. 266:4648-4653.

Chapter Two

MOLECULAR GENETIC TECHNIQUES APPLIED TO THE ANALYSIS OF ENZYMES OF ALKALOID BIOSYNTHESIS

Toni M. Kutchan

Laboratorium für Molekulare Biologie, Universität München
Karlstrasse 29, 80333 München, Germany

INTRODUCTION

The use of indigenous plants in traditional medicine is a practice which is older than even that of agriculture. In much of the world of today, these medicines are the only ones which are readily available. The search for the active principles in folk remedies led to the discovery of alkaloids such as morphine, ajmalicine, quinine, scopolamine and a host of other structures which are still in use as pharmaceuticals today. The complex structures of the alkaloids spurred chemists in the earlier part of this century to speculate on how plants were synthesizing these multi-ring carbon skeletons in a stereospecific manner. By the late 1950s and throughout the 1960s, these biogenetic hypotheses were being tested by radioisotope feeding experiments with plants. The results from these

Genetic Engineering of Plant Secondary Metabolism,
Edited by B.E. Ellis *et al.*, Plenum Press, New York, 1994

types of experiments outlined the biosynthetic pathways, but left many ambiguities, usually due to low incorporation rates of the radioactive precursors. With the expanded use of plant cell culture in the 1970s, it became possible to begin to identify the enzymes involved in the individual transformation steps leading to the alkaloids. In many cases, discovery of the alkaloid forming enzymes clarified portions of pathways that had remained ambiguous for many years. Once information on the alkaloid biosynthetic enzymes is obtained, the control mechanisms governing alkaloid biosynthesis in plants and in plant cell culture could be approached. In addition, the techniques of molecular genetics could be applied to these alkaloid biosynthetic systems. By isolating the cDNAs and genes that encode these enzymes, the potential exists to learn more about the enzyme reaction mechanisms and about the underlying regulation controlling the synthesis of these enzymes in the plant cell. There are presently several examples of these latter approaches being taken with alkaloid biosynthesis. To date, these include benzophenanthridine alkaloid biosynthesis in *Eschscholtzia californica*, monoterpenoid indole alkaloid biosynthesis in *Rauvolfia serpentina* and *Catharanthus roseus* and tropane alkaloid biosynthesis in *Hyoscyamus niger*. These three systems will be discussed here.

BENZO[c]PHENANTHRIDINE ALKALOIDS

The benzo[c]phenanthridine alkaloids are the active principles of the extracts of *Chelidonium majus*, (used in Europe for treatment of gastric cancer), *Sanguinaria canadensis* (used in Europe against warts, papillomas and condylomas)[1] and *Eschscholtzia californica* (used by the American Indian as a sedative).[2] The structures of selected benzophenanthridine alkaloids are given in Figure 1.

Robert Robinson had suggested that benzophenanthridine alkaloids arose from protopine by fission at C_6-N_7 followed by rearrangement and cyclization at C_6-C_{13}. Since the time of these early predictions about the biosynthesis of benzophenanthridine alkaloids, the biosynthetic pathway from the primary metabolite, *L*-tyrosine, through to the benzophenanthridine alkaloid, sanguinarine, has been completely elucidated at the level of each individual enzyme involved in the pathway. As was suggested by Robinson, the benzophenanthridine nucleus is derived from the benzylisoquinoline nucleus. The biosynthetic pathway can be divided into two portions: the pathway from *L*-tyrosine to (*S*)-reticuline and the pathway from (*S*)-reticuline to sanguinarine.

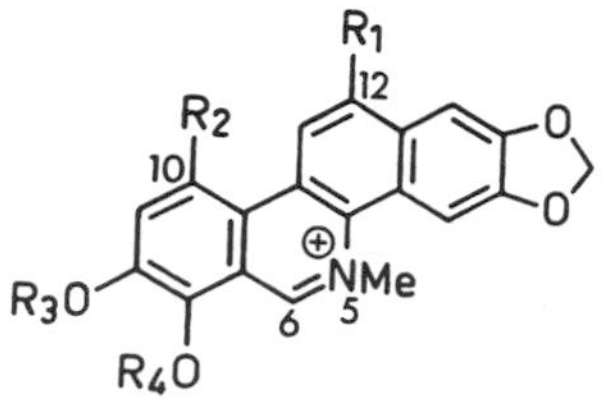

Fig. 1. Structures of select benzophenanthridine alkaloids. Sanguinarine ($R_1 = R_2 = H$, $R_3 + R_4 = -CH_2-$); chelerythrine ($R_1 = R_2 = H$, $R_3 = R_4 = CH_3$), chelirubine ($R_1 = H$, $R_2 = OCH_3$, $R_3 + R_4 = -CH_2-$); macarpine $R_1 = R_2 = OCH_3$, $R_3 + R_4 = -CH_2-$); 5,6-dihydrochelirubine ($R_1 = H$, $R_2 = OCH_3$, $R_3 + R_4 = -CH_2-$, C_5-C_6 saturated).

The pathway from tyrosine to the building blocks of the benzylisoquinoline alkaloids, dopamine and 4-hydroxyphenylacetaldehyde, remains a complex metabolic grid,[3] but can be pictured in a simplified form as shown in Figure 2, giving rise to (*S*)-norcolaurine.

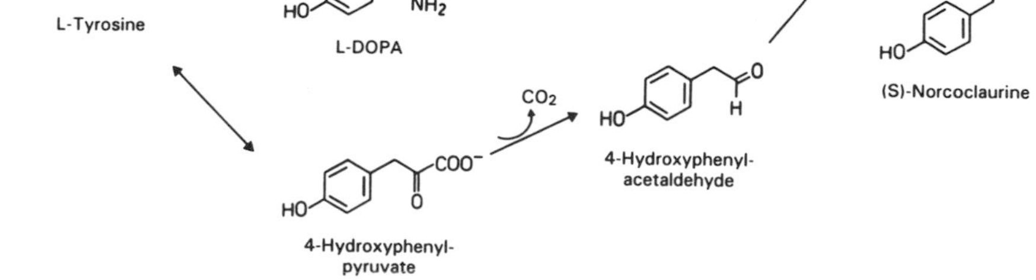

Fig. 2. Pathway from *L*-tyrosine to the benzylisoquinoline alkaloid (S)-norcoclaurine via building blocks, dopamine and 4-hydroxyphenyl-acetaldehyde.

The pathway that leads from dopamine and 4-hydroxyphenylacetaldehyde to form the isoquinoline and the benzyl moieties of the tetrahydroisoquinoline nucleus was revised[4] in recent years when it was discovered that (S)-norcoclaurine was the first alkaloidal intermediate in the pathway and not (S)-norlaudanosoline as was previously thought.[5] This cleared up ambiguities in results obtained in early feeding experiments where it could not be satisfactorily explained why tyramine and dopamine did not label both halves of the isoquinoline nucleus. It also provides good reasoning why norlaudanosoline has never been found to occur in nature. The revised pathway leading to (S)-reticuline is depicted in Figure 3.

The enzymes that catalyze each of these transformations have been described.[4]. The alkaloid, (S)-reticuline, occupies a pivotal point in alkaloid biosynthesis. It is the precursor to a myriad of different alkaloid families such as the morphinans, the protoberberines, the phthalideisoquinolines, the protopines

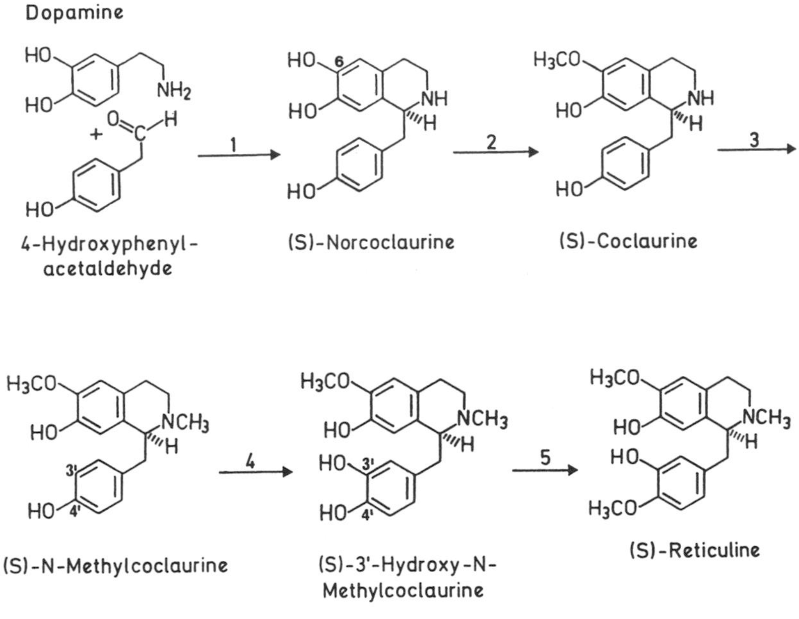

Fig. 3. Revised enzymatic pathway leading to the tetrahydrobenzylisoquinoline alkaloid (S)-reticuline. 1) (S)-Norcoclaurine synthase, 2) (S)-Adenosyl-L-methionine: (S)-norcoclaurine-6-O-methyltransferase, 3) (S)-Adenosyl-L-methionine: (S)-coclaurine-N-methyltransferase, 4) Phenolase, 5) (S)-Adenosyl-L-methionine: (S)-3'-hydroxy-N-methycoclaurine-4'-O-methyltransferase.

as well as the benzophenanthridines. The pathway that leads to the benzo-phenanthridine nucleus has been completely identified at the enzymatic level. There are eight transformation steps involving seven enzymes and one spontaneous rearrangement in this portion of the pathway (Fig. 4).[6] We will describe in detail step 1, the formation of (S)-scoulerine from (S)-reticuline by action of the berberine bridge enzyme.

The Berberine Bridge Enzyme

The branch-point enzyme that catalyzes the formation of (S)-scoulerine from (S)-reticuline, the berberine bridge enzyme [(S)-reticuline:oxygen oxidoreductase (methylene-bridge-forming), EC 1.5.3.9][7,8] is of particular interest in this pathway (Fig. 5).

The formation of the berberine bridge arises from oxidative cyclization of the N-methyl group of (S)-reticuline to form the berberine bridge carbon, C_8, of (S)-scoulerine. This is a reaction which is not found elsewhere in nature and cannot be mimicked chemically. For these reasons, the berberine bridge enzyme becomes mechanistically very interesting to examine. The conversion can be viewed as proceeding by two possible reaction mechanisms, ionic or radical (Fig. 6).

The stereochemistry with which the enzymatic reaction proceeds has been analyzed by ^3H NMR with a berberine bridge enzyme preparation from *Berberis*. and(S)-reticuline containing a chiral methyl on the nitrogen was found to be stereospecific. The enzyme replaces an N-methyl hydrogen with the phenyl group in an inversion mode.[9] The berberine bridge enzyme was originally purified from cell suspension cultures of *Berberis beaniana*[8], which showed no qualitative or quantitative changes in alkaloid content when treated with a yeast cell wall elicitor preparation.[10] The homogeneous enzyme was a single polypeptide with a molecular mass of 52 ± 4 kD as determined by SDS-PAGE. The enzyme converted (S)-reticuline, (S)-protosinomenine and (S)-laudanosoline to the corresponding (S)-tetrahydroprotoberberines in the presence of oxygen, stoichiometrically releasing H_2O_2. In the plant cell, the enzyme is located within a particle with the density p = 1.14 g/ml.[8, 11] Only 11 μg of the enzyme could be purified from the *Berberis* cell cultures; therefore, the interesting question of the identity of the cofactor which would be necessary for this oxidoreductase to function could not be approached. Benzophenanthridine alkaloids function in some species of the *Papaveraceae* and the *Fumariaceae* in an adaptive response to fungal infection. Of the benzophenanthridine class of alkaloids, at least, sanguinarine has been shown to have antimicrobial

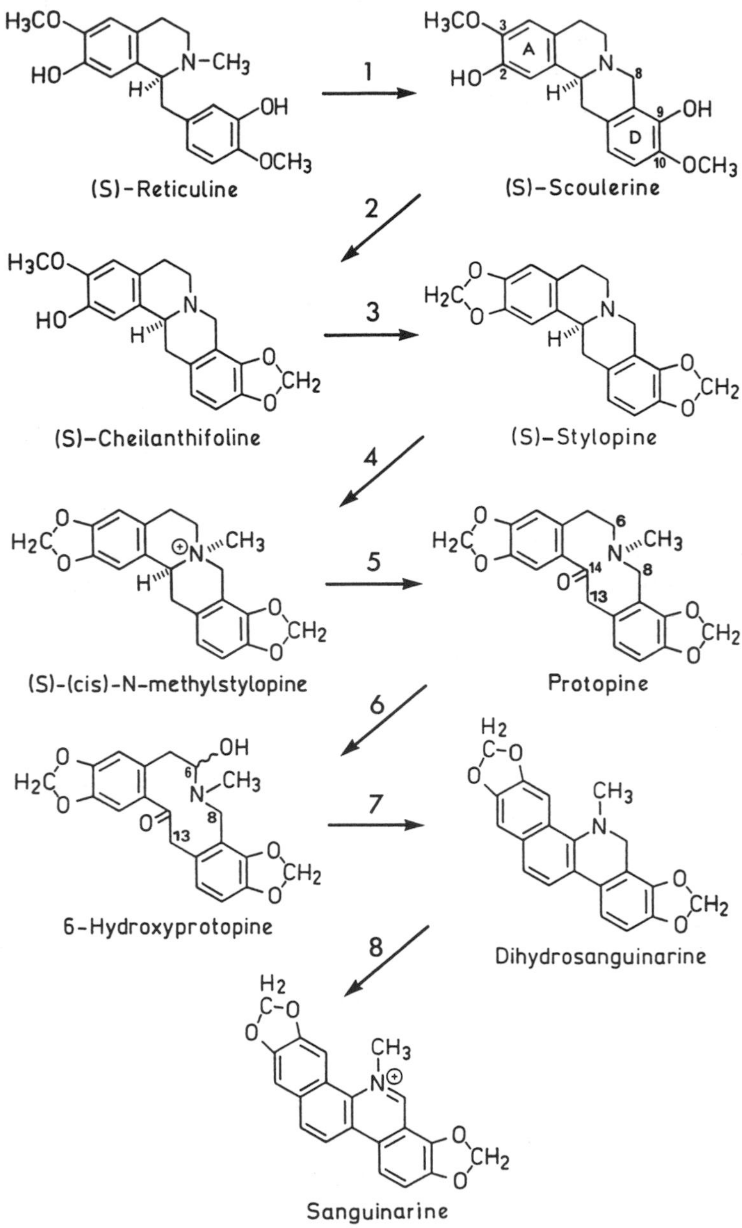

(S)-Reticuline   1   (S)-Scoulerine

2

(S)-Cheilanthifoline   3   (S)-Stylopine

4

(S)-(cis)-N-methylstylopine   5   Protopine

6

6-Hydroxyprotopine   7   Dihydrosanguinarine

8

Sanguinarine

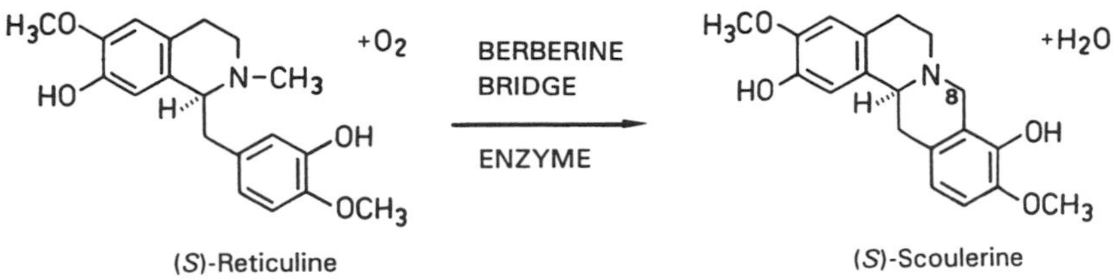

Fig. 5. Reaction catalyzed by the berberine bridge enzyme.

activity.[12,13] In cell suspension cultures of the California poppy, *Eschscholtzia californica*, the level of benzophenanthridine alkaloids as well as the level of the berberine bridge enzyme were increased after treatment of the cultures with a fungal cell wall preparation.[10] The berberine bridge enzyme is, therefore, a likely integral component of the defense system of plants which can produce benzophenanthridine alkaloids. Benzophenanthridine alkaloids are not only of use to the plant by functioning as phytoalexins, but also find pharmacological use in oral hygiene. Due to a commercial need for sanguinarine, much effort has been invested in producing this alkaloid by large scale cell suspension culture fermentation.[14] When all these factors are considered, the berberine bridge enzyme is involved in an ecochemical function in certain plants, in the production of sanguinarine as a pharmaceutical and in the catalysis of a reaction which is unique in nature and not chemically reproducible.

Fig. 4. Enzymatic pathway leading from (*S*)-reticuline to the benzophenanthridine alkaloid sanguinarine. 1. Berberine bridge enzyme, 2. (*S*)-Cheilanthifoline synthase, a cytochrome P-450 dependent monooxygenase, 3. (*S*)-Stylopine synthase, a cytochrome P-450 dependent monooxygenase, 4. (*S*)-Adenosyl-*L*-methionine: (*S*)-tetrahydroprotoberberine-*cis*-N-methyltransferase, 5. (*S*)-*Cis*-N-methyltetrahydroprotoberberine 14-hydroxylase, a cytochrome P-450 dependent monooxygenase, 6. Protopine 6-hydroxylase, a cytochrome P-450 dependent monooxygenase, 7. Spontaneous, 8. Dihydrobenzophenanthridine oxidase.

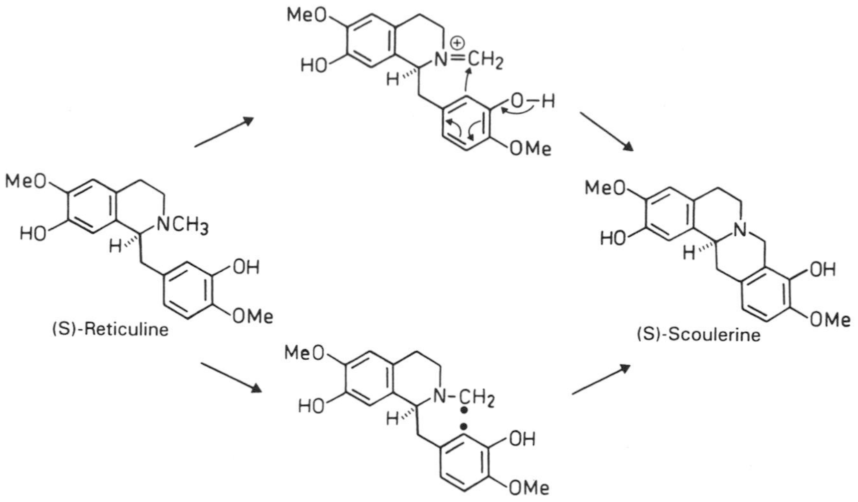

Fig. 6. Ionic versus radical mechanism of conversion of (S)-reticuline to (S)-scoulerine as catalyzed by the berberine bridge enzyme.

In order to study this interesting enzyme in more detail, larger quantities of the protein had to be made available. By far the best way to do this was to isolate the cDNA encoding the berberine bridge enzyme. E. californica was chosen as the experimental system because the elicitation characteristics[10] not only made it possible to increase the level of the enzyme for purification, but also provided a physiologically interesting system with which to work. The berberine bridge enzyme was, therefore, purified to homogeneity from elicited E. californica cell suspension cultures, a partial amino acid sequence was determined, and with the information thus obtained, the corresponding cDNA was isolated from an E. californica cDNA bank prepared from poly (A)+ RNA isolated from elicited cell cultures.[15] From the translation of the nucleotide sequence of this cDNA, several facts were learned about the berberine bridge enzyme. First, the enzyme is synthesized as a pre-protein. The first twenty-three amino acids constitute a signal peptide. It was known from the amino acid sequence that the amino terminal position was occupied by a glycine residue,which is in agreement with the predicted cleavage site between Gly 23 and Gly 24. This signal peptide is necessary for guiding the berberine bridge enzyme into the endoplasmic reticulum; this concurs with the finding that the enzyme is localized within a vesicle in the plant cell.[11] The amino acid sequence contains three consensus sequences for N-linked carbohydrate. One of the three was shown by amino acid sequence analysis to contain a modified Asn residue.[15] Probably the most important information obtained from the sequence was a 25% sequence homology to a bacterial enzyme, 6-hydroxy-D-nicotine oxidase, from *Arthrobacter oxidans*.[16] The overall homology was not strikingly high, but 6-hydroxy-D-nicotine oxidase is a flavoprotein which contains a covalently attached flavin.[17] The berberine bridge enzyme contained the consensus sequence necessary for covalent attachment of a flavin. Since the exact nature of the coenzyme that must be present on the berberine bridge enzyme in order for it to catalyze a redox reaction was not known, this provided lead information. The next goal was to obtain large enough quantities of the heterologously expressed enzyme to perform a biochemical analysis on the cofactor. The berberine bridge enzyme was not expressed in an enzymatically active form in *E. coli*. It was produced enzymatically active in yeast, although the overall level of accumulation was quite low.[15] The heterologous expression system which, by far, has been the most successful for the production of eukaryotic enzymes is that of the insect cell culture, *Spodoptera frugiperda* Sf9,[18] using the very powerful transcriptional promoter, *polh*, of the baculovirus *Autographa californica* nuclear polyhedrosis virus (AcNPV).[19] With this expression system, 4 mg of homogeneous berberine bridge enzyme could be obtained from one liter of spent culture medium.[20] Although the Sf9 cells are quite sensitive to temperature changes and shear force as well as to infection, the cells will perform post translational modifications on eukaryotic proteins similar to the original host organism. In the case of the berberine bridge enzyme, the signal peptide was properly cleaved and cofactor was attached such that the protein was produced in an enzymatically active form. The insect cell culture medium used for protein production was serum-free; this was advantageous because the berberine bridge enzyme produced by the Sf9 cells was secreted into the medium. Although within the plant cell the berberine bridge enzyme is sequestered in a specialized vesicle, the enzyme follows the default transport pathway in the insect cell, i.e. secretion. The use of a medium that contained minimal protein additives thus greatly simplified the purification procedure, and the enzyme was obtained in a homogeneous form after only three steps of purification.

Another interesting aspect of the berberine bridge enzyme is its regulation in cell suspension cultures of *E. californica*. It had been observed that lightly pastel-colored cells of *E. californica* turn dark red upon addition of a variety of abiotic and biotic elicitors.[10] This dark red pigment was identified as the benzophenanthridine alkaloid, macarpine. During this elicitation process, one of the enzymes that increased in total activity in the cell cultures was the berberine bridge enzyme. Given the cDNA clone for this enzyme, it was possible to investigate the nature of this induction. An analysis of the kinetics of induction of the berberine bridge enzyme in elicited cell suspension cultures of *E. californica* revealed that less than two hours after addition of a yeast cell wall elicitor preparation, an increase in the level of berberine bridge enzyme poly (A)$^+$ RNA could be detected.[15] This induction proceeded with a minimal lag period and likely reflected *de novo* transcription. Within four to six hours, the level of the berberine bridge enzyme had measurably increased. This time lag correlated well with *de novo* translation. After eight to ten hours, an increase in the amount of total benzophenanthridine alkaloids could be measured. Although the induction of transcript was transient, peaking at approximately six hours after elicitation, the enzyme level reached a maximum at sixteen hours and remained high throughout the 48 h period analyzed. Benzophenanthridine alkaloids continued to accumulate for several days. About the time that it was

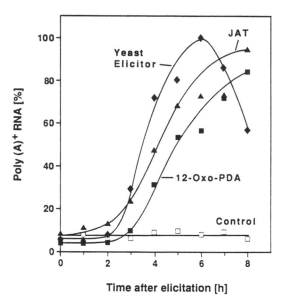

Fig. 7. Induction of berberine bridge enzyme poly (A)$^+$ RNA in *E. californica* cell suspension cultures in response to treatment with the indicated elicitor. 100% = maximal relative quantity of poly (A)$^+$ RNA in 50 µg total RNA; JAT = methyl jasmonate; 12-oxo-PDA = 12-oxo-phytodienoic acid.

found that addition of yeast elicitor induced *de novo* transcription of the berberine bridge enzyme gene, *bbe1*, in *E. californica*, it was discovered that methyl jasmonate was a potent inducer of protease inhibitor proteins in tomato. In a landmark paper, it was shown that airborne methyl jasmonate, as well, could induce this plant defense gene.[21] This effect of methyl jasmonate was found to be general, since it could induce accumulation of a wide variety of low molecular weight compounds, some of which were known defense-related compounds, in an equally wide range of plant genera and species.[22] The effects of methyl jasmonate and one of its biosynthetic prescursors, 12-oxo-phytodienoic acid, were compared to the yeast cell wall elicitor preparation with respect to the ability to induce the berberine bridge enzyme transcript in cell suspension cultures of *E. californica*.[23] It was found that both methyl jasmonate and 12-oxo-phytodienoic acid could induce *de novo* transcription of *bbe1* in a manner analogous to the yeast elicitor (Fig. 7).

This finding suggests that jasmonic acid and its octadecanoid pentacyclic biosynthetic precursor participate in a membrane lipid-based signalling system that activates defense genes in alkaloid-producing plants, as has been proposed for the activation of proteinase inhibitor genes in tomato,[24] for the genes of the phenylpropanoid pathway in parsley cell cultures[25] and for the induction of tendril coiling in *Bryonia*.[26] This also forms a basis for further study of the elements that regulate alkaloid gene activation.

MONOTERPENOID INDOLE ALKALOIDS

The use of plant genera such as *Rauvolfia* and *Vinca* in traditional medicine led chemists and pharmacists to investigate in more detail the active principles in these folk preparations. The earlier investigations of the 1950's resulted in the isolation of vincamine from *Vinca minor* L., which is used in the pharmaceutical industry as a vasodilator. Reports appeared later of the isolation and characterization of vindoline, vincaleukoblastine and catharanthine from the pantropical species *C. roseus*. The physical characteristics and chemotherapeutic activity of vincaleukoblastine were proposed and the isolation and structure of vincristine were reported. The interelationship between vincaleukoblastine and catharanthine and vindoline was also recognized. Soon after the reports of the isolation and characterization of these physiologically active indole alkaloids, the race to elucidate the biosynthetic pathways began; these efforts to clarify completely the pathways leading to the monoterpenoid indole alkaloids continue today. This work has been excellently reviewed.[27]

Much progress has also been made on the enzymatic verification of the proposed pathway that leads from strictosidine to the *Rauvolfia* alkaloid, ajmaline.[28] The enzymatic transformations discovered are given in Figure 8.

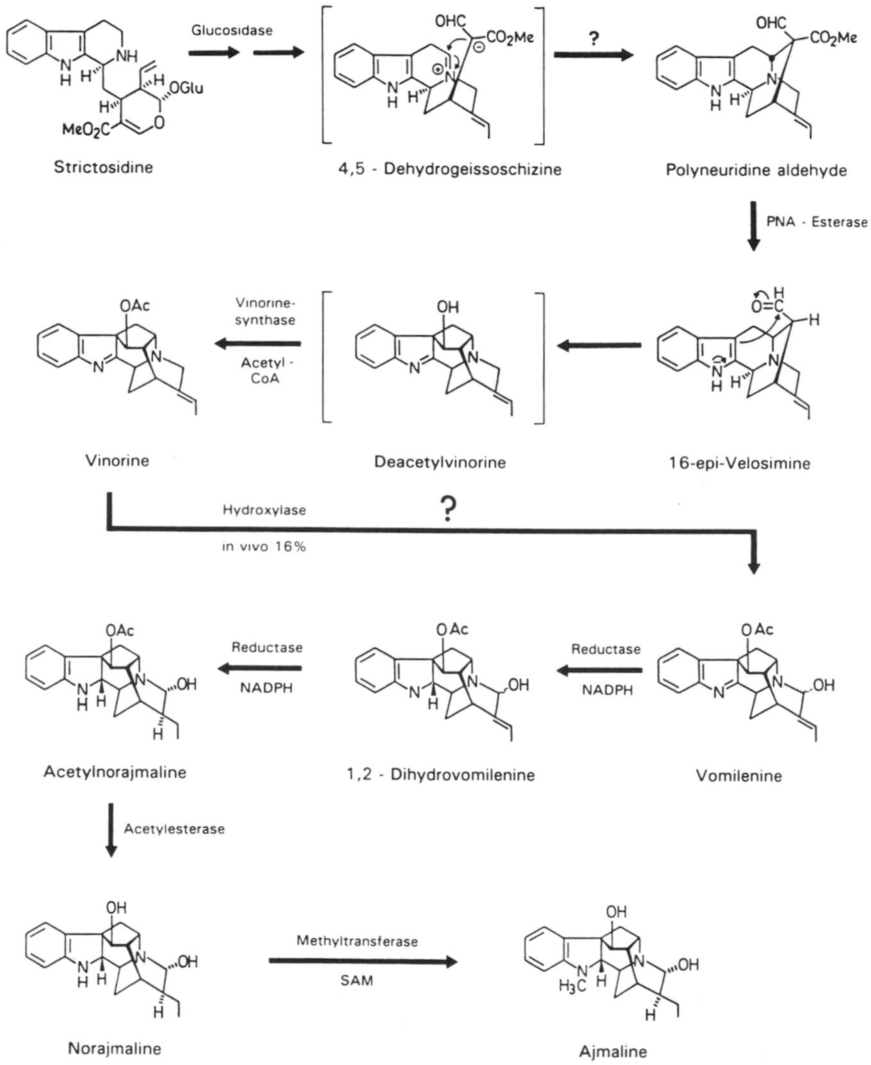

Fig. 8. Biosynthetic pathway leading to ajmaline in *Rauvolfia.*

Likewise, progress has been made in elucidating the enzymes involved in vindoline biosynthesis, in particular the later stages of the pathway to vindoline.[29-31] Some conflict exists in the literature as to the exact sequence of methylations and hydroxylations and, therefore, both enzymatic schemes are given here in Figure 9.

The enzymes involved in the very early steps of monoterpenoid indole alkaloid biosynthesis have caused less controversy in recent years and have provided us with the very first successful cDNA cloning experiments in alkaloid biosynthesis. These two enzymes, strictosidine synthase [EC 4.3.3.2] and tryptophan decarboxylase [EC 4.1.1.28], will be discussed in more detail here.

Strictosidine Synthase

The first enzyme in the biosynthesis of the alkaloid strictosidine [32] to yield to modern molecular genetic analysis was strictosidine synthase[33] from *R. serpentina* and, as such, opened a new type of study in the field of natural products (Fig. 10). The literature concerning strictosidine and strictosidine synthase has been very recently reviewed.[34]

Strictosidine synthase from *R. serpentina* was the first cDNA of alkaloid biosynthesis to be isolated[35] and heterologously expressed in an enzymatically active form.[36] The cDNA from *C. roseus* was later obtained using the sequences from the *Rauvolfia* clone[37] and was heterologously expressed in tobacco[38] and bacteria.[39] The cDNA clone encoding the *Rauvolfia* enzyme has now been expressed in the insect cell culture, *S. frugiperda* Sf9.[20] These expression systems can be summarized as follows in Table 1.

Heterologous expression in *Spodoptera* cell culture indicated that the *Rauvolfia* enzyme is synthesized as a pre-protein, although the cleavage site varied over a three amino acid range and did not correspond exactly to that predicted. The development of the various heterologous expression systems opens the way for a more detailed analysis of the enzyme reaction mechanism and for chemo-enzymatic synthesis of monoterpenoid indole alkaloids.

Some progress has been made on the regulation of monoterpenoid indole alkaloid biosynthesis using the strictosidine synthase clones. The gene, *str1*, for the *Rauvolfia* enzyme has been isolated.[43] *Str1* from ten different *Rauvolfia* species were then analyzed by PCR and found to be identical within the reading frame.[44] Unlike the only other known genes of secondary metabolism in plants, those of the general phenylpropanoid pathway, *str1* contained no introns and occurs as a single copy gene in the *Rauvolfia* genome

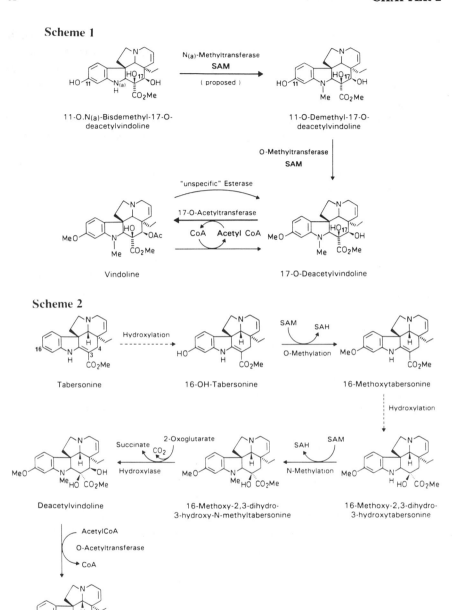

Fig. 9. Two proposed enzymatic biosynthetic pathways in the late stages of vindoline formation in *Catharanthus*.

Table 1. Host systems used for the expression of strictosidine synthase

Organism	Yield of Pure Enzyme (nkat/l)	Specific Activity (nkat/mg)	K_m Tryptamine (mM)	Secologanin (mM)	pH Opt.
R. serpentina [a,40]	29	184	4	4	6.5
C. roseus [a,41]	5.8	18	2.3	3.4	6.8
C. roseus [a,42]					
Isoform: I	1.6	20	0.9	-	6.7
II	6.7	44	6.6	-	6.7
III	12.3	68	1.9	-	6.7
IV	3.4	28	2.2	-	6.7
E. coli SG935[36]	138 [b]	-	-	-	-
E. coli XA90[39]	~840 [c]	~30 [c]	-	-	-
Nicotiana tabacum[38]	-	0.006 [b]	-	-	-
S. frugiperda[20]	267	62	1	1	6.5

[a] untransformed homologous system.
[b] based on crude extract activity.
[c] values not quantitatively determined.
- values not given in the literature.

The distribution of the transcript, enzyme and alkaloids in the R. *serpentina* plant is given in Table 2.[43]

The distribution of strictosidine synthase transcript in the C. *roseus* plant has also been determined. Although not quantitated, most of the strictosidine synthase transcript appears to be in the C. *roseus* root.[45] The induction kinetics of hybridizable strictosidine synthase transcript in elicited cell suspension cultures of C. *roseus* have also been reported.[45, 46]

Tryptophan Decarboxylase

Tryptophan decarboxylase catalyzes the formation of tryptamine from L-tryptophan. Tryptamine serves as substrate, together with secologanin, for strictosidine synthase in the formation of the first monoterpenoid indole alkaloid, strictosidine (Fig. 10). The cDNA encoding tryptophan decarboxylase in C. *roseus* has been isolated and the heterologous enzyme has been expressed in bacteria.[47] The subsequent report that tryptophan decarboxylase from C. *roseus*

Table 2. Distribution of strictosidine synthase poly(A)$^+$ RNA, enzyme activity and indole alkaloids in the *Rauvolfia serpentina* plant.[a]

Plant Organ	Poly(A)$^+$ RNA [%]	Enzyme Activity [%]	Indole Alkaloid Content [%]
Flower	21	18	24
Leaf	35	123	36
Stem	ND[b]	17	14
Root	100 [c]	100 [d]	100 [e]

[a]*Rauvolfia serpentina* (L.) Benth. ex Kurz. Root values were arbitrarily set at 100%.

[b]ND - not detectable

[c]100% = relative quantity of poly(A)$^+$ RNA in 50 µg total RNA. Hybridizable RNA levels were determined by scanning of the northern blot with a Berthold Linear Analyzer LB 2821.

[d]100% = 450 pkatal/g dry weight.

[e]100% = 11.6 mg/g dry weight indole alkaloids (the result of the summation of the total reserpine, ajmalicine, rescinnamine, vomilenine, yohimbine, serpentine, sarpagine, ajmaline and raucaffricine present in the tissue).

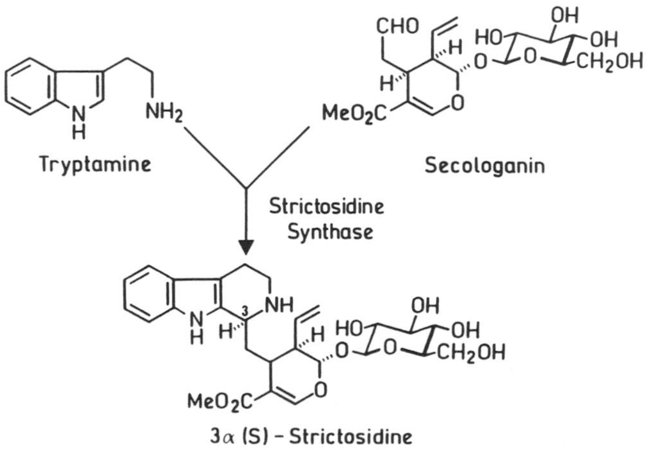

Fig. 10. Reaction catalyzed by strictosidine synthase.

is a pyridoxoquinoprotein containing two molecules of pyridoxal phosphate and two molecules of covalently bound pyrroloquinoline quinone per molecule of protein[48] poses interesting mechanistic questions about the correct covalent attachment of cofactor that would be necessary for expression of the enzyme in a catalytically active form in bacteria.[47] The kinetics of induction of the tryptophan decarboxylase transcript in elicited cell suspension cultures of *C. roseus*, as well as the effect of auxin, have recently been investigated.[45, 46, 49]

TROPANE ALKALOIDS

Hyoscyamine and scopolamine are major alkaloids of the genera *Duboisia, Hyoscyamus, Datura* and *Atropa*. Scopolamine is currently produced by the pharmaceutical industry for use in the prevention of motion sickness. The tropane alkaloids are esters of tropic acid and tropine derivatives. It has been quite rigorously demonstrated that tropic acid arises from phenylalanine and that tropine is derived from ornithine and/or arginine. The elucidation of several of the enzymes involved in tropane alkaloid biosynthesis[50-52] has helped to clarify results obtained by early feeding experiments with nicotine and tropane alkaloid producing plants. The proposed biosynthetic pathways are given in Figure 11.

Hyoscyamine 6β-hydroxylase

The best characterized step in scopolamine biosynthesis is the pentultimate step, the hydroxylation of hyoscyamine to 6β-hydroxy-hyoscyamine. In order to gain information on the enzymes of scopolamine biosynthesis, a tissue culture system had to be developed which would produce the tropane alkaloids. This was achieved by establishing root cultures of *Hyoscyamus niger*.[53] Undifferentiated cell suspension cultures did not possess the capacity for tropane alkaloid biosynthesis. In root cultures, however, sufficient quantities of alkaloid were produced to permit purification of the 2-oxo-glutarate-dependent dioxygenase, hyoscyamine 6β-hydroxylase [EC 1.14.11.11][54] (Fig. 12). The cDNA encoding hyoscyamine 6β-hydroxylase in *H. niger* has been isolated[55] and it has been elegantly demonstrated by both immunolocalization[56] and northern analysis that the enzyme and its transcript are accumulating exclusively in root tissue of both the intact plant and the cultured root, but are absent from leaf, stem, calyx, cultured cells and cultured shoots. More specifically, the protein was localized to the pericycle cells of the young root. Mature roots that had undergone secondary growth and lacked the

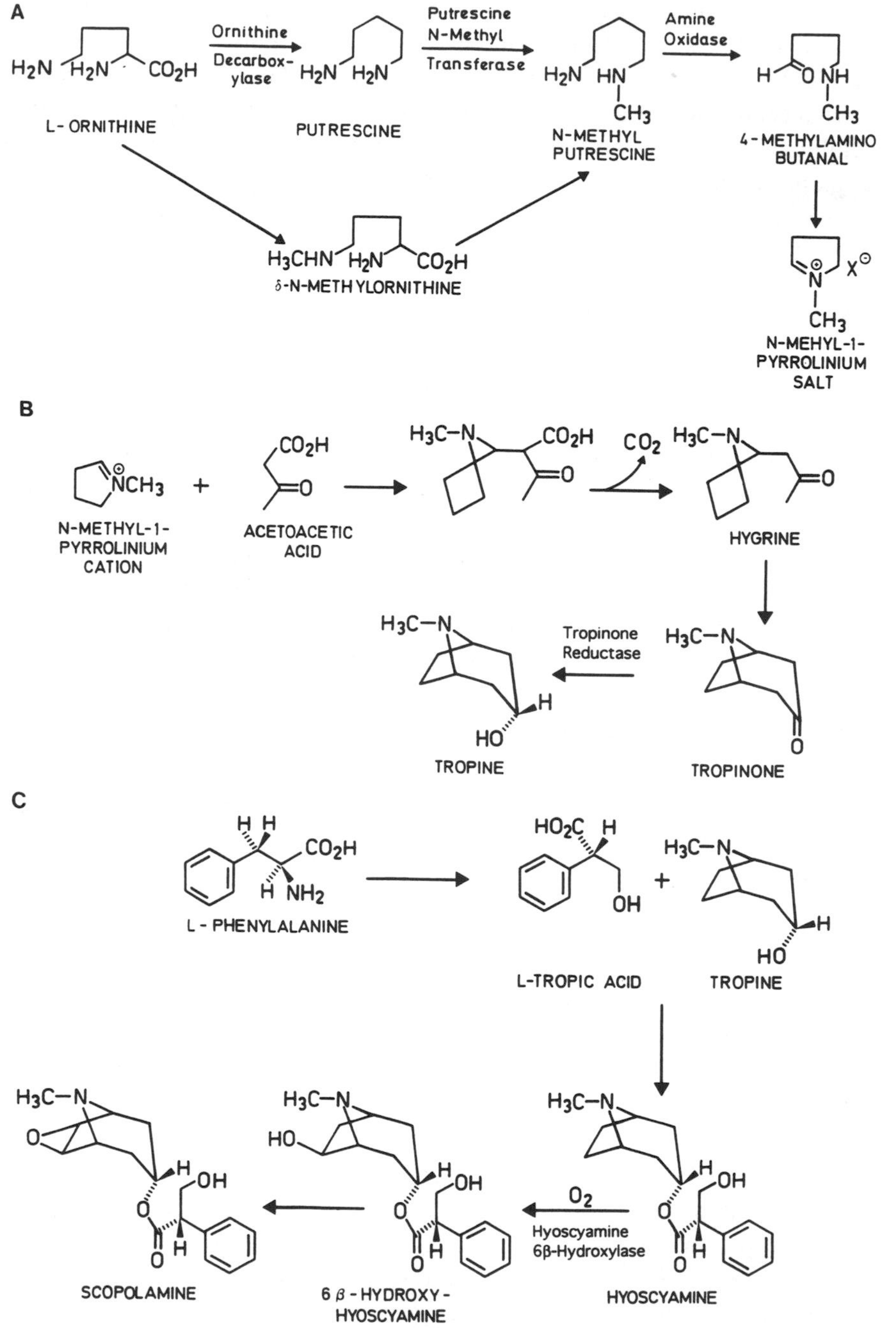

A

L-ORNITHINE → (Ornithine Decarboxylase) → PUTRESCINE → (Putrescine N-Methyl Transferase) → N-METHYL PUTRESCINE → (Amine Oxidase) → 4-METHYLAMINO BUTANAL → N-MEHYL-1-PYRROLINIUM SALT

δ-N-METHYLORNITHINE

B

N-METHYL-1-PYRROLINIUM CATION + ACETOACETIC ACID → (– CO_2) → HYGRINE → TROPINONE → (Tropinone Reductase) → TROPINE

C

L-PHENYLALANINE → L-TROPIC ACID + TROPINE

HYOSCYAMINE → (O_2, Hyoscyamine 6β-Hydroxylase) → 6 β-HYDROXY-HYOSCYAMINE → SCOPOLAMINE

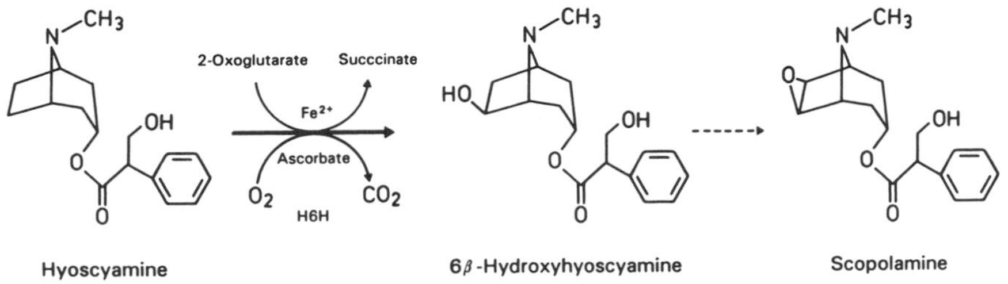

Fig. 12. Reaction catalyzed by hyoscyamine 6β-hydroxylase.

pericycle also lacked the enzyme. This type of tissue-specific expression provides solid anatomical evidence that explains the absence of scopolamine production in cell suspension culture and adds to our understanding of how alkaloid gene expression is regulated in plant tissue culture. *Atropa* can now be transformed to yield scopolamine, thereby providing an elegant example of a new approach to modifying medicinal plants.[57]

CONCLUSION

The application of molecular genetic technology to the topic of alkaloid biosynthesis has led, in recent years, to an increase in our understanding of how select alkaloid biosynthetic pathways are regulated and to an increase in our knowledge concerning the biocatalysts that form these complex molecules in the plant. The current evidence for this comes in the tropane, monoterpenoid indole and protoberberine / benzophenanthridine classes of alkaloids. Although still at a very early stage, continued molecular genetic investigations on natural product pathways will eventually lead to the capability of altering these metabolic pathways *in planta* as well as to use of the heterologously produced enzymes in chemo-enzymatic production of alkaloids.

Fig. 11. Biosynthesis of scopolamine. (A) Two proposed pathways leading to the N-methyl-1-pyrrolinium salt. (B) Proposed pathway leading to tropine. (C) Formation of tropic acid and condensation with tropine.

ACKNOWLEDGEMENT

The portion of the work presented here that was done in the Munich laboratory was supported by a grant from the Bundesminister für Forschung und Technologie, Bonn.

REFERENCES

1. CORDELL, G.A. 1981. Introduction to Alkaloids: A Biogenetic Approach, Wiley Interscience, New York, 1055pp.

2. CHENEY, R.H. 1964. Therapeutic potential of *Eschscholtziae californicae* herba. Quart. J. Crude Drugs 3:413-416.

3. RUEFFER, M., ZENK, M.H. 1987. Distant precursors of benzylisoquinoline alkaloids and their enzymatic formation. Z. Naturforsch. 42c:319-332.

4. STADLER, R., ZENK, M.H. 1990. A revision of the generally accepted pathway for the biosynthesis of the benzyltetrahydroisoquinoline alkaloid reticuline. Liebigs Ann. Chem. 555-562.

5. SPENSER, I.D. 1968. Biosynthesis of alkaloids and of other nitrogenous secondary metabolites. Compr. Biochem. 20:231-413.

6. KUTCHAN, T.M., DITTRICH, H., BRACHER, D., ZENK, M.H. 1991. Enzymology and molecular biology of alkaloid biosynthesis. Tetrahedron 47:5945-5954.

7. RINK, E., BÖHM, H. 1975. Conversion of reticuline into scoulerine by a cell free preparation from *Macleaya microcarpa* cell suspension cultures. FEBS Lett. 49:396-399.

8. STEFFENS, P., NAGAKURA, N., ZENK, M.H. 1985. Purification and characterization of the berberine bridge enzyme from *Berberis beaniana* cell cultures. Phytochemistry 24:2577-2583.

9. FRENZEL, T., BEALE, J.M., KOBAYASHI, M., ZENK, M.H., FLOSS, H.G. 1988. Stereochemistry of the enzymatic formation of the berberine bridge in protoberberine alkaloids. J. Am. Chem. Soc. 110:7878-7880.

10. SCHUMACHER, H.-M., GUNDLACH, H., FIEDLER, F., ZENK, M.H. 1987. Elicitation of benzophenanthridine alkaloid biosynthesis in *Eschscholtzia* cell cultures. Plant Cell Reports 6:410-413.

11. AMANN, M., WANNER, G., ZENK, M.H. 1986. Intracellular compartmentation of two enzymes of berberine biosynthesis in plant cell cultures. Planta 167: 310-320.

12. DZINK, J.L., SOCRANSKY, S.S. 1985. Comparative *in vitro* activity of sanguinarine against oral microbial isolates. Antimicrob. Agents Chemother. 27:663-665.

13. CLINE, S.D., COSCIA, C.J. 1988. Stimulation of sanguinarine production by combined fungal elicitation and hormonal deprivation in cell suspension cultures of *Papaver bracteatum*. Plant Physiol. 86:161-165.

14. PARK, J.M., YOON, S.Y., GILES, K.L., SONGSTAD, D.D., EPPSTEIN, D., NOVAKOVSKI, D., FRIESEN, L., ROEWER, I. 1992. Production of sanguinarine by suspension culture of *Papaver somniferum* in bioreactors. J. Ferment. Bioeng. 74:292-296.

15. DITTRICH, H., KUTCHAN, T.M. 1991. Molecular cloning, expression, and induction of berberine bridge enzyme, an enzyme essential to the formation of benzophenanthridine alkaloids in the response of plants to pathogenic attack. Proc. Natl. Acad. Sci. USA 88:9969-9973.

16. BRANDSCH, R., HINKKANEN, A.E., MAUCH, L., NAGURSKY, H., DECKER, K. 1987. 6-Hydroxy-D-nicotine oxidase of *Arthrobacter oxidans*. Gene structure of the flavoenzyme and its relationship to 6-hydroxy-L-nicotine oxidase. Eur. J. Biochem. 167:315-320.

17. BRANDSCH, R., BICHLER, V. 1985. *In vivo* and *in vitro* expression of the 6-hydroxy-D-nicotine oxidase gene of *Arthrobacter oxidans*, cloned into *Escherichia coli*, as an enzymatically active, covalently flavinylated polypeptide. FEBS Lett. 192:204-208.

18. VAUGHN, J.L., GOODWIN, R.H., TOMPKINS, G.J., MCCAWLEY, P. 1977. The establishment of two cell lines from the insect *Spodoptera frugiperda* (Lepidoptera: Noctuidae) In Vitro 13:213-217.

19. MILLER, L.K. 1989. Insect baculoviruses - powerful gene expression vectors. Bioessays 11:91-95.

20. KUTCHAN, T.M., BOCK, A., DITTRICH, H. 1993. Heterologous expression of the plant proteins strictosidine synthase and berberine bridge enzyme in insect cell culture. Phytochemistry (in press).

21. FARMER, E.E., RYAN, C.A. 1991. Interplant communication: Airborne methyl jasmonate induces synthesis of proteinase inhibitors in plant leaves. Proc. Natl. Acad. Sci. USA 87:7713-7716.

22. GUNDLACH, H., MÜLLER, M.J., KUTCHAN, T.M., ZENK, M.H. 1992. Jasmonic acid is a signal transducer in elicitor-induced plant cell cultures. Proc. Natl. Acad. Sci. USA 89:2389-2393.

23. KUTCHAN, T.M. 1993. 12-Oxo-phytodienoic acid induces accumulation of berberine bridge enzyme transcript in a manner analogous to methyl jasmonate. J. Plant Physiol., in press.

24. FARMER, E.E., RYAN, C.A. 1992. Octadecanoid precursors of jasmonic acid activate the synthesis of wound-inducible proteinase inhibitors. Plant Cell 4: 129-134.

25. DITTRICH, H., KUTCHAN, T.M., ZENK, M.H. 1992. The jasmonate precursor, 12-oxo-phytodienoic acid, induces phytoalexin synthesis in *Petroselinum crispum* cell cultures. FEBS Lett. 309:33-36.

26. WEILER, E.W., ALBRECHT, B., GROTH, B., XIA, Z.-Q., LUXEM, M., LIß, H., ANDERT, L., SPENGLER, P. 1993. Evidence for the involvement of jasmonates and their octadecanoid precursors in the tendril coiling response of *Bryonia dioica*. Phytochemistry 32:591-600.

27. CORDELL, G.A. 1974. The biosynthesis of indole alkaloids. Lloydia 37:219-298.

28. STÖCKIGT, J., SCHÜBEL, H. 1988. Cultivated plant cells: An enzyme source for alkaloid formation. NATO ASI Series H18 Plant Cell Biotechnology (M.S.S. Pais. ed.) Springer-Verlag, Berlin Heidelberg, pp. 251-264.

29. FAHN, W., GUNDLACH, H., DEUS-NEUMANN, B., STÖCKIGT, J. 1985. Late enzymes of vindoline biosynthesis. Acetyl-CoA: 17-*O*-deacetylvindoline 17-*O*-acetyltransferase. Plant Cell Reports 4:333-336.

30. FAHN, W., LAUßERMAIR, E., DEUS-NEUMANN, B., STÖCKIGT,

J. 1985. Late enzymes of vindoline biosynthesis. *S*-Adenosyl-*L*-methionine: 11-*O*-demethyl-17-*O*-deacetylvindoline 11-*O*-methyltransferase and unspecific acetylesterase. Plant Cell Reports 4:337-340.

31. DETHIER, M., DE LUCA, V. 1993. Partial purification of an N-methyltransferase involved in vindoline biosynthesis in *Catharanthus roseus*. Phytochemistry 32:673-678 and references therein.

32. SMITH, G.N. 1968. Strictosidine: A key intermediate in the biogenesis of indole alkaloids. J. Chem. Soc. Chem. Commun. 912-914.

33. STÖCKIGT, J., ZENK, M.H. 1977. Strictosidine (Isovincoside): The key intermediate in the biosynthesis of monoterpenoid indole alkaloids. J. Chem. Soc. Chem. Commun. 646.

34. KUTCHAN, T.M. 1993. Strictosidine: From alkaloid to enzyme to gene. Phytochemistry 32:493-506.

35. KUTCHAN, T.M., HAMPP, N., LOTTSPEICH, F., BEYREUTHER, K., ZENK, M.H. 1988. The cDNA clone for strictosidine synthase from *Rauvolfia serpentina*: DNA sequence determination and expression in *Escherichia coli*. FEBS Lett. 237:40-44.

36. KUTCHAN, T.M. 1989. Expression of enzymatically active cloned strictosidine synthase from the higher plant *Rauvolfia serpentina* in *Escherichia coli*. FEBS Lett. 257:127-130.

37. MCKNIGHT, T.D., ROESSNER, C.A., DEVAGUPTA, R., SCOTT, A.I., NESSLER, C.L. 1990. Nucleotide sequence of a cDNA encoding the vacuolar protein strictosidine synthase from *Catharanthus roseus*. Nucl. Acids Res. 18:4939.

38. MCKNIGHT, T.D., BERGEY, D.R., BURNETT, R.J., NESSLER, C.L. 1991. Expression of enzymatically active and correctly targeted strictosidine synthase in transgenic tobacco plants. Planta 185:148-152.

39. ROESSNER, C.A., DEVAGUPTA, R., HASAN, M., WILLIAMS, H.J., SCOTT, A.I. 1992. Purification of an indole alkaloid biosynthetic enzyme, strictosidine synthase, from a recombinant strain of *Escherichia coli*. Protein Expression and Purif. 3:295-300.

40. HAMPP, N., ZENK, M.H. 1988. Homogeneous strictosidine synthase

from cell suspension cultures of *Rauvolfia serpentina*. Phytochemistry 27:3811-3815.

41. TREIMER, J.F., ZENK, M.H. 1979. Purification and properties of strictosidine synthase, the key enzyme in indole alkaloid formation. Eur. J. Biochem. 101:225-233.

42. PFITZNER, U., ZENK, M.H. 1989. Homogeneous strictosidine synthase isozymes from cell suspension cultures of *Catharanthus roseus*. Planta Med. 55:525-530.

43. BRACHER, D., KUTCHAN, T.M. 1992. Strictosidine synthase from *Rauvolfia serpentina*: Analysis of a gene involved in indole alkaloid biosynthesis. Arch. Biochem. Biophys. 294:717-723.

44. BRACHER, D., KUTCHAN, T.M. 1992. Polymerase chain reaction comparison of the gene for strictosidine synthase from ten *Rauvolfia* species. Plant Cell Reports 11:179-182.

45. PASQUALI, G., GODDIJN, O.J.M., DE WAAL, A., VERPOORTE, R., SCHILPEROORT, R.A., HOGE, J.H.C., MEMELINK, J. 1992. Coordinated regulation of two indole alkaloid biosynthetic genes from *Catharanthus roseus* by auxin and elicitors. Plant Mol. Biol. 18:1121-1131.

46. ROEWER, I.A., CLOUTIER, N., NESSLER, C.L., DE LUCA, V. 1992. Transient induction of tryptophan decarboxylase (TDC) and strictosidine synthase (SS) genes in cell suspension cultures of *Catharanthus roseus*. Plant Cell Reports 11:86-89.

47. DE LUCA, V., MARINEAU, C., BRISSON, N. 1989. Molecular cloning and analysis of cDNA encoding a plant tryptophan decarboxylase: Comparison with animal DOPA decarboxylases. Proc. Natl. Acad. Sci. USA 86:2582-2586.

48. PENNINGS, E.J.M., GROEN, B.W., DUINE, J.A., VERPOORTE, R. 1989. Tryptophan decarboxylase from *Catharanthus roseus* is a pyridoxoquinoprotein. FEBS Lett. 255:97-100.

49. GODDIJN, O.J.M., DE KAM, R.J., ZANETTI, A., SCHILPEROORT, R.A., HOGE, J.H.C. 1992. Auxin rapidly down regulates transcription of the tryptophan decarboxylase gene from *Catharanthus roseus*. Plant Mol. Biol. 18:1113-1120.

50. YAMADA, Y., HASHIMOTO, T. 1988. Biosynthesis of tropane alkaloids. In: Applications of Plant Cell and Tissue Culture. (Ciba Foundation Symposium 137) Wiley, Chichester, pp. 199-212.

51. NARUHIRO, H., FUJITA, T., HATANO, M., HASHIMOTO, T., YAMADA, Y. 1992. Putrescine *N*-methyltransferase in cultured roots of *Hyoscyamus albus*. n-Butylamine as a potent inhibitor of the transferase both *in vitro* and *in vivo*. Plant Physiol. 100:826-835.

52. HASHIMOTO, T., NAKAJIMA, K., ONGENA, G., YAMADA, Y. 1992. Two tropinone reductases with distinct stereospecificities from cultured roots of *Hyoscyamus niger*. Plant Physiol. 100:836-845.

53. HASHIMOTO, T., YAMADA, Y. 1986. Hyoscyamine 6β-hydroxylase, a 2-oxoglutarate-dependent dioxygenase, in alkaloid-producing root cultures. Plant Physiol. 81:619-625.

54. HASHIMOTO, T., YAMADA, Y. 1987. Purification and characterization of hyoscyamine 6β-hydroxylase from root cultures of *Hyoscyamus niger* L. Eur. J. Biochem. 164:277-285.

55. MATSUDA, J., OKABE, S., HASHIMOTO, T., YAMADA, Y. 1991. Molecular cloning of hyoscyamine 6β-hydroxylase, a 2-oxoglutarate-dependent dioxygenase, from cultured roots of *Hyoscyamus niger*. J. Biol. Chem. 266:9460-9464.

56. HASHIMOTO, T., HAYASHI, A., AMANO, Y., KOHNO, J., IWANARI, H., USUDA, S., YAMADA, Y. 1991. Hyoscyamine 6b-hydroxylase, an enzyme involved in tropane alkaloid biosynthesis, is localized at the pericycle of the root. J. Biol. Chem. 266:4648-4653.

57. HASHIMOTO, TO, YUN, D.-J, YAMADA, Y. 1993. Production of tropane alkaloids in genetically engineered root cultures. Phytochemistry 32: 713-718.

Chapter Three

POLYKETIDE BIOSYNTHESIS: ANTIBIOTICS IN *STREPTOMYCES*

Richard Plater and William R. Strohl*

Department of Microbiology, The Ohio State University
484 West 12th Avenue, Columbus, Ohio 43210 USA

INTRODUCTION

The polyketides comprise a class of secondary metabolites found mainly in bacteria, filamentous fungi and plants. They are defined as molecules assembled by the head to tail condensation of thioesters derived from short chain carboxylic acids, most commonly acetyl-SCoA, malonyl-SCoA, methylmalonyl-SCoA, and ethylmalonyl-SCoA.[1] Despite this common biosynthetic origin, members of this class of natural products include a wide variety of structures ranging from small molecules such as the seven-carbon

---

*Corresponding author

Genetic Engineering of Plant Secondary Metabolism,
Edited by B.E. Ellis *et al.*, Plenum Press, New York, 1994

furan structure, furanomycin,[2] to large macrocycles such as the avermectins, which have as many as 33 carbons of polyketide origin[3] (Fig. 1).

A similar diversity is found in the vast array of biological activities displayed by various polyketides. Among the plant polyketides are several phytoalexins such as 6-hydroxymellein and stilbenes that function to protect the producing species against fungal pathogens.[4-6] Plant polyketides are becoming increasingly recognized for their potential pharmacological significance.[7] Recent examples of such plant-derived pharmacologically active polyketides, 8-hydroxyannonacin, 10,13-*trans*-13,14-*erythro*- and 10,13-*trans*-13,14-*threo*-densicomacin isolated from the stem bark of the Peruvian plant *Annona*

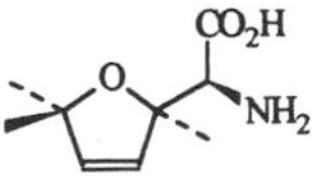

Furanomycin

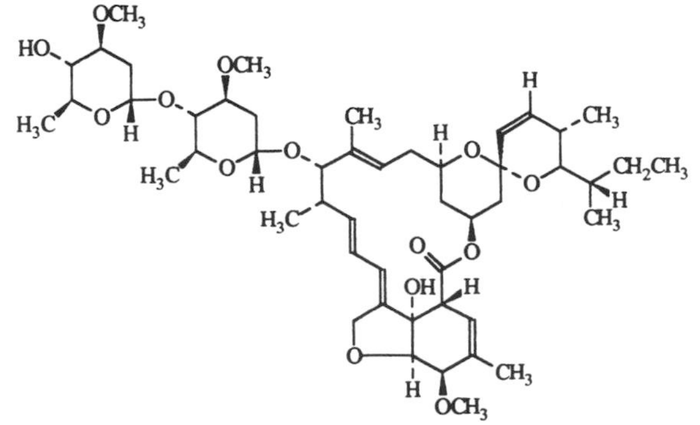

Avermectin A1

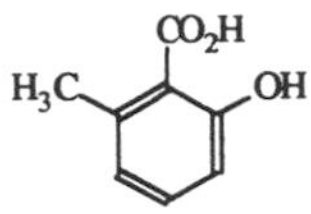

6-Methylsalicylic acid

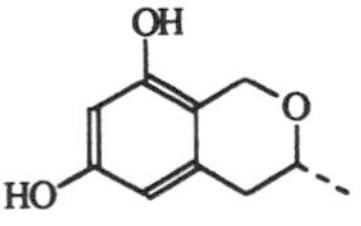

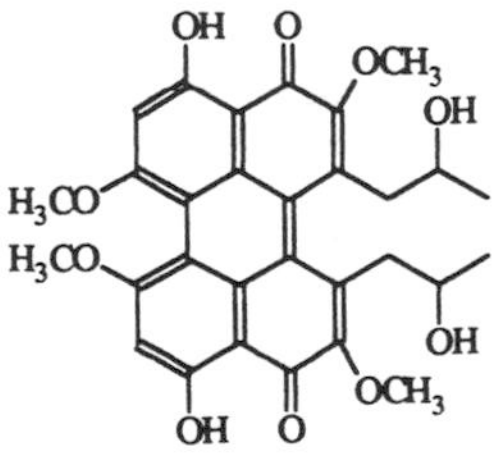

6-Hydroxymellein

Ent-isophleichrome

Fig. 1. Structures of several diverse compounds produced by polyketide synthase mechanisms.

densicoma, have significant cytotoxic activity against human tumor cells.[8]

Phytopathogenic fungi also produce polyketides, such as *ent*-isophleichrome[9] and certain polyketide-derived melanin-like compounds[10] which are important in the pathogenicity of the crop disease causing fungi, *Cladosporium herbarum* and various *Pyricularia* and *Colletotrichum* species. Moreover, fungal-derived aflatoxins, which are economically important because of their toxic properties in food products, are also polyketides.[11] Other fungal polyketides, such as griseofulvin, produced by *Penicillium* species are clinically important antifungal antibiotics.[12]

An impressive array of polyketide natural metabolites are produced by the streptomycetes, which are mycelial, primarily soil-borne bacteria of the actinomycete group that undergo a morphological development from substrate mycelium to aerial mycelium and aerial spore formation as part of their natural life cycle. Concomitant with the morphological differentiation process, these organisms also undergo a biochemical differentiation to produce secondary metabolites. Streptomycetes produce approximately seventy percent of the naturally-derived antibiotics known today, making them among the most important producers of clinically-used natural products. Among the metabolites produced by streptomycetes are several polyketides which possess remarkable antimicrobial activity and thus probably function to impede competing soil inhabitants.[13] At least one such metabolite, pamamycin, however, has been shown to act as a signal molecule that stimulates cellular differentiation.[14] Whatever their function in the physio-ecological context, the biological activities that the streptomycete polyketides possess make them undoubtedly the most commercially important class of bacterially-produced natural products. Polyketides produced by streptomycetes and related organisms include clinically important anti-cancer drugs (e.g., doxorubicin, daunorubicin, aclarubicin[15,16]), anti-bacterial agents (e.g., erythromycin A[17] and the tetracyclines[18]), immunosuppressive agents (e.g., FK506 and rapamycin[19,20]), antihelmithic agents (e.g., the avermectins[21]), insecticides (e.g., tetranactin[22]), livestock feed additives (e.g., tylosin[23]), and herbicides (e.g., herboxidiene[24]). It is clear, therefore, that a fundamental understanding, at both the enzymic and genetic levels, of how the naturally occurring polyketides are produced can be of considerable utility. Such an understanding, coupled with the technology required to manipulate the relevant biosynthetic pathways, should permit the metabolic engineering of microorganisms to produce polyketide metabolites with enhanced biological and pharmacological activities, particularly in the areas of crop/pathogen interactions and drug development. In this article, we endeavor to show that for the streptomycetes, such a detailed understanding of polyketide biosynthesis and its molecular basis is rapidly emerging. Indeed, the first examples of the use of genetic manipulations to engineer biosynthetic pathways leading to predicted and/or desired structures have already been reported.[25-28]

CLASSIFICATION OF POLYKETIDE SYNTHASES

Polyketides are produced by polyketide synthases which, analogous to fatty acid synthases,[1,29] fall into two groups, Type I and Type II. Recent reviews by David Hopwood,[29] Leonard Katz,[30] Peter Leadlay,[31] and John Robinson[32] and their respective collaborators describe in detail the characteristics of both Type I and Type II polyketide synthases and the proposed mechanisms by which they catalyze the synthesis of various polyketides.

Type I polyketide synthases, similar to Type I fatty acid synthases of yeast,[33] consist of very large (subunits of ca. 200,000 M_r), multifunctional enzymes that contain several enzymatic domains, including the distinctive acyl carrier peptide domain. Type I polyketide synthases have been suggested, and in some cases proven, to catalyze the biosynthesis of erythromycin,[34-36] spiromycin,[37] oleandomycin (D.G. Swan et al., Unpublished; GenBank accession no. L09654), avermectin,[38] and tylosin[39] in streptomycetes, 6-hydroxymellein in plants,[4] and 6-methylsalicylic acid[40] and a spore coat pigment[41] in filamentous fungi.

Type II polyketide synthases, on the other hand, like fatty acid synthase of *Escherichia coli*,[42] consist of several distinct proteins, each containing a single or, at most, perhaps a few active domains. The definitive characteristic of both Type II fatty acid synthases and polyketide synthases is that the acyl carrier protein (ACP), which carries the functionally critical 4'-phosphopantetheine prosthetic group, is produced by a separate open reading frame (ORF) and is usually a small (M_r ca. 7,000 to 10,000), acidic protein.[42] Type II polyketide synthases have been shown to catalyze the biosynthesis of actinorhodin,[43] dihydrogranaticin,[44] tetracenomycin C,[45] oxytetracycline,[29] daunorubicin (J.-s. Ye and W.R. Strohl, unpublished results), as well as other molecules currently of unknown structure produced by *Streptomyces coelicolor* (the WhiE product),[46] *Streptomyces halstedii* (a spore pigment),[47] *Streptomyces cinnamonensis*,[48] and *Streptomyces curacoi* (possibly curamycin).[49] The chalcone and resveratrol synthases of plants consist of multiple, distinct proteins, each containing single or a few enzymic domains similar to Type II polyketide synthases, but are apparently evolutionarily distinct, based on amino acid sequence alignments and the location of the active-site cysteines.[50]

COMPARISON OF POLYKETIDE BIOSYNTHESIS TO FATTY ACID BIOSYNTHESIS

When considering the manner in which the carbon backbone of polyketides is believed to be assembled, an informative comparison can be drawn with the well characterized mechanism of fatty acid biosynthesis catalyzed by the fatty acid synthase system.[29,33,42] Fatty acid synthases catalyze the head-to-tail condensation of C_2 units, comprising a starter unit derived from acetyl-SCoA and subsequent extender units derived from malonyl-SCoA, to form long (usually 16 or 18 carbon), often fully-saturated fatty acids. The catalytic cycle of this complex enzyme (Fig. 2) begins with the activity of acetyl-SCoA:ACP S-acetyltransferase (E.C. 2.3.1.38) which loads an acetyl unit onto the prosthetic group thiol of an ACP. This starter unit is then transferred to the active site thiol of a β-ketoacylsynthase domain or protein, leaving the ACP-thiol free to accept the first malonyl extender unit, a reaction catalyzed by malonyl-SCoA:ACP S-acyltransferase (E.C. 2.3.1.39). With the enzyme thus loaded, the chain assembly is initiated by a decarboxylative condensation of the malonyl extender group to the acetyl starter unit, catalyzed by the β-ketoacylsynthase activity, leaving a C_4 moiety attached via thioester linkage to the ACP. The distal carbonyl group of this C_4 chain is then reduced by the successive action of β-ketoacylreductase, dehydratase, and enoylreductase activities.[42] Transfer of the fully saturated carbon chain back to the β-ketoacylsynthase thiol allows the ACP to accept another malonyl moiety, so that a new extension and reduction cycle can be initiated.

Termination of fatty acid chain extension typically occurs after seven or eight catalytic cycles such that C_{16} or C_{18} fatty acyl chains are produced. For both Type I and the Type II fatty acid synthases, termination at this point appears to be governed by a kinetic mechanism. In each case the specificity of the β-ketoacylsynthase activity is such that the rate of chain extension falls progressively and rapidly with each C_2 addition once a C_{16} chain length has been reached.[51-53] The result of this is that a chain terminating event becomes favored over further elongation. For mammalian Type I fatty acid synthases, this termination event is catalyzed by a thioesterase domain within the multifunctional enzyme which releases the assembled chain as the free acid.[33,53] The $E.$ $coli$ Type II and the yeast Type I synthases, on the other hand, employ a palmitoyl-SACP:SCoA S-acyltransferase activity which transfers the completed chain from the ACP-thiol to a coenzyme A thioester.[33,42,51-52] Alternatively, in $E.$ $Coli$ the acyl-ACP moieties are substrates for glycerol phosphate acyltransferases, resulting in direct incorporation of the acyl moieties into phospholipids.[42]

Polyketide assembly, catalyzed by the polyketide synthases, involves a complement of catalytic activities directly analogous to that described above for fatty acid biosynthesis. Polyketide biosynthesis, as a whole, is more complex in three important respects (Fig. 2). First, the various polyketide synthases employ a wide range of starter units. For example, the starter units for actinorhodin, dihydrogranaticin, tetracenomycin C, steffimycin, and monensin are derived from acetyl-SCoA, the starter units for the antitumor drugs, doxorubicin, daunomycin, and aclarubicin are derived from propionyl-SCoA, the starter unit for oxytetracycline is malonamyl-SCoA, the starter unit for avermectin is derived from 2-methylbutyryl-SCoA, and 4-coumaryl-SCoA is the precursor for the starter used by chalcone synthase.

Secondly, polyketide synthase complexes can accept a variety of extender units. For polyketides produced by Type I enzymes, the extender units may be different within different parts of the polyketide chain, each extender assembly being encoded by a series of repeated domains. Malonyl-SCoA, methylmalonyl-SCoA, and ethylmalonyl-SCoA are most often used as the precursors for the extender units,[29-30] although more unusual extender moieties are also employed, such as that derived from succinyl-SCoA used in nonactin biosynthesis.[54] Incorporation of methyl or ethylmalonyl moieties results in a two carbon extension to the polyketide chain with the remaining carbon atoms appearing as alkyl substituent groups. Concomitant with any alkyl substitution of the polyketide chain is the generation of a chiral center. Unlike the fatty acid synthases, therefore, certain polyketide synthase complexes must exercise stereochemical control if the correct (i.e., bioactive) end product is to be reached.

The third major difference between polyketide synthase and fatty acid synthase systems is that polyketide synthases are able to curtail at any point the series of reactions that normally lead to full saturation in the fatty acid synthase catalytic cycle (Fig. 2). By varying the degree of post-condensation processing, polyketide synthases are able to leave keto (when only condensation occurs), hydroxyl (condensation and β-ketoacylreduction), enoyl (condensation, β-ketoacylreduction, and dehydration) or alkyl (condensation, β-ketoacylreduction, dehydration, and enoyl reduction) functionality in the growing polyketide chain. Unlike the inherently iterative process of fatty acid assembly, therefore, the biosynthesis of any given polyketide requires that the catalyzing polyketide synthase system choose correctly from a number of options with respect to starter and extender unit selection and the precise nature, often including stereochemistry, of the processing catalyzed after each condensation event. The relevant information or "programming" must be encoded in the structure of each individual polyketide synthase. A detailed understanding of how this is achieved

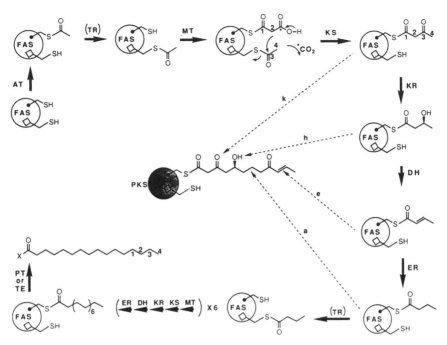

Fig. 2. Comparison of fatty acid synthase mechanism with that proposed for polyketide synthase. The dotted arrows labelled k, h, e, and a indicate the points at which polyketide synthases can curtail the catalytic cycle to permit different functionalities in the carbon chain. Abbreviations: FAS, fatty acid synthase; AT, S-acyltransferase; TR, transferase; MT, malonyl-SCoA:S-ACP S-acyltransferase; KS, β-ketoacyl-synthase; KR, β-ketoacylreductase; DH, dehydrase; ER, enoyl-reductase; PT, palmitoylacyl-transferase; TE, thioesterase; PKS, polyketide synthase. Reproduced, with permission, from the Annual Review of Genetics, Volume 24, ©1990 by Annual Reviews Inc.

has been the long-standing goal of research into polyketide biosynthesis. Furthermore, such an understanding is an absolute prerequisite to any attempts to engineer, by genetic means, polyketide synthases that catalyze specific reaction sequences and thus lead to desired novel structures.

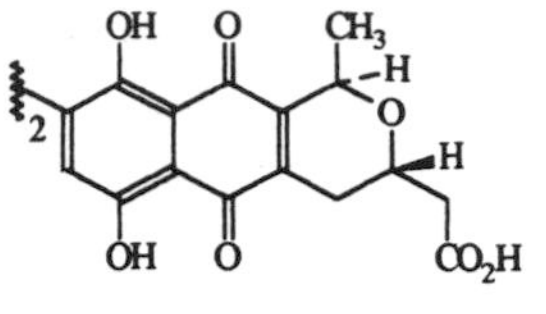

Actinorhodin

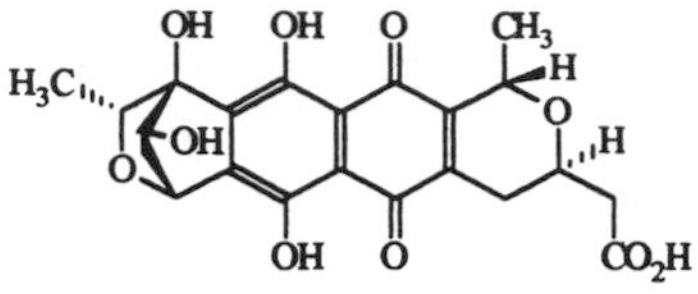

Dihydrogranaticin

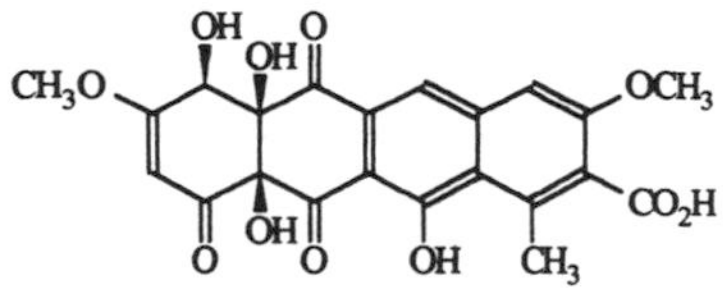

Tetracenomycin C

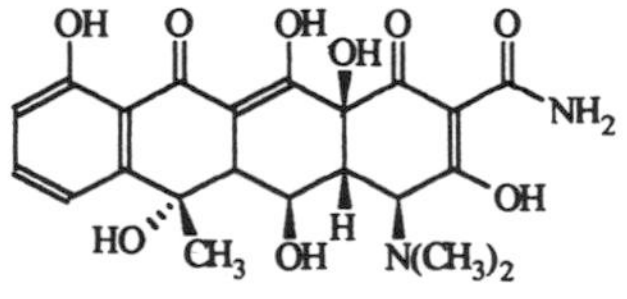

Oxytetracycline

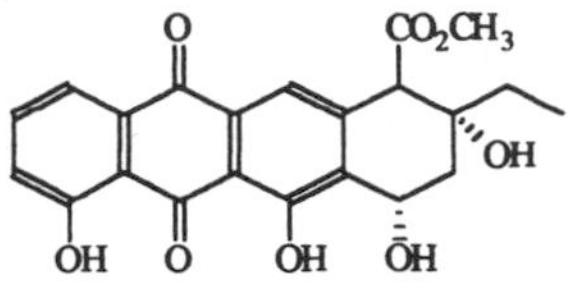

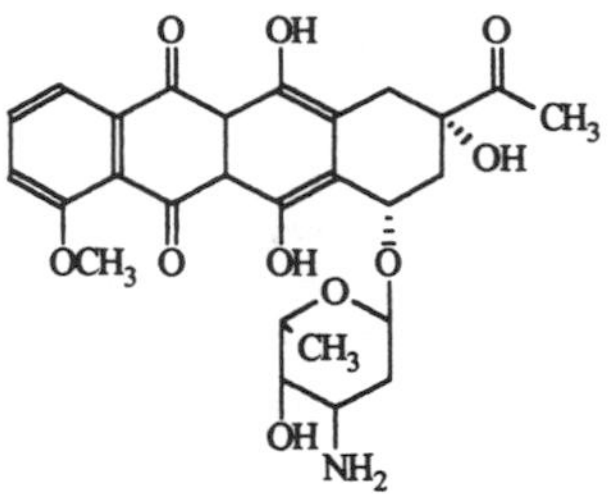

Daunomycin

Aklavinone

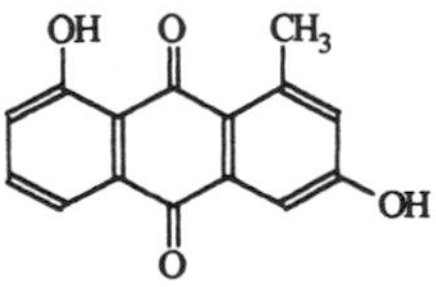

Aloesaponarin II

Fig. 3. Aromatic antibiotic compounds produced by Type II polyketide synthases in streptomycetes.

BACTERIAL TYPE II POLYKETIDE SYNTHASES

The streptomycete aromatic polyketides, actinorhodin, dihydro-granaticin, and tetracenomycin C, (Fig. 3) are each assembled using an acetyl starter unit and malonyl extender units, seven extender moieties for actinorhodin and dihydrogranaticin and nine for tetracenomycin C. In the actinorhodin and dihydrogranaticin systems, the carbonyl group derived from the fourth extender unit must be processed by β-ketoacylreduction and dehydration to give the double bond which is required for a subsequent ring closure event to proceed correctly. In tetracenomycin C biosynthesis no such processing of the polyketide chain is required.

The genes encoding the polyketide synthases for actinorhodin, dihydrogranaticin, and tetracenomycin C have been fully characterized by nucleotide sequencing[43-45] and in each case the complement of activities encoded is entirely consistent with the respective biosynthetic profiles. The actinorhodin and dihydrogranaticin gene regions include ORFs which apparently encode a heterodimeric β-ketoacylsynthase (ORF1/ORF2), an ACP (ORF3), a single (for actinorhodin) or two (for dihydrogranaticin) β-ketoacylreductase(s) and a dehydratase activity[55] (Fig. 4). The latter is in fact encoded as part of a bifunctional cyclase/dehydratase, the cyclase portion of which is required for a subsequent ring closure.[55] The tetracenomycin C gene region contains an analogous arrangement of β-ketoacylsynthase and ACP ORFs (ORFs 1 to 3), but as predicted from the structure and assembly, no reductase or dehydratase-encoding ORFs are present.[45] A cyclase activity required for later ring closure events is present, in this case as part of a presumed bifunctional cyclase/O-methyltransferase[55] (Fig. 4).

Interestingly, in all three of these Type II streptomycete polyketide synthases, the ORF2 encoded component of the presumed β-ketoacylsynthase, although similar over its entire length to other known β-ketoacylsynthases, lacks a cysteine residue known to be required for condensing enzyme activity.[43-45] Despite this, in the case of the actinorhodin system, at least, the ORF2 gene product is an absolute requirement for polyketide synthase function,[56] leading investigators to believe that ORF2 has an important role in chain-length determination and possibly starter unit selection, as discussed later.

The primary structures of these Type II polyketide synthases confirm, in broad terms, a fatty acid synthase-like model for polyketide assembly as described in the previous section. Several key questions, however, particularly with respect to the issue of "programming", remain unclear as discussed in the following sections.

Programming of Starter and Extender Unit Selection

A key aspect of the "programming" of polyketide synthases is their ability to select the correct precursor molecules for use as the starter and extender units. For fatty acid synthases, where the starter unit is derived from acetyl-SCoA and all the extender units are derived from malonyl-SCoA, this may be achieved by the action of specific acetyl-SCoA:ACP and malonyl-SCoA:ACP S-acyltransferase activities. It is striking, therefore, that nowhere in the polyketide synthase gene clusters described above,[43-45] nor in any of the other regions encoding streptomycete Type II polyketide antibiotics sequenced to date (W.R.S., unpublished analysis of sequences available in GenBank), have any potential S-acyltransferase-encoding ORFs been found (Fig. 4). In this respect, it is important to note that for actinorhodin and tetracenomycin C, heterologous expression experiments have established that the sequenced regions encode all the proteins required to direct biosynthesis of these secondary metabolites.

Two separate resolutions to this apparent conundrum seem plausible. The first relates to the observation that each of the β-ketoacylsynthase ORF1 gene products includes in its C-terminal amino acid sequence a G-H-S-X-G motif which is characteristic of acyltransferases.[43] Thus, the ORF1 products may contain two functional domains, one catalyzing the loading of the ACP-thiol with either the starter or extender units, and the second catalyzing the subsequent condensation event. It seems clear that this putative ORF1 "acyltransferase domain" could not be responsible for both starter and extender selection. The high fidelity that all polyketide synthases must display in the selection of different starter and extender units is not consistent with a single acyltransferase active site being responsible for both starter and extender unit selection.

This argument can be extended further by consideration of the results of experiments in which heterologous DNA was used to complement mutants in the actinorhodin β-ketoacylsynthase ORF1.[30] Oxytetracycline is an aromatic polyketide product assembled by the condensation of eight malonyl extender moieties onto a malonamyl starter unit. Use of this starter unit results in an amide side chain on the final fused ring structure (Fig. 3). A complete characterization of the oxytetracycline polyketide synthase has not yet been reported; but initial sequence data have revealed an arrangement of β-ketoacylsynthase and ACP encoding ORFs (ORFs 1 to 3) analogous to that found in the actinorhodin, dihydrogranaticin, and tetracenomycin C systems (Fig. 4).[29] Significantly, a clone carrying the oxytetracycline ORF1 was able to restore actinorhodin production in a non-producing mutant specifically deficient for a functional copy of actinorhodin ORF1.[30] This clearly implies that the

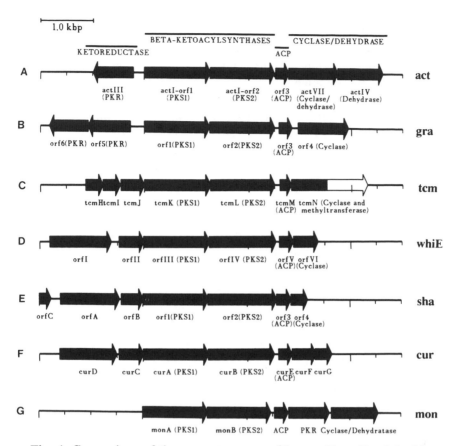

Fig. 4. Comparison of the gene structures of known Type II polyketide synthases based on the nucleotide sequences. A, (act) actinorhodin, produced by *S. coelicolor* strain A3(2) (GenBank accession no. X63449; reference 43); B, (gra) dihydrogranaticin, produced by *S. violaceoruber* strain Tü22 (GenBank accession no. X16144; reference 44); C, (tcm) tetracenomycin C, produced by *S. glaucescens* (GenBank accession no. M80674; reference 45); D, (whiE) *whiE* locus encoding spore pigment of *S. coelicolor* A3(2) (GenBank accession no. X55942; reference 46); E, (sha) *S. halstedii* locus putatively encoding spore pigment (GenBank accession no. L05390; reference 47); F, (cur) locus in *S. curacoi* believed to encode curamycin biosynthesis (GenBank accession no. X62518; reference 49); G, (mon) locus in *S. cinnamonensis* of unknown function (GenBank accession no. Z11511; reference 48).

proposed ORF1 "*S*-acyltransferase" domain, if functional at all, likely catalyzes the loading of the extender units rather than the starter unit. If the reverse were true, the complementation event would have resulted in a "re-programming" of the actinorhodin polyketide synthase with respect to starter unit choice, and thereby yielded a novel compound.

Thus, in this model, the *S*-acyltransferase responsible for starter unit loading remains unaccounted for. In considering this, note that β-ketoacylsynthase III (KAS III) involved in fatty acid synthesis in *E. coli* catalyzes the condensation of malonyl-SACP directly with acetyl-SCoA, rather than requiring that the acetate starter unit first be transacylated to the ACP-thiol.[57] If the polyketide synthases were to carry out condensation in this manner, no starter unit acyl:ACP *S*-acyltransferase would theoretically be required.

A second possible explanation for how precursors are selected and loaded onto the streptomycete Type II polyketide synthases is simply that the relevant activities from the fatty acid synthase complex of the producing organism are able to function in this capacity. This would presumably require that the fatty acid synthases of the polyketide-producing streptomycetes have a Type II organization. Although to date no streptomycete fatty acid synthase complexes have been characterized in detail,[58] the fatty acid synthase from the closely related bacterium *Saccharopolyspora erythraea* (formerly *Streptomyces erythreus*) appears to be a Type II enzyme complex.[59] This explanation can obviously only be satisfactory, however, for polyketides derived from an acetyl starter unit and malonyl extender units. All of the aromatic polyketides shown in Figure 3 are indeed assembled using exclusively malonyl extender units. As described previously, however, oxytetracycline biosynthesis requires a malonamyl starter unit while daunorubicin and other similar anthracyclines are assembled using a propionyl starter unit. For these metabolites, therefore, a specific acyl:ACP *S*-acyltransferase governing starter unit selection, presumably encoded within the polyketide biosynthesis gene cluster, would seem to be required. We have recently discovered in the daunorubicin biosynthesis pathway gene cluster of *Streptomyces* sp. strain C5 an ORF that encodes a protein having high sequence similarity with known acyltransferases (J.-s Ye and W.R. Strohl, unpublished observations). Whether this putative acyltransferase gene may encode a propionyl-SCoA:ACP acyltransferase is currently unknown. If such acyl:ACP *S*-acyltransferases are present, their role in starter unit selection, and therefore in the programming of the polyketide formed, could be crucial. The potential role of such starter unit-specific acyl:ACP *S*-acyltransferases in the programming of novel polyketide metabolites, however, can only be resolved by further analysis.

Programming of β-Ketoacylreduction and Dehydration Events

The ability to select and then modify correctly only the required carbonyl group(s) within a polyketide chain is the second key aspect of polyketide synthase "programming". As described in the introduction to this section, the biosynthesis of both actinorhodin and dihydrogranaticin requires the processing of one carbonyl group, that derived from the fourth extension cycle, by β-ketoacylreduction[60] followed by dehydration. Conceptually, two possibilities exist for how this is achieved. Either the reductive steps occur immediately after the fourth condensation event and before the polyketide chain is further extended (this corresponds to a processive model), or the entire carbon poly-β-keto chain is assembled, followed by the specific reductive step(s).

Traditionally it has been believed that polyketide synthases employ the above processive mode of chain assembly,[61] and certainly this is the most attractive hypothesis when complex polyketides requiring many "programming" choices are considered (discussed further in the section describing Type I polyketide synthases). Recently, however, experimental evidence has suggested that the Type II polyketide synthase systems responsible for actinorhodin and anthracycline (i.e., aklavinone) assembly do not conform to the processive model.[27,62]

The first key observation was made when the gene encoding the actinorhodin β-ketoacylreductase was cloned into a mutant strain of *Streptomyces galilaeus* that produces 2-hydroxy-aklavinone.[27] The recombinant strain produced aklavinone, indicating that the heterologous actinorhodin β-ketoacyl-reductase was responsible for removal of the 2-hydroxyl group from the molecule programmed by the chromosomally encoded polyketide synthase. The inverse experiment showed the corresponding result; i.e., cloning of the *actI-orfs1,2,3*, *actVII* (cyclase/dehydrase), and *actIV* (cyclase) genes (Fig. 4) into wild-type *S. galilaeus* 31133 resulted in formation of the anthraquinone, aloesaponarin II (Fig. 5), in which the molecule was initiated, extended, and cyclized on the basis of the plasmid-encoded enzymes, but β-ketoacylreduction was encoded by the resident chromosomal β-ketoacylreductase gene.[27,62] Since both the actinorhodin and aclarubicin β-ketoacylreductases functioned interchangeably on either polyketide chain (e.g., C_{16}-actinorhodin or C_{21}-aklavinone chains) at a fixed distance from the carboxy terminus of the molecules, it was postulated that the enzymes "measure" back from the carboxy-terminus of the respective polyketide precursors to determine where they must act[27,62] (Fig. 5). Significantly, for this to occur, the polyketide chain must be fully assembled, as the carboxy-terminus is formed last.[29] These data imply that there is low specificity for β-ketoacylreductase activity,[27,62] and that programming of this event requires only that a carbonyl group be present in the proper position with respect to the finally produced polyketide chain.

This non-processive hypothesis for production of polyketides by Type II polyketide synthases derived from the results described above,[62] is attractive in that it allows for the assembly of the carbon backbone of actinorhodin (and thus by deduction, also anthracyclines) by a purely iterative series of extension cycles. This is consistent with the data indicating the presence of only one β-ketoacylsynthase active site within the actinorhodin and dihydrogranaticin polyketide synthases.[43,44] As shall be seen later in this article, Type I polyketide synthases, which catalyze a series of dissimilar extension cycles, possess a separate active site for each condensation event.

Programming of Chain Length Determination

The final consideration with respect to polyketide "programming" is the issue of chain length determination. As described earlier, fatty acid synthase systems specify chain length by a combination of the reduced activity of β-ketoacylsynthase for extensions beyond C_{16} and a competing chain terminating activity (either a palmitoyl-SACP:SCoA S-acyltransferase or a thioesterase[33,42,51-53]). Again, experiments involving the heterologous expression of specific actinorhodin polyketide synthase components have given the first insights into this aspect of "programming" for the Type II polyketide synthases. *S. galilaeus* 31133 normally produces the anthracycline metabolite aklavinone, derived from a propionyl-SCoA starter unit and nine malonyl-derived extender units. When transformed with a plasmid clone carrying the genes for the two components of the actinorhodin β-ketoacylsynthase (ORFs 1 and 2), however, the major product, as mentioned above, was aloesaponarin II,[62] which was proven to be derived from an acetyl starter unit and seven malonyl extender units.[27]

Three important inferences can be drawn from these results with respect to starter unit specificity and chain length determination. Firstly, if, as suggested earlier, a specific acyl:ACP S-acyltransferase determines the choice of a propionyl moiety as the starter unit in aklavinone biosynthesis, then this S-acyltransferase is obviously unable to interact correctly with the hybrid polyketide synthase system (the recombinant ActI-Orfs 1,2 and the host anthracycline ACP, β-ketoacylreductase, ActVII- and ActIV-homologues; see Fig. 5) that is functioning in the recombinant strain. This is apparent from the fact that the hybrid polyketide synthase system does not utilize the available propionyl starter unit, thus indicating that some specificity in the choice of starter unit is encoded within ORFs 1 and/or 2 of the actinorhodin polyketide synthase. Whether this specificity is due to the requirement of certain protein-protein interactions ("docking") with a starter unit-specifying acyltransferase, or to the catalytic specificity of the β-ketoacylsynthase, is currently unknown.

Secondly, we hypothesize that in order for aloesaponarin II to be produced by the recombinant strain, the acetyl starter unit, which would not be provided by the anthracycline polyketide biosynthesis genes, may be provided by the acetyl:ACP S-acyltransferase that normally functions in fatty acid biosynthesis, since an appropriate enzyme would not be present for anthracycline biosynthesis.

Thirdly, the fact that the hybrid polyketide synthase complex, which included only the β-ketoacylsynthase from the actinorhodin polyketide synthase system, follows "actinorhodin-like programming" in the number of extender units added[27] clearly indicates that it is the specificity of the heterodimeric condensing enzyme that dictates chain length. This could possibly be by a mechanism similar to that described for fatty acid synthases, where the specificity of β-ketoacyl synthase causes the rate of chain extension to drop markedly once the required chain length is reached. How the completed polyketide chain is then removed from the "stalled" polyketide synthase is not known, but, this, rather than a role in precursor loading, may be the function of the putative "ORF1 acyltransferase" domain. It will be interesting to see the results of experiments initiated in the Khosla[63] and Sherman[56] laboratories in which the various polyketide synthase units (ORFs 1,2, and ACPs) are swapped for one another using molecular procedures.

Polyketide Chain Folding and Cyclization Events

One final aspect of polyketide biosynthesis that is not directly related to programming of the polyketide synthase systems is, nevertheless, very important with respect to the generation of the correct final molecule. Biosynthesis of the fused ring structures discussed in this section requires not only the accurate assembly of a polyketide chain but also the correct post-assembly folding of that chain. To effect proper cyclization of the assembled polyketide chain, it is expected that the appropriate carbon atoms within the chain must be brought into proximity to ensure that the correct ring closing carbon-carbon bonds may be formed.[62] The majority of the required ring closures occur by the joining of a methylene group and a carbonyl group, presumably via an aldol-type condensation. Subsequently, various dehydration events are required to introduce the required double bonds and aromaticity.[62]

Despite having no direct catalytic role, polyketide synthase functions (the polyketide synthase being defined here as those activities involving polyketide chain assembly and functionalization) may be very important in ensuring that these ring closure events proceed correctly. This assertion derives from studies of one particular actinorhodin non-producing mutant of *S. coelicolor* lacking a functional copy of the *actVII* gene, which putatively encodes a bifunctional cyclase/dehydratase[64] (Fig. 5). Within the actinorhodin

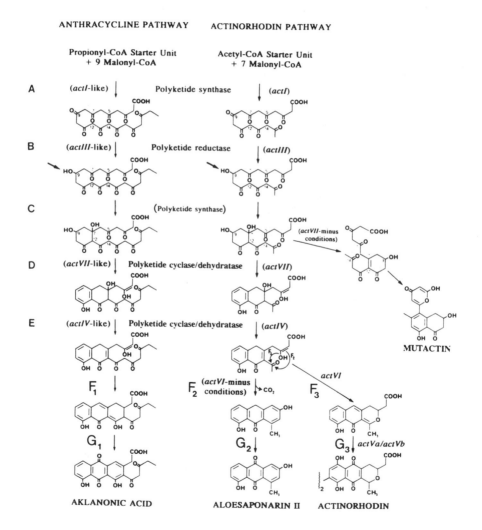

biosynthetic pathway, the entire ca. 22 kilobase gene cluster for which has been sequenced, the *actVII* gene product is the only enzyme that has been assigned a cyclase (aldolase) activity,[43] despite the fact that two ring closing condensations are required. Interestingly, the mutant lacking a functional copy of *actVII* accumulates the shunt product mutactin, the production of which appears to result from the incorrect folding of the polyketide chain after the completion of the first, but in the absence of the second, of two required aldol condensations.[64] After this incorrect folding, an alternative cyclization event, presumed to be spontaneous,[55,62,64] occurs to give mutactin.

These observations suggest that the first ring closure event only requires that the polyketide chain be held in the correct conformation such that it can occur in a spontaneous fashion.[62,64] By contrast the second ring closure must be enzymatically catalyzed, and in the absence of the required activity, the carbon chain folds erroneously, resulting in the observed shunt product. Thus, it seems clear that the conformation of the polyketide chain must be controlled tightly throughout its assembly to preclude the possibility of spontaneous and erroneous cyclization events. Precisely how this is achieved is not currently known, although it seems likely that the polyketide synthase components must contribute to this process by forming a kind of scaffolding to stabilize the correct

Fig. 5. Hypothetical pathways from acetyl-SCoA and propionyl-SCoA to aklanonic acid, a precursor in anthracycline biosynthesis in *S. galilaeus*, *S. peucetius*, and *Streptomyces* sp. C5, and from acetyl-SCoA to aloesaponarin II, mutactin, and actinorhodin in *S. galilaeus* 31133 containing recombinant *actI-orfs 1,2* (β-ketoacylsynthase) from *S. coelicolor*, *S. coelicolor* strain B40 (*actVII⁻*) and *S. coelicolor* A3(2), respectively.[62] The similarities between anthracycline and actinorhodin biosynthesis have been deduced from the ability of recombinant *S. galilaeus*, containing *actI-orfs 1,2* (β-ketoacylsynthase) from *S. coelicolor*, to produce aloesaponarin II. According to these hypothetical pathways, reaction A, generating the polyketide precursor, would be pathway-specific, whereas reactions B-E are carried out by essentially identical functions. Reactions F and G diverge, depending on the pathway, to produce different aromatic polyketide metabolites. Reproduced, with permission, from Molecular Microbiology 6:1723-1728 (1992).

conformation of the growing polyketide chain.[55]

Taken together with the evidence discussed earlier suggesting post-chain reduction by the polyketide reductase, these data indicate that the developing polyketide molecule remains protein bound (and perhaps is transferred from protein to protein without solubilization) throughout the first several processing events, suggesting strong protein-protein interaction not only among the polyketide synthase components, but also encompassing the reductase and cyclization enzymes of the pathway. This is consistent with the concept of "metabolite channelling"[65] as described for several other pathways, most notably for the metabolon of tricarboxylic acid cycle enzymes in yeast mitochondria.[66]

A final point of interest with respect to Type II polyketide synthases is that in addition to the β-ketoacylreductases, the dehydratases and cyclases (encoded by the *actVII* and *actIV* genes or their equivalents in anthracycline biosynthesis; see Fig. 5), as well as the acyl carrier proteins (for actinorhodin, encoded by *actI-orf3*), operative in anthracycline and actinorhodin biosynthesis were found to function interchangeably.[27,62] Cloning of only the *actI-orf*s 1 and 2 (putatively forming a functional heterodimeric β-ketoacylsynthase[62]) into wild-type anthracycline producers results in the formation of the anthraquinone, aloesaponarin II, while cloning of those genes into *S. galilaeus* 31671, lacking polyketide reductase activity, resulted in the formation of 3-hydroxy-aloesaponarin II (known as desoxyerythrolaccin[27,62]). In order for these products to be formed, the resident chromosomally encoded anthracycline dehydratases and cyclases must be able to function normally on a molecule produced and programmed by the actinorhodin β-ketoacylsynthase.[62] This interchangeability of Type II polyketide synthase functions becomes important when the prospects for engineering novel polyketides by generating hybrid polyketide synthases are considered.

BACTERIAL TYPE I POLYKETIDE SYNTHASES

The polyketide metabolites discussed in the last section, encoded by Type II polyketide synthases, require comparatively few programming choices to be made by the catalyzing enzyme systems. From the results obtained with Type II polyketide synthase systems thus far, it appears that there is built-in starter unit specificity, that the extender units required are all identical (i.e., derived from malonyl-SCoA), and that post-condensation, pre-folding processing occurs either just once or not at all.

Biosynthesis of large, macrolide polyketides, on the other hand,

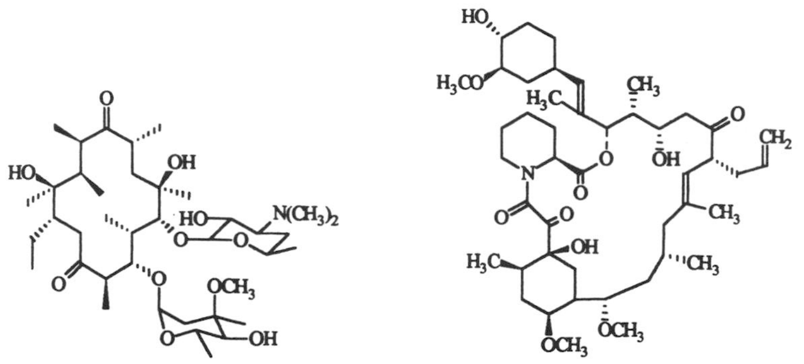

Erythromycin A   FK-506

Fig. 6. Structures of the broad-spectrum antibiotic, erythromycin
A, and the immunosuppressive agent, FK-506, produced by Type
I polyketide synthases in streptomycetes and similar organisms.

involves a large degree of processing, including the incorporation of a variety of
extender units as well as differential processing during assembly of the
molecules. The most complete data currently available describing such a system
are derived from the biosynthesis of 6-deoxyerythronolide-B, the polyketide core
of the clinically important antibiotic, erythromycin A[36,67] (Fig. 6). After
selection of a propionyl starter unit, six elongation cycles are required. The first
and third of these elongation cycles incorporate extender units derived from (2R)-
methylmalonyl-SCoA, while the second, fifth and sixth utilize the (2S)-isomer.
The stereochemistry of the fourth extender unit cannot be deduced from the final
structure and at this point is not known.[31] Furthermore, the first, second, fifth
and sixth condensation events are followed by reduction to leave hydroxyl
functionality, while processing to keto and fully saturated functionality follows
the third and fourth condensations, respectively[36,67] (Fig. 7).

The polyketide synthase catalyzing the assembly of this complex
polyketide has been cloned from the producing organism, S. erythraea, and the
encoding DNA sequenced in its entirety.[36,67,68] The startling observation made
was that, unlike the Type II polyketide synthase gene regions described earlier
that are made up almost exclusively of ORFs encoding monofunctional proteins,
the 6-deoxyerythronolide B polyketide synthase region encodes three large ORFs
for multifunctional polypeptides.[36,67] Computer assisted analysis of the deduced

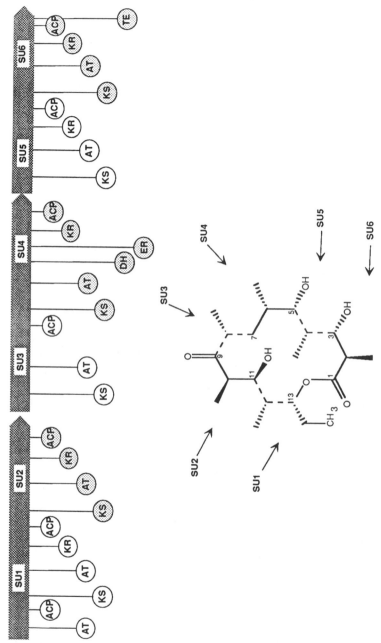

Fig. 7. Model for structure of three ORFS of Type I polyketide synthase encoding the biosynthesis of 6-deoxyerythronolide B, the precursor for erythromycin A. Each ORF shown by an arrow is ca. 370,000 M_r. AT, acyltransferase; ACP, acyl carrier protein; KS, β-ketoacylsynthase; KR, β-ketoacylreductase; DH, dehydrase; ER, enoylreductase; TE, thioesterase. The structures within 6-deoxyerythronolide B for which each subunit is responsible are denoted by arrows. Reproduced with permission.[67]

primary structure of the three polyketide synthase components revealed that each could be divided into two synthase units, so that six units are present in all.[36,67] Each of these synthase units includes the correct complement of acyl:ACP S-acyltransferase, β-ketoacylsynthase and ACP domains required for one of the six condensation events, as well as the relevant β-ketoacylreductase, enoyl reductase and dehydratase domains required for the correct post condensation processing. These multidomain units can thus be named synthase units one through six in accordance with the elongation cycle they putatively catalyze.[30,36,67] In addition, the order in which the units are encoded within the polyketide synthase gene region appears to mirror precisely the order in which the extension steps they catalyze must occur (Fig. 7). An additional acyltransferase domain is encoded within the N-terminal region of the ORF believed to encode synthase units one and two, while the far C-terminal region of the third multifunctional ORF, that is presumed to encode synthase units five and six, contains a deduced thioesterase domain[30,36,67] (Fig. 7).

The clear inference of this analysis is that programming in this Type I polyketide synthase is achieved by its organization into functional modules, each containing an acyl:ACP S-acyltransferase able to select and load the correct precursor, all the activities required to catalyze chain extension with that extender unit, and all the relevant post-condensation processing required for each round of extension. Transfer of the growing polyketide chain between these modules in the correct order would, therefore, theoretically ensure that the required product is reached. The additional acyltransferase domain at the N-terminal end of the first multifunctional protein functions to select the starter unit while the thioesterase domain at the C-terminal end of the third multifunctional protein releases the completed polyketide chain from the enzyme complex.[36,67]

The validity of the model described above was demonstrated by the generation of an in-frame deletion removing the β-ketoacylreductase encoding region from the fifth synthase unit.[36] The product predicted, in which no β-ketoacylreduction would be expected to occur in the fifth extension cycle, namely 5,6-dideoxy-3-α-mycarosyl-5-oxoerythronolide B, was indeed produced by the mutant strain.[36] Thus, not only was a modified drug produced through recombinant genetic engineering, but strong evidence for the processive synthesis of 6-erythronolide B via the Type I polyketide synthase was obtained. From this, it is expected that investigators will be able to determine what changes can be incorporated, via metabolic engineering, in the design of polyketides produced by Type I enzymes.

To date, full sequence analysis has not been reported for any other gene region encoding the polyketide synthase for a complex polyketide. The polyketide synthase for the large macrocylic polyketide avermectin (Fig. 1), produced by *Streptomyces avermitilis*, however, has been extensively characterized by a more targeted sequencing approach.[38] Avermectin biosynthesis requires twelve elongation cycles, seven incorporating extender units derived from malonyl-SCoA and five incorporating methylmalonyl moieties.[3] Two extension cycles require no post condensation processing, five require β-ketoacylreduction, and the remaining five require both β-ketoacylreduction and dehydration.[3] Short segments from throughout the polyketide synthase region for this metabolite have been analyzed and the overall nature of the catalytic domains that they partially encode could thus be deduced.[38] This analysis has revealed that the avermectin polyketide synthase consists of twelve synthase unit encoding modules, consistent with the biosynthetic profile outlined above.[38] The analysis allowed the deduction that the modules are organized into two convergently transcribed groups of six. How the modules are arranged into multifunctional proteins has not, however, been determined.

The picture emerging, therefore, is that complex polyketides such as the macrolides, rather like non-ribosomally synthesized peptides, are assembled by the use of a complex protein template. Programming is achieved by the correct channelling of the extending polyketide between specifically designed synthase units. This channelling is presumably based upon a precise spatial arrangement of the synthase units and on specific protein-protein interactions between them.[30] A detailed understanding of these factors and their influence on the fidelity of complex polyketide assembly will require the physical and biochemical characterization of the these Type I polyketide synthases. Despite this lack of understanding, however, it has already been shown that when the precise event catalyzed by each of the encoded activities within a Type I polyketide synthase can be assigned, then the generation of novel structures by targeted polyketide synthase modification is possible. In the next section, the prospects for rational design of molecules by polyketide synthase engineering, and problems that may be encountered in attempts to do this, will be described further.

PROSPECTS FOR METABOLIC ENGINEERING OF HYBRID POLKETIDE METABOLITES

In the introduction to this review it was stated that the genetic engineering of polyketide synthase systems with the ability to assemble desired and predicted novel structures is rapidly becoming possible. The Type I and Type

II polyketide synthase systems present very different challenges in attempts to achieve these goals. When considering Type II polyketide synthases, two important points must be made. Firstly, it appears that the Type II polyketide synthases are only capable of a limited range of programming choices. In all the examples currently known, an iterative series of condensation cycles using malonyl-SACP extender units is employed and post-condensation processing, when it occurs, is confined to β-ketoacylreduction at one point in the polyketide chain.[29,30,62] Thus, the possibilities for reprogramming may well be limited to starter unit choice and polyketide chain length. Secondly, as has been described, our understanding of how the required programming is encoded within the structure of these Type II polyketide synthases is still far from complete. This is particularly true with respect to how starter unit choice is determined and how polyketide chain folding and cyclization is governed.

Despite these problems, one important observation holds promise for the engineering of Type II polyketide synthases. Numerous experiments, some of which were described earlier, have demonstrated that polyketide synthase components from one *Streptomyces* strain, when introduced into a second strain, are able to form a functional hybrid polyketide synthase with indigenous polyketide synthase components. Thus, once the precise contribution made to polyketide synthase programming by each of the individual components is fully understood, the engineering of novel polyketide synthases by the mixing of polyketide synthase components from different organisms could hopefully proceed with few problems. In this respect it is important to note that methodology enabling the replacement, at the chromosomal level, of specific polyketide synthase components by analogous components from other strains has already been developed.[56,63]

The pertinent question is: how might a more detailed understanding of Type II polyketide synthase programming be achieved? Clearly, the gene clusters encoding a wider range of the known aromatic polyketides must be fully characterized at the genetic level so that the precise role played by each of the encoded ORFs can be investigated by the type of "ORF-swapping" experiments that have already proved so informative. Eventually it will be possible to build up a library of polyketide synthase components, each specifying a known programming choice from which the required activities could be selected and combined. Examples of what this library might contain include a set of *S*-acyltransferases specifying different starter unit choices and a set of β-ketoacylsynthases each specifying a different chain length.

However, two cautionary points must be noted. Firstly, it is not clear that all combinations of polyketide synthase components will be compatible. As described earlier, the actinorhodin β-ketoacylsynthase, when cloned into anthracycline producing strains, did not employ the starter unit specified by the host polyketide synthase. Whether the problem lies in the catalytic specificity of the β-ketoacylsynthase, or in the ability of the host and heterologous polyketide synthase components to interact correctly, is not known.

Secondly, the manner in which polyketide chain folding is dictated is poorly understood. Subtle interactions between the growing polyketide chain and the various polyketide synthase components may be vitally important in this respect. Thus, it is entirely possible that when new combinations of polyketide synthase functions are generated, the polyketide chain assembled will not fold as predicted.

A different set of concerns are relevant when considering the prospects for engineering reprogrammed Type I polyketide synthases. Little doubt can remain concerning how the relevant programming is encoded within these polyketide synthases (namely by the presence of a separate synthase unit for each extension cycle). What is not clear, however, is precisely how it is ensured that the extending polyketide chain passes between these different synthase units in the required order. This channeling of the extending polyketide chain must depend on a precise spatial arrangement of the synthase units.[30] In addition, it would seem clear that certain residues within the synthase units serve to make the contacts with the polyketide chain that channel it in the required direction. Clearly, therefore, any attempts to engineer novel synthase units, by the deletion, addition, or swapping of certain catalytic domains, must preserve those regions of the polypeptides responsible for the required interactions. A further related point is that even if the correct residues are retained it is not clear that those acting "downstream" of any modified synthase unit would be able to interact correctly with the modified chain produced.

As with the Type II polyketide synthases, initial observations with the Type I polyketide synthases suggest that despite the potential pitfalls, successful engineering of Type I polyketide synthases can be achieved. One example of the reprogramming of the erythromycin polyketide synthase has already been reported, i.e., the deletion of the β-ketoacylreductase domain of the fifth synthase unit resulting in the expected novel polyketide.

The recent expression in *E. coli* of the three multifunctional proteins of the erythromycin polyketide synthase should allow a detailed study of the interactions between them by traditional biochemical means.[31] In addition, a systematic program of domain swapping and/or altering experiments should allow a set of ground rules to be drawn up defining what is, and what is not, possible with respect to the reprogramming of these Type I polyketide synthases. It will be particularly interesting to see if synthase units from different organisms can be combined to catalyze novel series of extension cycles, or if duplicates of existing synthase units can be added to specify a longer chain length.

Theoretically at least, the power of this type of technology to produce novel structures is almost limitless. Where the practical limits lie can currently only be hypothesized.

ACKNOWLEDGEMENTS

We thank Dr. Jingsong Ye for sharing unpublished information with us. This work was supported by a National Science Foundation grant MCB-90-19585.

REFERENCES

1.   BIRCH, A.J. 1967. Biosynthesis of polyketides and related compounds. Science 156:202-206.
2.   PARRY, R.J., YANG, N. 1992. Isolation and characterization of furanomycin nonproducing *Streptomyces threomyceticus* mutants. J.Antibiot. 45:1161-1166.
3.   CANE, D.E., LIANG, T.-C., KAPLAN, L., NALLIN, M.K., SCHULMAN, M.D., HENSENS, O.D., DOUGLAS, A.W., ALBERS-SCHÖNBERG, G. 1983. Biosynthetic origin of the carbon skeleton and oxygen atoms of the avermectins. J. Am. Chem. Soc. 105:4110-4112.
4.   KUROSAKI, F., ITOH, M., TAMADA, M., NISHI, A. 1991. 6-Hydroxymellein synthetase as a multifunctional enzyme complex in elicitor-treated carrot root extract. FEBS Lett. 288:219-221.
5.   KUROSAKI, F., KIZAWA, Y., NISHI, A. 1989. Biosynthesis of dihydroisocoumarin by extracts of elicitor-treated carrot root. Phytochem. 28:1843-1845.
6.   SCHÖPPNER, A., KINDL, H. 1984. Purification and properties of a stilbene synthase from induced cell suspension cultures of peanut. J. Biol. Chem. 259:6806-6811.
7.   COLLINGE, M. 1986. Ways and means to plant secondary metabolites. Trends Biotechnol. 4:299-301.

8.   YU, J.G., HO, D.K., CASSADY, J.M., XU, L., CHANG, C.-j. 1992. Cytotoxic polyketides from *Annona densicoma* (Annonaceae): 10,13-*trans*-13,14-*erythro*-densicomacin, 10,13-*trans*-13,14-*threo*densi-coma-cin, and 8-hydroxyannonacin. J. Org. Chem. 57:6198-6202.

9.   ROBESON, D.J., JALAL, M.A.F. 1992. Formation of *ent*-isophleichrome by *Cladosporium herbarum* isolated from sugar beet. Biosci. Biotech. Biochem. 56:949-952.

10. BELL, A.A., WHEELER, M.H. 1986. Biosynthesis and functions of fungal melanins. Annu. Rev. Phytopathol. 24:411-451.

11. DUTTON, M.F. 1988. Enzymes and aflatoxin biosynthesis. Microbiol. Rev. 52:274-295.

12. GROVE, J.F. 1963. Griseofulvin. Quart. Rev. 17:1-19.

13. CHATER, K.F. 1984. Morphological and physiological differentiation in *Streptomyces*. In: Microbial Development, (R. Losick and L. Shapiro, eds.). Cold Spring Harbor Laboratory, Cold Spring Harbor, NY, pp. 89-115.

14. KONDO, S., YASUI, K., NATSUME, M., KATAYAMA, M., MARUMO, S. 1988. Isolation, physico-chemical properties and biological activity of pamamycin-607, an aerial mycelium-inducing substance from *Streptomyces alboniger*. J. Antibiot. 41:1196-1204.

15. STROHL, W.R., BARTEL, P.L., CONNORS, N.C., ZHU, C.-B., DOSCH, D.C., BEALE, J.M. JR., FLOSS, H.G., STUTZMAN-ENGWALL, K., OTTEN, S.L., HUTCHINSON, C.R. 1989. Biosynthesis of natural and hybrid polyketides by anthracycline-producing streptomycetes. In: Genetics and Molecular Biology of Industrial Microorganisms. (C.L. Hershberger, S.W. Queener, and G. Hegeman, eds.), American Society for Microbiology, Washington, D.C., pp. 68-84.

16. HUTCHINSON, C.R. 1993. Anthracycline antibiotics. In: Genetics and Biochemistry of Antibiotic Production. (L.C. Vining and C. Stuttard, eds.), Butterworth-Heinemann Publ. Co., Stoneham, MA, pp. 000-000. (in press).

17. WEBER, J.M., LEUNG, J.O., MAINE, G.T., POTENZ, R.H.B., PAULUS, T.J., DEWITT, J.P. 1990. Organization of a cluster of erythromycin genes in *Saccharopolyspora erythraea*. J. Bacteriol. 172:2372-2383.

18. BUTLER, M.J., FRIEND, E.J., HUNTER, I.S., KACZMAREK, F.S., SUGDEN, D.A., WARREN, M. 1989. Molecular cloning of resistance genes and architecture of a linked gene cluster involved in

the biosynthesis of oxytetracycline by *Streptomyces rimosus*. Mol. Gen. Genet. 215:231-238.

19. KINO, T., HATANAKA, H., HASHIMOTO, M., NISHIYAMA, M., GOTO, T., OKUHARA, M., KOHSAKA, M., AOKI, H., IMANAKA, H. 1987. FK-506, a novel immunosuppressant isolated from a *Streptomyces*. I. Fermentation, isolation, and physico-chemical and biological characteristics. J. Antibiot. 40:1249-1255.

20. PAIVA, N.L., DEMAIN, A.L. 1991. Incorporation of acetate, propionate, and methionine into rapamycin by *Streptomyces hygroscopicus*. J. Nat. Prod. 54:167-177.

21. BURG, R.W., MILLER, B.M., BAKER, E.E., BIRNBAUM, J., CURRIE, S.A., HARTMAN, R., KONG, Y.-L., MONAGHAN, R.L. OLSON, G., PUTTER, I., TUNAC, J.B., WALLICK, H., STAPLEY, E.O., ŌIWA, R., ŌMURA, S. 1979. Avermectins, new family of potent anthelmintic agents: Producing organism and fermentation. Antimicrob. Agents Chemother. 15:361-367.

22. DEMAIN, A.L. 1983. New applications of microbial products. Science 219:709-714.

23. MCGUIRE, J.M., BONIECE, W.S., HIGGENS, C.E., HOEHN, M.M., STARK, W.M., WESTHEAD, J. WOLFE, R.N. 1961. Tylosin, a new antibiotic: I. Microbiological studies. Antibiot. Chemother. 11:320-327.

24. MILLER-WIDEMAN, M., MAKKAR, N., TRAN, M., ISAAC, B., BIEST, N., STONARD, R. 1992. Herboxidiene, a new herbicidal substance from *Streptomyces chromofuscus* A7847.Taxonomy, fermentation, isolation, physico-chemical and biological properties. J. Antibiot. 45:914-921.

25. HOPWOOD, D.A., MALPARTIDA, F., KIESER, H.M., IKEDA, H., DUNCAN, J., FUJII, I., RUDD, B.A.M., FLOSS, H.G., ŌMURA, S. 1985. Production of 'hybrid' antibiotics by genetic engineering. Nature (London) 314:642-644.

26. ŌMURA, S., IKEDA, H., MALPARTIDA, F., KIESER, H.M., HOPWOOD, D.A. 1986. Production of new hybrid antibiotics, mederrhodins A and B, by a genetically engineered strain. Antimicrob. Agents Chemother. 29:13-19.

27. BARTEL, P.L., ZHU, C.-B., LAMPEL, J.S., DOSCH, D.C., CONNORS, N.C., STROHL, W.R., BEALE, J.M. JR., FLOSS, H.G. 1990. Biosynthesis of anthraquinones by interspecies cloning of actinorhodin biosynthesis genes in streptomycetes: Clarification of

actinorhodin gene functions. J. Bacteriol. 172:4816-4826.

28.   EPP, J.K., HUBER, M.L.B, TURNER, J.R., GOODSON, T., SHONER, B.E. 1989. Production of a hybrid macrolide antibiotic in *Streptomyces ambofaciens* and *Streptomyces lividans* by introduction of a cloned carbomycin biosynthetic gene from *Streptomyces thermotolerans*. Gene 85:293-301.

29.   HOPWOOD, D.A., SHERMAN, D.H. 1990. Molecular genetics of polyketides and its comparison to fatty acid biosynthesis. Annu. Rev. Genet. 24:37-66.

30.   KATZ, L., DONADIO, S. 1993. Polyketide synthesis: Prospects for hybrid antibiotics. Annu. Rev. Microbiol. 47:875-912.

31.   LEADLAY, P.F., STAUNTON, J. APARICIO, J.F., BEVITT, D.J., CAFFREY, P., CORTES, J., MARSDEN, A., ROBERTS, G.A. 1993. The erythromycin-producing polyketide synthase. Biochem. Soc. Trans. 21:218-222.

32.   ROBINSON, J.A. 1991. Polyketide synthase complexes: Their structure and function in antibiotic biosynthesis. Phil. Trans. R. Soc. Lond. B 332:107-114.

33.   WAKIL, S.J., STOOPS, J.K., JOSHI, V.C. 1983. Fatty acid synthesis and its regulation. Annu. Rev. Biochem. 52:537-579.

34.   CORTEZ, J., HAYDOCK, S.H., ROBERTS, G.A., BEVITT, D.J., LEADLAY, P.F. 1990. An unusually large multifunctional polypeptide in the erythromycin-producing polyketide synthases of *Saccharopolyspora erythraea*. Nature 348:176-178.

35.   DONADIO, S., STAVER, M.J., McALPINE, J.B., SWANSON, S.J., KATZ, L. 1991. Modular organization of genes required for complex polyketide biosynthesis. Science 252:675-679.

36.   DONADIO, S., KATZ, L. 1992. Organization of the enzymatic domains in the multifunctional polyketide synthase involved in erythromycin formation in *Saccharopolyspora erythraea*. Gene 111:51-60.

37.   GEISTLICH, M., LOSICK, R., TURNER, J.R., RAO, R.N. 1992. Characterization of a novel regulatory gene governing the expression of a polyketide synthase gene in *Streptomyces ambofaciens*. Mol. Microbiol. 6:2019-2029.

38.   MACNEIL, D.J., OCCI, J.L., GEWAIN, K.M., MACNEIL, T., GIBBONS, P.H., RUBY, C.L., DANIS, S.J. 1992. Complex organization of the *Streptomyces avermitilis* genes encoding the avermectin polyketide synthase. Gene 115:119-125.

39.   BALTZ, R.H., SENO, E.T. 1988. Genetics of *Streptomyces fradiae* and

tylosin biosynthesis. Annu. Rev. Microbiol. 42:547-574.

40. BECK, J., RIPKA, S., SIEGNER, A., SCHLITZ, E., SCHWEIZER, E. 1990. The multifunctional 6-methylsalicylic acid synthase gene of *Penicillium patulum*. Its gene structure relative to that of other polyketide synthases. Eur. J. Biochem. 192:487-498.

41. MAYORGA, M.E., TIMBERLAKE, W.E. 1992. The developmentally regulated *Aspergillus nidulans wA* gene encodes a polypeptide homologous to polyketide and fatty acid synthesis. Mol. Gen. Genet. 235:205-212.

42. MAGNUSON, K, JACKOWSKI, S ,ROCK, C.O., CRONAN, J.E., Jr. 1993. Regulation of fatty acid biosynthesis in *Escherichia coli*. Microbiol. Rev. 57:522-542.

43. FERNÁNDEZ-MORENO, M.A., MARTÍNEZ, E., BOTO, L., HOPWOOD, D.A., MALPARTIDA, F. 1992. Nucleotide sequence and deduced functions of a set of cotranscribed genes of *Streptomyces coelicolor* A3(2) including the polyketide synthase for the antibiotic actinorhodin. J. Biol. Chem. 267:19278-19290.

44. SHERMAN, D.H., MALPARTIDA, F., BIBB, M.J., KIESER, H.M., BIBB, M.J., HOPWOOD, D.A. 1989. Structure and deduced function of the granaticin-producing polyketide synthase gene cluster of *Streptomyces violaceoruber* Tü22. EMBO J. 9:2717-2725.

45. BIBB, M.J., BIRO, S., MOTAMEDI, H., COLLINS, J.F., HUTCHINSON, C.R. 1989. Analysis of the nucleotide sequence of the *Streptomyces glaucescens tcmI* genes provides key information about the enzymology of polyketide antibiotic biosynthesis. EMBO J. 8:2727-2736.

46. DAVIS, N.K., CHATER, K.F. 1990. Spore colour in *Streptomyces coelicolor* A3(2) involves the developmentally regulated synthesis of a compound biosynthetically related to polyketide antibiotics. Mol. Microbiol. 4:1679-1691.

47. BLANCO, G., PEREDA, A., MENDEZ, C., SALAS, J.A. 1992. Cloning and disruption of a fragment of *Streptomyces halstedii* DNA involved in the biosynthesis of a spore pigment. Gene 112:59-65.

48. ARROWSMITH, T.J., MALPARTIDA, F., SHERMAN, D.H., BIRCH, A., HOPWOOD, D.A., ROBINSON, J.A. 1992. Characterisation of *actI*-homologous DNA encoding polyketide synthase genes from the monensin producer *Streptomyces cinnamonensis*. Mol. Gen. Genet. 234:254-264.

49. BERGH, S., UHLEN, M. 1992. Analysis of a polyketide synthesis-

encoding gene cluster of *Streptomyces curacoi*. Gene 117:131-136.

50.   LANZ, T., TOPF, S., MARNER, F.-J., SCHRÖDER, J., SCHRÖDER, G. 1991. The role of cysteines in polyketide synthases. Site-directed mutagenesis of resveratrol and chalcone synthases, two key enzymes in different plant-specific pathways. J. Biol. Chem. 266:9971-9976.

51.   GREENSPAN, M.D., BIRGE, C.H., POWELL, G., HANCOCK, W.S., VAGELOS, P.R. 1970. Enzyme specificity as a factor in regulation of fatty acid chain length in *Escherichia coli*. Science 170:1203-1204.

52.   SUMPER, M., OESTERHELT, D., RIEPERTINGER, C., LYNEN, F. 1969. Die Synthese verschiedener Karbonsauren durch den Multienzymkomplex der Fettsauresynthese aus Hefe und die Erklarung ihrer Bildung. European J. Biochem. 10:377-387.

53.   LIBERTINI, L.J., SMITH, S., 1979. Synthesis of long chain acyl-enzyme thioesters by modified fatty acid synthases and their hydrolysis by a mammary gland thioesterase. Arch. Biochem. Biophys. 192:47-60.

54.   ASHWORTH, D.M., ROBINSON, J.A., TURNER, D.L. 1988. Biosynthesis of the macrotetrolide antibiotics: Incorporation of carbon-13 and oxygen-18 labelled acetate, propionate, and succinate. J. Chem. Soc. Perkin Trans. 1:1719-1727.

55.   SHERMAN, D.H., BIBB, M.J., SIMPSON, T.J., JOHNSON, D., MALPARTIDA, F., FERNANDEZ-MORENO, M., MARTINEZ, E., HUTCHINSON, C.R., HOPWOOD, D.A. 1991. Molecular genetic analysis reveals a putative bifunctional polyketide cyclase/dehydrase gene from *Streptomyces coelicolor* and *Streptomyces violaceoruber*, and a cyclase/O-methyltransferase from *Streptomyces glaucescens*. Tetrahedron 47:6029-6043.

56.   SHERMAN, D.H., KIM, E.-S., BIBB, M.J., HOPWOOD, D.A. 1992. Functional replacement of genes for individual polyketide synthase components in *Streptomyces coelicolor* A3(2) by heterologous genes from a different polyketide pathway. J. Bacteriol. 174:6184-6190.

57.   TSAY, J. -T., OH, W., LARSON, T.J., JACKOWSKI, S., ROCK, C.O. 1992 Isolation and characterization of the β-ketoacyl carrier protein synthase III gene (*fabH*) from *Eschericia coli* K-12. J. Biol. Chem. 267:6807-6814.

58.   FLATMAN, S., PACKTER, N.M. 1983. Partial purification of fatty acid synthetase from *Streptomyces coelicolor*. Biochem. Soc. Trans. 11:597.

59. REVILL, W.P., LEADLAY, P.F. 1991. Cloning, characterization, and high-level expression in *Escherichia coli* of the *Saccharopolyspora erythraea* gene encoding an acyl carrier protein potentially involved in fatty acid biosynthesis. J. Bacteriol. 173:4379-4385.

60. HALLAM, S.E., MALPARTIDA, F., HOPWOOD, D.A. 1988. DNA sequence, transcription and deduced function of a gene involved in polyketide antibiotic biosynthesis in *Streptomyces coelicolor*. Gene 74:305-320.

61. CANE, D.E., YANG, C.-C. 1987. Macrolide biosynthesis, 4. Intact incorporation of a chain-elongation intermediate into erythromycin. J. Am. Chem. Soc. 109:1255-1257.

62. STROHL, W.R., CONNORS, N.C. 1992. Significance of anthraquinone formation resulting from the cloning of actinorhodin genes in heterologous streptomycetes. Molec. Microbiol. 6:147-152.

63. KHOSLA, C., EBERT-KHOSLA, S., HOPWOOD, D.A. 1992. Targeted gene replacements in a *Streptomyces* polyketide synthase gene cluster: role for the acyl carrier protein. Molec. Microbiol. 6:3237-3249.

64. ZHANG, H.-I., HE, X.-G., ADEFARATI, A., GALLUCCI, J., COLE, S.P., BEALE, J.M., KELLER, P.J., CHANG, C.-J., FLOSS, H.G. 1990. Mutactin, a novel polyketide from *Streptomyces coelicolor*. Structure and biosynthetic relationship to actinorhodin. J. Org. Chem. 55:1682-1684.

65. EASTERBY, J.S. 1989. The analysis of metabolite channelling in multienzyme complexes and multifunctional proteins. Biochem. J. 264:605-608.

66. SUMEGI, B., MACCAMMON, M.T., SHERRY, A.D., KEYS, D.A., MACALISTER, H.L., SRERE, P.A. 1992. Metabolism of (3-carbon-13) pyruvate in TCA cycle mutants of yeast. Biochem. (USA) 31:8720-8725.

67. DONADIO, S., STAVER, M.J., MCALPINE, J.B., SWANSON, S.J., KATZ, L. 1992. Biosynthesis of the erythromycin macrolactone and a rational approach for producing hybrid macrolides. Gene 115:97-103.

68. BEVITT, D.J., CORTES, J., HAYDOCK, S.F., LEADLAY, P.F. 1992. 6-Deoxyerythronolide-B synthase 2 from *Saccharopolyspora erythraea*. Cloning of the structural gene, sequence analysis and inferred domain structure of the multifunctional enzyme. Eur. J. Biochem. 204:39-49.

Chapter Four

THE BIOLOGIST'S PALETTE: GENETIC ENGINEERING OF ANTHOCYANIN BIOSYNTHESIS AND FLOWER COLOR

Neal Courtney-Gutterson

DNA Plant Technology Corporation
6701 San Pablo Avenue
Oakland, California 94608

INTRODUCTION

Anthocyanin chemistry and biochemistry have been studied for many decades, largely out of a desire to understand the basis of colors produced in a variety of plant parts. The color of flowers has received particular attention. Interest in flower color was focused not only in chemistry and biochemistry, but also in the genetics of flower color in a number of amenable plants, including *Petunia hybrida, Antirrhinum majus* (snapdragon), *Dianthus caryophyllus* (carnation), and *Matthiola incana* (stock). Interest in the genetics of anthocyanin biosynthesis was driven not only by a desire to understand the basis of flower

Genetic Engineering of Plant Secondary Metabolism,
Edited by B.E. Ellis *et al.*, Plenum Press, New York, 1994

color, but also out of an interest to manipulate flower color for commercial purposes. The subject of this discussion is similarly two-fold; fundamental understanding of the factors determining flower color, and control of flower color for commercial purposes can be achieved through genetic engineering.

Although formally the subject of this review is genetic engineering of anthocyanins per se, the focus will be broader, to include the engineering of any factor that can impact flower color produced by anthocyanins. For example, the factors determining blue flower color produced with anthocyanins include the anthocyanidin base, its subsequent chemical modification, interaction with other flavonoids and interaction with ions, particularly protons.[2] Genes which control the synthesis of any molecules which can impact flower color, their isolation and engineering, and the impact of altered expression in transgenic plants are the subject of this review.

An enormous variety of flower color is attributable to flavonoid compounds, from creamy yellow to bright red, from pale pink to blue and violet.[3] Why use genetic engineering to modulate flower color when this range of variation already exists? The reasons are two-fold: 1) to introduce into specific ornamental crops a color type not available through natural variation, and 2) to modify flower color in a specific cultivar which has a commercially valuable combination of agronomic and consumer characteristics. The former approach is one not available to the classical breeder, as the genes necessary for particular colors do not exist within the gene pool of a particular ornamental crop. The latter approach is available to the classical breeder, but the approach is extremely difficult to carry out due to the high degree of heterozygosity carried by many commercially important varieties. Thus, the work of the breeder has become also the work of the molecular geneticist. A molecular genetic approach is not likely to replace classical breeding in standard commercial practice, but for specific purposes this approach can be superior. For example, in a crop such as *Rosa hybrida*, it will be much simpler to produce a pink-flowering rose with desirable commercial characteristics from an existing red-flowering line that already possesses those characteristics than it will be to breed for a pink-flowering derivative with the same desired characteristics. As will be shown below, it will be possible to create a range of flower color intensities in a uniform background; the best flower color can then be selected by the breeder in the desired agronomic and consumer-trait background.

To understand the approaches available to plant molecular geneticists for manipulation of flower color, it is necessary to understand the biochemistry and genetics of pathways for anthocyanin biosynthesis as worked out in several plants.[4-6] This will be the subject of the next section. Two general approaches to modulating flower color--changing the amount of pigments produced or changing the specific pigments and/or copigments produced--are being made currently. These approaches, achieved largely in *Petunia hybrida,* will be described, followed by applications of these approaches to important cut flower crops. In addition to reporting on genetic engineering of anthocyanin biosynthesis to produce plants with modified flower colors, I will present work on the use of anthocyanin biosynthesis in *Petunia* as a model system for studying a recently developed method to control gene expression in plants.

BIOCHEMISTRY AND MOLECULAR GENETICS OF ANTHOCYANIN BIOSYNTHESIS

Two very useful and relevant reviews have appeared recently which summarize the genetics and molecular biology of anthocyanin biosynthesis and color expression.[7,8] Consequently, only a brief overview of genes encoding enzymes of the flavonoid biosynthetic pathway, or which effect expression of color by anthocyanins, will be discussed here. The principal reactions of anthocyanin biosynthesis, as elucidated in *Petunia hybrida, Antrrhinum majus,* and *Zea mays,* are summarized in the schematic of Figure 1 with reference to the production of one anthocyanin, pelargonidin-3-glucoside. Here the pathway is considered as three distinct blocks which can produce overlapping sets of specialized products. The first four steps in the pathway, block 1, are common to a range of different secondary metabolites, and they comprise the core phenylpropanoid biosynthetic pathway. The following step, chalcone synthase, is the first step committed to flavonoid biosynthesis; it initiates the second block of the overall pathway. This portion of the pathway terminates with the synthesis of dihydrokaempferol. A range of side reactions can convert these intermediates into flavonoid products, such as the flavones and flavonols, which can interact with anthocyanins to modify flower color.[9] The final portion of the pathway is specific to anthocyanin biosynthesis, consists of a series of four reactions, and terminates with the addition of glucose to the 3-hydroxyl of an anthocyanidin. With the addition of glucose to the 3-hydroxyl group, the first anthocyanin is produced, as the glycosylation of anthocyanidins is essential for stable expression of color. It should be noted that the anthocyanins accumulate in vacuoles, although most biosynthetic reactions occur in the cytosol.[9]

Three additional sets of reactions are important in determining the final color expressed by an anthocyanin. First, additional hydroxyl groups can be added

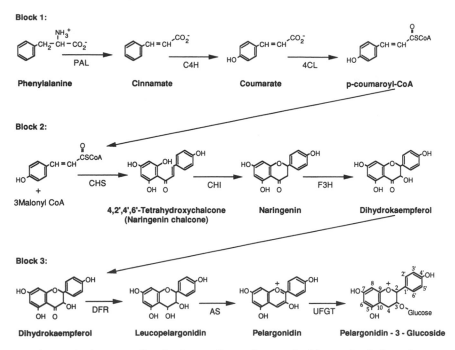

Figure 1. A generalized scheme for anthocyanin biosynthesis based on results obtained from many plants. The individual blocks are discussed in the text. Block 1 includes functions required for the synthesis of the phenylpropanoids; block 2 includes functions required only for the synthesis of flavonoids; block 3 includes functions required for the synthesis of anthocyanins. The numbering system used to refer to positions of the anthocyanin A-ring (2-10) or B-ring (1'-6') is indicated in the final structure of the figure.

to the 3' or 5' positions of the anthocyanidin B-ring. The additional hydroxyl groups cause a shift in maximum absorption, from 520 nm for pelargonidin (monohydroxylated) to 535 nm for cyanidin (3',4'-hydroxylated) to 546 nm for delphinidin (3',4',5'-trihydroxylated). Flavonols can also be synthesized with the same set of different hydroxylations, resulting in kaempferol, quercetin or myricetin. Second, additional glycosyl or acyl groups can be added to the anthocyanin, generally at the 3 and/or 5 positions. Among the sugars which are found as part of anthocyanins are rhamnose, xylose, glucose and galactose.[10] Additions of caffeic, coumaric, malonic, malic, succinic or acetic acids to a sugar moiety added at the 3-position or 5-position have been reported.[11-13] The glycosylations and acylations can alter color as well, as exemplified by the difference in flower color observed in the *rt* and *gf* mutants of *Petunia hybrida* .[4] Finally, the anthocyanin B-ring hydroxyls can be methylated at the 3' and/or 5' positions, with relatively minor effects on flower color.

The anthocyanins present in flowers can produce a range of colors depending upon secondary components within flower vacuoles. These differences in perceived flower color are most significant when delphinidin derivatives are present. For example, delphinidin is associated with the blue color of most ornamental plants; however, blue color is only found when an appropriate copigment (such as glycosylated flavonols) and an appropriate pH are present.[1] At a pH near neutrality, delphinidin gives a blue color, whereas a red color is found with a pH of 5.5 or less. This has been confirmed *in vivo*, using flower color mutants of *Petunia hybrida* .[14]

A number of the genes encoding enzymes of the biosynthetic pathway, or encoding proteins which regulate the expression of these structural genes, have been identified now.[7] Of most relevance to the genetic engineering work presented below are chalcone synthase (CHS) and dihydroflavonol reductase (DFR) genes. The first chalcone synthase gene was isolated in 1985; similar genes have now been isolated from more than a dozen different plant species.[15,16] Often plants are found to have a multiple gene family for chalcone synthase, with as many as 8 different genes in petunia and 2 different genes in maize.[17-19] The first dihydroflavonol synthase gene was isolated also in 1985, and a number of DFR genes have since been isolated from different plant sources.[20-22] There appears to be only a single DFR gene in each of maize, snapdragon and petunia. Several other genes encoding pathway enzymes have also been isolated, including chalcone isomerase and flavonol-3-hydroxyl transferase. Genes encoding enzymes of the phenylpropanoid pathway were also cloned in the 1980's.

MODULATION OF COLOR INTENSITY

Chalcone Synthase, Flavanone-3-hydroxylase, and Dihydroflavonol Reductase

Creating plants that produce flowers of reduced color intensity requires a disruption in a step of the biosynthetic pathway. If the goal is to obtain varieties producing a range of flower colors, then the disrupted step could be virtually any of the steps of the anthocyanin biosynthetic pathway which are general with respect to late modifications (e.g., hydroxylations). Candidate steps are chalcone synthase (CHS), flavanone-3-hydroxylase (F3H), and dihydroflavonol reductase (DFR). Disruption of the chalcone isomerase step is not a good target, since accumulation of chalcone intermediates can result in a yellow color. Thus, blockage of anthocyanin biosynthesis has been achieved using either block 2 or block 3 enzymes as targets. The use of block 1 enzymes as targets might be expected to result in plants with pleiotropic modifications, perhaps in plant responses to pathogen infection. Disruption of expression of the first enzyme of phenylpropanoid biosynthesis (PAL) has resulted in such plants.[23]

If the goal of the approach is to ensure that white-flowering transgenic plants can be produced, it is best to disrupt a step of block 2. The accumulation of flavonols or flavanones, which are synthesized using block 2 intermediates as substrates, can result in cream colored flowers, rather than pure white flowers. Most of the work on altered flower color intensity has focused on disruption of expression of the CHS gene, although work has also been done using the DFR gene. Both sense suppression and antisense suppression have been used.

Initial Demonstration of Antisense Technology in *Petunia hybrida*

During the middle of the 1980's, constructs designed to express anti-sense RNA were found to reduce the expression of endogenous genes with a sequence related to the inroduced gene (e.g. [24,25]). *Petunia hybrida* was selected by Mol and his colleagues to evaluate antisense suppression of anthocyanin biosynthesis in flowers. They had demonstrated earlier[17,26] that petunia contains 8 different CHS genes, only 2 of which are expressed in flowers (CHS-A and CHS-J, at relative levels of approximately 10:1). Van der Krol et al. reported in 1988 that the introduction of a full-length antisense CHS gene construct resulted in suppression of anthocyanin biosynthesis.[27] A range of phenotypes was obtained, including flowers with chaotic white sectors, flowers with uniformly reduced pigmentation, and flowers with nearly white color throughout the corolla. In these experiments, expression of the full-length coding sequence was programmed for expression in the antisense orientation by the 35S promoter from cauliflower mosais virus, a strong constitutive promoter. The alteration in flower pigmentation was attributed to a reduction in chalcone synthase gene expression, by demonstrating that CHS protein and mRNA were reduced in white corolla tissue, by demonstrating that white tissue lacked all flavonoids normally present, and by demonstrating that pigment synthesis could be restored by supplementing with naringenin-chalcone, the product of the CHS reaction. More detailed analysis of pigment levels and gene expression indicated a correlation between the extent of suppression of anthocyanin biosynthesis and the residual level of CHS message (Fig. 2).[28] Co-segregation of altered flower color phenotype and the antisense gene construct was shown for one transgenic individual in backcrosses to the two different parents contributing to the hybrid line used for these studies. All 10 plants with a modified phenotype carried the anti-sense construct, while all 10 plants with normal flower color lacked the anti-sense construct. In these initial experiments, suppression of anthocyanin biosynthesis was also demonstrated in tobacco, which is closely related to petunia, using the same anti-sense constuct.

Although the flower phenotype varied between independent transgenic plants, flower phenotype on an individual plant was usually fairly uniform. The variation observed was in the degree of coloration of flowers expressing a consistent pattern of coloration.[28] Individual branches with different extent of suppression were propagated independently to determine whether a consistent flower color pattern would be produced. However, the range of variation found in the primary transformant was found in the vegetatively-propagated plants, indicating that the variation was due to uncontrollable internal or environmental factors impacting anthocyanin biosynthesis or the antisense suppression phenomenon itself. The products of anti-sense suppression of CHS in petunia would appear to have limited commercial utility, as most transgenic plants produced flowers with asymmetric, irregular flower color patterns. White-flowering transgenic plants, perhaps obtained by screening larger numbers of CHS antisense transformants, could be used for commercial purposes in an appropriate setting.

Sense Suppression of Anthocyanin Biosynthesis in Petunia

Experiments published two years after the above antisense suppression work described an alternative approach to suppressing the expression of chalcone synthase[29] or dihydroflavonol reductase[30] in the production of plants with novel flower color patterns or white flowers. Napoli et al.[29] attempted to enhance the production of anthocyanin pigments in three different petunia lines which ranged from pink to deep purple. A full-length chalcone synthase A gene coding sequence was programmed for expression by the constitutive 35S promoter, and introduced into each of the target lines. No increase in flower color was found; in contrast, some transgenic plants from each parental line had white sectors, a result similar to that obtained by van der Krol et al.[27] using an antisense

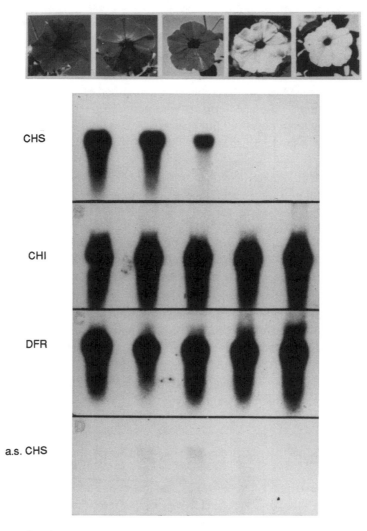

Figure 2. The accumulation of chalcone synthase message correlates with the amount of anthocyanin synthesized in petunia flowers for a set of independent CHS transformants (Reprinted with permission of Springer Productions, van der Krol et al. [28]

Figure 3. The range of flower color phenotypes obtained via sense suppression of chalcone synthase gene expression in petunia

construct. In addition, 3 of 37 transgenic plants derived from one parent, produced only pure white flowers (see Figure 3). As this result was so unexpected, Napoli *et al.*[29] tested the construct which was used in these experiments to determine whether a subtle mutation introduced into the CHS coding sequence could be responsible for the production of white tissue. The coding sequence introduced was identical to the published coding sequence. In addition, when the coding sequence was expressed in *E. coli*, a protein of the expected size was detected immunologically. Suppression of CHS expression was not due to an artifact of the experiment, suggesting the possibility that the presence of an additional, functional copy of the CHS-A gene was able to suppress expression of the endogenous CHS-A gene.

In light of these results, van der Krol et al.[28] introduced a chimeric DFR gene into petunia to obtain suppression of anthocyanin biosynthesis. They succeeded in producing transgenic plants with a range of floral patterns similar to those obtained upon the introduction of a CHS gene, and they reported a reduction in expression of the endogenous DFR gene in lines with reduced pigment synthesis. We have introduced a chimeric DFR gene into petunia line V26 and obtained transgenic plants producing a) pale purple flowers through partial suppression, 2) nearly white flowers, and 3) patterned flowers with purple and pale purple color (unpublished observations). Anthocyanin biosynthesis can thus be modulated by targetting either an enzyme of block 2 (CHS) which is required for synthesis of flavonoids, or an enzyme of block 3 which is required only for anthocyanin biosynthesis.

Since we have used sense suppression to modify flower color in chrysanthemum and rose (see below), it is necessary to describe briefly some features of sense suppression itself. The white-flowering transgenic plants produced through introduction of a CHS transgene into petunia have been analyzed to provide insight into mechanism and utility. After many months of growth, each of the three white-flowering transgenic plants produced by sense suppression was found to give off branches producing only purple flowers identical to the flowers produced in the parental line. Expression of the introduced, chimeric CHS-A gene and the endogenous CHS-A gene was evaluated in white flowers and purple revertant flowers of one primary transformant. The result, shown in Figure 4, indicates that both the transgene and the endogenous gene are expressed at high levels in purple revertant flowers, and at very low levels in white flowers. Similar results have since been obtained with the two other white-flowering lines reported by Napoli *et al.* Thus, the transgene and the endogenous gene are coordinately suppressed (co-suppressed) in the white-flowering, transgenic plants. These data suggested either 1) an interaction between the endogenous gene sequence and the transgene sequence in the genome, or 2) an interaction between the transcripts of these genes, leading to a reduction in cytoplasmic accumulation of each message.

In petunia flowers, two different CHS genes are expressed, with CHS-A contributing approximately 10-fold more total message than CHS-J. Since pure white-flowers were produced by CHS-suppressed plants, it seemed likely that both CHS-A and CHS-J were suppressed. Van der Krol *et al.*[30] demonstrated that both CHS-A and CHS-J were suppressed in a white-flowering plant which they examined. We have measured accumulation of CHS-J message, using RNAse protections, in all four lines obtained using a CHS-A coding sequence in

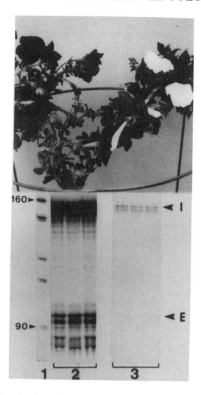

Figure 4. Comparison of steady-state CHS message levels in violet and white flowers derived from a CHS sense-suppressed line. RNase protection assays for violet and white flowers are shown under the corresponding branches of the transgenic individual. Lane 1 is the end-labeled pBR322 molecular weight standard; the arrows indicate 90 and 160 bases. Lanes labeled "2" are RNase protections of RNA isolated from three separate 40-mm-long purple corollas. Lanes labeled "3" are RNase protections of RNA isolated from three separate 40-mm-long white corollas. "E" indicates the position of the protected fragment for the endogenous CHS transcript and "I" indicates the position of the protected fragment for the introduced CHS transcript. (Reprinted with permission of American Society of Plant Physiology, Napoli et al.).[29]

the introduced chimeric gene. In each case, CHS-J message accumulation was reduced by approximately the same extent as the endogenous CHS-A message (Courtney-Gutterson et al., manuscript in preparation). Thus, sense suppression can be obtained for at least two different endogenous genes using a single chimeric gene, even though the CHS-J coding sequence is only 85% identical to the CHS-A coding sequence.[18]

Using petunia as a model, we have posed the question, could transgenic plants with modified phenotypes produced through sense suppression be used reliably for experimental and commercial purposes? Two criteria must be satisfied: 1) uniformity of flower phenotype during prolonged growth; 2) uniformity of flower phenotype amongst progeny, either produced sexually or vegetatively. In other words, sense suppression must be stably transmitted through mitosis as well as through meiosis. Flower color in petunia offers a particularly good assay for changes in suppression, because a) flower color is easily visualized, b) petunia flowers over an extended period of time, and c) partial suppression of CHS in petunia is manifested not as a uniform reduction in flower color but as white and colored sectors. The latter point is important, since a low level of suppression probably would not be detected if expressed as uniform reduction of flower color, but it is easily detected here as a small amount of white tissue on a purple background.

Analysis of the genetic behavior of suppressing transgenes was difficult in the white-flowering plants because suppression of CHS expression resulted in conditional male sterility.[31,32] Homozygous plants could not be obtained initially. The three white-flowering transgenic lines differed considerably in their genetic behavior. When back-crossed with the parental line, line 218.38 reproducibly produced approximately 50% white-flowering progeny. In contrast, progeny of a backcross of line 218.41 with the parental gave consistently 30-50% progeny exhibiting suppressed flowers. Many of the progeny did not give fully white flowers, however. For the third line, 975, an even lower percentage of plants gave fully white flowers. For line 218.38, PCR analysis was performed in one experiment to determine whether the transgene and the phenotype cosegregated. Of 35 progeny, 18 white-flowering progeny all had inherited the transgene and 17 purple-flowering progeny all lacked the transgene. These results indicated that although genetic behavior was variable amongst suppressed transformants, lines could be identified which give good inheritance characteristics. Because homozygous plants were not available, it was not feasible to determine whether inheritance of the modified phenotype was sufficiently reproducible for commercial purposes.

In order to obtain data on a larger population, we worked with primary transformants from which we could obtain homozygotes. When flower color is only partially suppressed, male fertility is partially retained. Two CHS-suppressed primary transformants were self-pollinated, resulting in three classes

Figure 5. Gene dosage effect of sense suppression for two different transgenic individuals. A). The primary transformant 10129.11 (left) and its monozygous progeny (right); B) The primary transformant 10129.31 (top) and its progeny: hemizygous (right); homozygous (left).

of progeny, with either parental flowers or one of two different patterns of flower color (see Figure 5) (Courtney-Gutterson *et al.*, ms in preparation). The two different flower color classes were shown to be due to a gene dosage effect for each line, with homozygous plants having a more extreme suppression phenotype than hemizygous plants. This conclusion was supported by Southern hybridization analysis which indicated no change in DNA structure, and by crosses with the parental line "V26". Using homozygous lines developed from each primary transformant, we obtained progeny from a backcross to the parental line. Over 12,000 progeny were found to produce flowers with very similar phenotypes (see Figure 6), indicating that sense suppression is expressed uniformly from each locus. The expected flower color phenotype was lost from only a few progeny from each line. Southern hybridization analysis revealed that loss of the suppression phenotype was associated with changes in T-DNA

A   B

Figure 6. Uniform expression of sense suppression phenotypes in a large scale inheritance trial. Homozygous individuals derived from two different primary transformants were backcrossed with the parental line "V26" to produce a genetically uniform population. The photographs represent a portion of the plants in flower: a) progeny derived from 10129.11; b) progeny derived from 10129.31.

content. When individual homozygous or hemizygous plants derived from each of the primary transformants were grown for an extended period of time (over a year now), no revertant branches were obtained for either line. These results indicate that sense suppression phenotypes can be maintained well through either mitosis or meiosis. The demonstrated stability is sufficient for many commercial applications in the ornamentals industry.

Applications to Ornamental Species

Chrysanthemum is a commercially important cut-flower crop, ranking third in world-wide sales.[33] Having demonstrated that sense suppression of chalcone synthase could result in transgenic plants producing pure white flowers, we decided to use this approach to produce white-flowering chrysanthemum derivatives. Anti-sense suppression was also used, in order to compare the two methods. The motivation for these experiments was to provide a commercial breeder with a white-flowering relative of a pink-flowering cultivar ("Moneymaker") which has a number of desirable characteristics, including a very short time to produce flowers. A gene introduction method was developed for *Dendranthema morifolium*, which proved to be applicable to a wide variety of chrysanthemum cultivars (Robinson et al., submitted for publication). A cDNA library was prepared using floral mRNA from a commercial chrysanthemum cultivar producing red flowers. Several cDNA clones were identified by hybridization to an oligonucleotide synthesized to match a consensus sequence identified in a number of CHS coding sequences.[16] Several of these were sequenced partially, indicating the presence of at least three alleles (consistent with the hexaploid nature of Florist's chrysanthemum). One of these was sequenced completely, revealing a sequence having a single open reading frame of the expected length, approximately 1100-1150 base pairs.

As was done in petunia, the full-length CHS coding sequence was engineered for expression in plants using the CaMV 35S promoter and a polyadenylation sequence from the nopaline synthase gene. Constructs were prepared with the coding sequence either in the sense or anti-sense orientation relative to the promoter, in order to compare the anti-sense and sense methods of gene suppression. Approximately 100 transgenic plants were produced using each of these constructs, and an additional population of plants was produced using the progenitor vector. Yellow-flowering and dark pink-flowering individuals were obtained with all three constructs, suggesting that these derivatives were variants produced through *in vitro* culture. No white-flowering transgenics were obtained using the control vector; however, white-flowering transgenics were obtained using both the anti-sense and sense vectors (3/83 and

2/133, respectively). An additional pale pink-flowering line was produced using sense suppression. The frequency of obtaining CHS suppression was comparable for the two suppression methods.

If the white-flowering transgenics were created through CHS suppression, the accumulation of CHS precursors would be expected. This was tested by thin layer chromatography of flower extracts, which revealed that both anti-sense and sense-suppressed white-flowering varieties accumulated increased levels of caffeic acid, presumably by hydroxylation of the accumulating coumaric acid. Analysis of endogenous CHS message level was performed by RNAse protection analysis for the 2 white-flowering lines suppressed using sense constructs. In each case, a dramatic reduction in endogenous CHS message was observed (Courtney-Gutterson et al., submitted for publication).

A key requirement for the commercial use of either anti-sense or sense suppression to produce ornamental varieties with modified traits is sufficient stability of the modified trait through either vegetative or sexual propagation. To test this, one anti-sense white-flowering transgenic individual and one sense white-flowering transgenic individual were propagated vegetatively at a moderate scale to test for stability through propagation and during growth of the resulting plants. The tests were conducted at three different sites throughout the United States--in Florida, South Carolina and in California. Approximately 200 cuttings were planted at each site. The results on propagative stability were similar at the three sites, with all plants producing flowers with a suppressed phenotype. However, the number of plants producing flowers with some pigment synthesis varied considerably between the sites, ranging from 2% up to 12%. This variation in anthocyanin biosynthesis has been seen in other chrysanthemum culivars (Ans Otten, personal communication), presumably reflecting a sensitivity to environmental conditions such as temperature and light intensity. The stability of the phenotype produced through sense suppression was comparable to the stability of the phenotype produced through anti-sense suppression.

Gerbera hybrida is another commercially important cut-flower crop, achieving sixth in sales in 1991 through the Dutch auctions.[33] Groups from the University of Helsinki and the company Kemira Oy collaborated to demonstrate the ability to control flower color using anti-sense suppression as a rapid method for introducing novel flower colors into elite Gerbera varieties.[34] Two non-allelic chalcone synthase cDNA clones were isolated from "Terra Regina", a red-flowering variety; good expression was found in petals and sepals. A nearly full length CHS cDNA clone (gchs1), programmed for expression in the anti-sense orientation by the constitutive 35S promoter, was introduced into "Terra Regina". Due to the inefficiency of the transformation system, only 4 mature plants containing the anti-sense construct were obtained. Two transgenic individuals retained the parental flower color, and two had reduced pigment synthesis, resulting in one pink-flowering variety and one cream-flowering variety. Both transformants with modified flower color had significantly reduced CHS message accumulation as determined by northern analysis. CHS enzyme assays of control and the two lines with altered flower color gave results similar to those of the northern analysis.

A sense suppression approach was taken to modify flower color in the most important commercial cut-flower crop, *Rosa hybrida*. Whereas mutation breeding is a useful tool in some plants (e.g., chrysanthemum) for producing color variants with otherwise unaltered plant characters, such an approach is not possible with rose varieties. Consequently, it is currently not possible to produce a family of varieties with different colors but with otherwise the same plant characters. Such families of plant varieties can be produced using a genetic engineering approach which produces plants altered in only a single character. We tested this approach in the rose variety "Royalty", an important commercial variety grown and sold in the United States. Royalty produces very deep red flowers on stems giving generally good performance through storage and giving a fairly good consumer vase life. Our goal was to produce a range of pink-flowering derivatives and a white-flowering derivative. A rose CHS gene coding sequence was isolated from petal total RNA using RNA-based PCR and primers based on conserved sequences of CHS genes from several dicotyledonous plants. A fragment of approximately 850 bp was produced, with the full-length coding sequence expected to be approximately 1150 bp based on known sequences from other plants. The rose CHS sequence was manipulated so that expression programmed by the 35S promoter would produce a transcript containing the coding sequence in-frame for translation.

The CHS construct was introduced into "Royalty" embryogenic calli (Firoozabady et al., ms. in preparation); transgenic calli were regenerated into plants through an embryogenesis step.[35] Over 100 plants were produced, mostly with a T-DNA element containing a neomycin phosphotransferase gene (nptII), the chimeric CHS construct, and the *uidA* gene of *E. coli* as a marker of transformation (GUS, beta-glucuronidase; pFLG10371). A second construct was used to produce a small number of plants which did not carry the *uidA* gene, and which had a different orientation of the nptII gene and the CHS gene (pFLG10381). Fifteen transgenic plants were obtained which produced flowers of

Table No. 1. Anthocyanin content in CHS-suppressed rose. A chimeric CHS gene constructed with approximately 75% of a rose petal chalcone synthase coding sequence was introduced into the rose cultivar "Royalty" using two different plasmid vectors, pFLG10371 and pFLG10381. These vectors differed in that pFLG10381 has an additional gene as a marker of transformation (*beta*-glucuronidase), and in the orientation of the genes within the construct. The anthocyanin level was measured by extracting frozen petal tissue into methanolic hydrochloric acid, and then measuring absorbance at 540 nm.

CLONE NUMBER	% OF PARENTAL
10371.53	41
10371.81	24
10371.140	25
10371.184	20
10371.229	38
10371.281	20
10371.577	17
10381.703	4
10381.711	17
10381.718	3
10381.719	8

reduced color intensity (i.e., pink rather than red). A range of pink colors was obtained, with reductions in anthocyanin level ranging from 59% to 97%. (Table 1) The plants which had the most dramatic reduction in flower color intensity were obtained using the vector pFLG10381. The cause of the difference in results obtained with the two vectors is not presently understood. However, it does indicate that details of T-DNA constructs used for suppression can markedly impact the results quantitatively. Thin layer chromatography was used to demonstrate increased accumulation of *p*-coumaric acid in two of the transgenic plants producing pink flowers. Rose CHS message accumulation was measured in flowers of several pink-flowering and several red-flowering transgenic

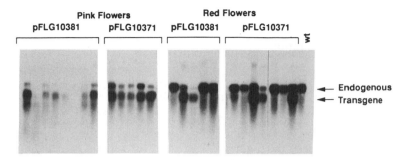

Figure 7. Northern analysis of CHS RNA in transgenic "Royalty" plants. RNA was isolated from corolla tissue, and probed with the partial rose CHS coding sequence introduced into the transgenic plants. The transgenic CHS message and the endogenous CHS message can be distinguished, since a coding sequence fragment was used for the construction of the transgene. RNA isolated from an untransformed "Royalty" flower is shown on the far right.

individuals using northern analysis; reduced accumulation was associated with the pink flowers.(Fig. 7) The plants producing lower levels of anthocyanin also accumulated reduced endogenous CHS message.

The flower color of the transgenic rose lines has been stable for well over a year, and through several flushes of flower production. The production characteristics of these lines are being evaluated further. Overall, this approach appears to be capable of producing a range of flower color, and even some patterning, from which a grower could select the most desirable plants to produce.

No white-flowering transgenic individual was obtained in these experiments, in contrast to the results obtained for CHS suppression in petunia and chrysanthemum. There are a number of possible explantations for this result: a) it could be due to the use of a fragment of the CHS coding sequence, as only a full-length CHS sequence was capable of complete suppression in a series of experiments in petunia (Courtney-Gutterson et al., ms. in preparation); b) the frequency of obtaining white-flowering lines may be low, and an insufficient number of plants may have been screened; c) another CHS gene is expressed in rose flowers with sufficient sequence divergence from the introduced sequence that it is not suppressed.

MODULATION OF FLOWER COLOR

Pelargonidin Production in Petunia

Petunia flowers produce a limited set of anthocyanin pigments. Only dihydroxylated and trihydroxylated B-ring derivatives are produced, with cyanidin or delphinidin as the base anthocyanidins. A very wide range of flower colors can be produced, based on these two pigments, the production of copigments, and covalent modification of the pigments. However, the salmon or orange colors which are found in rose flowers or in pelargonium flowers are not found in any petunia lines.

The basis for the absence of pelargonidin derivatives in petunia was shown by Forkmann and Ruhnau[36] to be the substrate specificity of the dihydroflavonol reductase (DFR) of petunia. The DFR enzyme converts dihyroflavonols into leucoanthocyanidins. When dihydrokaempferol (DHK) was added to petunia flower extracts, no pelargonidin was produced. However, when leucopelargonidin, produced from DHK by DFR, was added to extracts, pelargonidin was produced. A similar result could be obtained by feeding these precursors to intact flower buds. These results suggested that introducing a DFR gene encoding an enzyme able to convert DHK to leucopelargonidin would enable petunia to produce pelargonidin.

The maize DFR gene that encodes an enzyme with the necessary specificity was isolated and introduced into petunia by Meyer et al.[37] The maize DFR coding sequence was programmed to be expressed constitutively using the 35S promoter. The chimeric construct was introduced into a line that has neither 3'-hydroxylase nor 3',5'-hydroxylase activity, so that DHK accumulates in flowers. The normal flower color of this line is nearly white, with a small amount of anthocyanin produced due, apparently, to leakiness of the biosynthetic mutations. A number of the transgenic petunias produced flowers identical to those of the parent; others produced uniformly colored flowers with a new color; and still others produced flowers with white sectors and sectors corresponding to the new color. Flowers with the new color accumulated a pelargonidin derivative with the same type of structural modifications found in the parental line. As predicted, the effective alteration of DFR specificity in petunia resulted in the production of an anthocyanin pigment not normally produced by these genotypes.

Although color modification was successful, some of the transgenic plants produced surprisingly complex flower color patterns. The basis for this variation in flower color was found to lie in the structural complexity of the introduced sequences, and in the methylation state of the introduced sequences.[38-40] An inverse correlation was found between the number of copies of the introduced gene and the production of a new flower color. Methylation of a specific site in the 35S promoter correlated with the number of maize DFR gene copies, suggesting that reduced expression of the maize DFR gene was related to methylation of the 35S promoter. This is similar to results found in a wide range of other transgenic plants, in which promoter methylation correlated inversely with expression level.[41-43] Meyer *et al.* further demonstrated that even a single copy of the introduced gene could fail to express well if it had integrated into a region of the genome that is normally hypermethylated.

Meyer *et al.* identified a transgenic individual that displayed good Mendelian behavior for the new flower color phenotype.[40] This individual was made homozygous, and all backcross progeny in a small test produced uniformly pigmented flowers. A single plant was then propagated vegetatively to allow sufficient amount of seed to be produced for a large scale experiment. Approximately 40,000 hemizygous progeny were sown in the field, and allowed to flower throughout the summer in Germany. Flower color in this population was found to vary, with both genetic and environmental factors appearing to play a role. For example, approximately 5% of the plants produced flowers which were not uniformly colored at any time during growth. In addition, more than 50% of the flowers produced immediately following a period of high temperature and high light intensity were either white or very pale, much more like the parental line than the transgenic line. This result sounded a note of caution in terms of the use of transgenic plants for commercial purpose. Perhaps the inconsistency of the transgenic phenotype is due to normal effects on anthocyanin biosynthesis in the petunia backgound selected. There is significant variation for tolerance to environmental conditions amongst ornamental crops.

The novel flower color created in petunia by the maize DFR gene has been introduced into commercial petunia lines by scientists at Zaadunie, B.V. They have shown that salmon color flowers are produced only in specific genetic backgrounds (Mart van Grinsven, personal communication). In particular, a line producing cyanidin or delphinidin normally produces little pelargonidin when transformed with the maize DFR gene, and there is little impact on flower color. It appears that the maize DFR gene as expressed in the transgenic plants does not produce sufficient activity to compete effectively for DHK as substrate. A hydroxylase enzyme activity is sufficient to channel DHK into cyanidin or delphinidin precursors rather than pelargonidin precursors.

Genes for Creating Blue Flower color

The creation of rose varieties producing blue flowers has been a goal long pursued by rose breeders. Rudyard Kipling even memorialized this pursuit in a poem in which he concluded that the search "was but an idle quest--roses white and red are best".[44] However, the occasional development of new varieties with novel flower colors presumably kept alive the hope that a blue flowering variety would be produced through breeding. For example, before 1929 modern rose plants only produced flowers in a range of red shades, as cyanidin was the only anthocyanidin pigment produced in rose flowers. A flower color variant was obtained in 1929 which produced orange-scarlet flowers due to the synthesis of pelargonidin[45] The new line presumably lacked the 3'-hydroxylase necessary to produce dihydroquercetin from dihydrokaempferol. It is apparent with our current understanding of anthocyanin biosynthesis that a gain of function or functions is required to produce blue color in roses, indicating that a traditional breeding approach will not be successful. For example, Eugster and Marki-Fischer[45] concluded that the "absence of delphinidin derivatives is most important, since the hopes of a blue rose vanishes with them." Not only the rose breeder, but florists as well, would welcome the development of good cut flower crops with blue colors. Most ornamental crops producing blue flowers are not long-lasting, and they are not available year-round. It seems appropriate to consider what genetic engineering steps would be required to create rose or carnation varieties producing blue flowers, and what progress has been made toward this end.

Based on the biochemistry of blue flowers, there are several requirements for producing blue color, assuming that only structurally simple anthocyanins are produced. First, delphinidin must be produced through the activity of a 3',5'-hydroxylase enzyme. Second, appropriate covalent modification of the anthocyanidin base is necessary. It appears that 3,5-bisglucoside is a minimum for blue color, although the shade of blue will be different with different modifications (such as the presence of a coumaryl or cinnamyl moiety added to the sugar at the 3 position of the A-ring). Third, a co-pigment such as a glycosylated flavonol or flavone must be present. Finally, the pH of the vacuolar compartment in which anthocyanins are stored must be near 6.0 or higher. In petunia flowers having appropriate conditions, red flowers are produced when the combined vacuolar and cellular pH is 5.5. However, when the combined vacuolar and cellular pH rises to 5.9, blue flowers result. (Note that vacuolar volume is approximately 90% of the total cellular volume in petunia petals.)

In many rose varieties, two of these four requirements are met. The anthocyanins are bisglucosylated, and large amounts of flavonols are produced.[45]

In such backgrounds, flower vacuolar pH will need to be raised (rose petals have combined cellular pH of 4.0-5.0) and the synthesis of delphinidin will be needed, through expression of a 3',5'-hydroxylase gene. Genes encoding a function required to maintain vacuolar pH and encoding a 3',5'-hydroxylase enzyme have been isolated recently in different laboratories. The genes have been isolated from petunia, which was selected as a model system because of the combination of biochemical understanding of anthocyanin biosynthesis, genetic definition of factors influencing anthocyanin biosynthesis, and the facility of molecular genetic manipulations.

3',5'-hydroxylase. A research group from Calgene Pacific (in Victoria, Australia) has recently reported the isolation of 2 genes encoding 3',5'-hydroxylase activity from petunia (patent filing). They relied on the observation that 3',5'-hydroxylase is a cytochrome P450 enzyme, based on inhibitor studies. An elegant combination of biochemistry, molecular biology and genetics was used in their approach. A series of PCR amplifications and library screenings were performed, using sequences conserved amongst eukaryotic cytochrome P450 genes.[46,47] To evaluate initially whether individual PCR-based cDNA clones could correspond to the 3',5'-hydroxylase, the expression pattern of the corresponding genes was compared to the known expression pattern of the hydroxylase based on enzyme assays. After a couple of rounds of screening and amplification, two candidate clones were identified. Support for the identity of the clones was obtained first by mapping studies. One clone was closely linked to the Hf1 gene, and the other was closely linked to the Hf2 gene. (The Hf1 and Hf2 genes are known to program 3',5'-hydroxylase activity in an overlapping set of plant tissues). Each clone was then programmed for expression in yeast, relying on the ability of yeast to provide the reductase activity necessary for complete function of the cytochrome P450. Extracts prepared from yeast cells expressing each clone had 3',5'-hydroxylase activity, confirming the isolation of these genes. As a final test, the candidate Hf1 clone was shown to enhance 3',5'-hydroxylase activity in a petunia line which normally produced very low amounts .

Ph6 gene. A group working at DNA Plant Technology Corp.[48] has recently reported the isolation of a gene from petunia which is capable of regulating the pH of flowers to shift flower color from blue to red to. The approach selected here was heterologous transposon tagging.[49] The Ac element from maize was introduced into petunia, and lines were developed with a high frequency of transposition. A large number of progeny were produced from such lines, either by self-pollination or after crossing with a petunia line that carries a number of mutations in flower color genes. A new line was derived which, when self-pollinated, produced flowers with a variegated color pattern. Because this line is a hybrid, the variegation appeared in a range of flower color backgrounds, giving either red sectors on a blue background or dark sectors on a paler background. Revertant flowers were identifed during the growth of this line. The pH of revertant flowers was compared with that of the variegated flowers, indicating that the mutation resulted in an increase of approximately 0.4 pH units in petal cell extracts. Genetic analysis indicated a perfect correlation between the variegated phenotype and the presence of a new Ac element in the derived line. The sequences adjacent to the Ac element were cloned by making use of Ac homology. Using only the adjacent sequence as a probe, bands of the same size were obtained in Southern hybridizations as those obtained using the Ac element when probing variegated progeny. Allelism tests were carried out using four different known ph mutants of petunia, ph1, ph2, ph4 and ph6. Restoration of normal pH was obtained in these crosses, except for the cross with the ph6 mutant, indicating that the variegated phenotype and pH change were created by an Ac insertion in the Ph6 gene.

In order to produce blue flower color in a plant such as rose, the pH of the vacuole could be raised by eliminating expression of a Ph6-homologous gene. To do this, the Ph6 gene of petunia must be used as a probe to identify the rose homolog. Additional experiments are now in progress to determine how the Ph6 gene regulates vacuolar pH.

CONCLUSION

The long-term goal of much of the work described here is to develop tools which enable a molecular breeder to create in plants of a given flower color, a flower color which is more desirable for any number of reasons. To date, only a few tools have been assembled, and only a few examples of color conversions have been obtained. In addition to the genes encoding functions determining flower color and the tools for regulating gene expression, an understanding of the pigment biochemistry of any given ornamental species--and in fact, any particular cultivar--is essential to approach color modification as an *engineering* enterprise and not as an experimental exercise.

Flower color intensity. The work presented above demonstrates that modulation of flower color intensity, particularly from highly pigmented flowers to less pigmented flowers, is of general utility. Four different ornamental crops have been manipulated in this way--petunia, chrysanthemum, gerbera and rose.

The work reported in rose has been particularly important in that it indicates that flower color intensity can be varied over a wide range in a single experiment. This approach should be fairly straightforward to apply in other systems as well, since a) chalcone synthase gene sequences are highly conserved in a number of evolutionary diverse plants, and b) modulation of a single enzyme activity is sufficient to regulate flower intensity. Molecular breeding can deliver a range of flower color intensities in identical genetic backgrounds, from which a producer can select the most desirable color, or a set of varieties producing a range of colors. Similar work has been performed using dihydroflavonol reductase suppression in petunia, in which both uniform reduction of flower color intensity and flower color patterns have been produced.

One complication with this approach is that the plant can possess multiple chalcone synthase genes, with more than one copy expressed in the corolla. This is the case in petunia, but it has been shown that both expressed copies are suppressed when a single CHS gene sequence is used to suppress anthocyanin biosynthesis. In the case of rose, the failure to obtain completely white flowers could be due to the expression of a second CHS gene which is sufficiently different from the introduced gene sequence that it is not suppressed. If this were the case, rose varieties producing white flowers could be obtained by introducing different sequences targeting the specific genes. Since there is generally only a single DFR gene expressed in corolla tissue, suppression with an isolated DFR gene is more likely to give complete suppression of anthocyanin biosynthesis. However, DFR gene sequences are less highly conserved than CHS sequences, making it potentially more difficult to isolate a new DFR gene. In addition, the complete suppression of DFR still may not produce a completely white flower, as flavonols can still accumulate and give rise to a creamy yellow color.

From the work presented here, it can be seen that either antisense or sense suppression methods can be used to suppress expression of genes required for anthocyanin biosynthesis. The more recently developed sense suppression method has been shown to be useful in a range of other plants and for a range of other genes as well.[50-53]

Another aspect of controlling flower color intensity is the conversion of white or pale flowers to more intense flower color of essentially the same hue. This approach requires that the step responsible for the failure to produce anthocyanins is identified, and that a construct that expresses the necessary gene is introduced. This should result in a range of flower color intensities, as for suppression of anthocyanin biosynthesis; such an approach has not been successfully demonstrated yet.

Flower color hue. Alteration of the hue of flowers can be achieved either by suppression of a step of the biosynthetic pathway or by extending the pathway existing in a given plant by producing a pigment or copigment not produced in the original variety. In either case, specific elaborations of the general anthocyanin biosynthetic pathway present in any given species (and cultivar) should be understood prior to attempting manipulation. This is not a requirement for reduction of flower color intensity, as chalcone synthase and dihydroflavonol reductase are common to all pathways. Suppression of a step to modify flower hue has not been demonstrated yet, but with the tools being developed, such approaches can be developed easily. For example, the synthesis of dihydroquercetin (DHQ) from dihydrokaempferol (DHK) requires the action of a 3'-hydroxylase. DHQ is the precursor of cyanidin (which is responsible for red and pink color) whereas DHK is the precursor of pelargonidin (which is responsible for orange colors). By suppressing completely the expression of 3'-hydroxylase, an orange-flowering variety can be produced from a red-flowering, cyanidin--producing variety. Partial suppression would lead to orange-red flowers with a range of possible colors.

Alteration of flower color hue through extension of the normal pathway found in a plant was demonstrated already several years ago through the production of pelargonidin in petunia. This is an excellent example of the requirements for this approach, as the biochemistry of anthocyanin biosynthesis was well understood in petunia. It is also an illustrative example, since the genetic background of the particular petunia variety into which a maize DFR gene was introduced significantly impacted the final color obtained. The initial experiment was performed with a petunia line mutated for both 3'-hydroxylase and 3',5'-hydroxylase, and which therefore accumulated dihydrokaempferol. Since dihydrokaempferol was still a precursor for the synthesis of pelargonidin, high levels of pelaragonidin production were achieved, and a good orange/salmon color was obtained. However, when the maize DFR gene was introduced into a variety having intact 3' and 3',5'- hydroxylase functions, very little pelargonidin was produced and there was little detectable difference in flower hue. Clearly, alteration of anthocyanidin biosynthesis was more difficult when the change required competition for a common precursor.

A very specific example of flower hue modification would be the creation of a rose variety producing blue flowers from one producing orange or red flowers. The range of possible colors exhibited by flowers containing delphinidin is large, depending upon the secondary modifications of delphinidin (e.g., glycosylation and acylation), the presence of copigments and the vacuolar pH. Metal ions can also complex with anthocyanins to influence final color. Thus, although the essential molecular tools for creating blue-flowering roses or carnations are in hand, the specific shade of blue produced when delphinidin is produced in a flower containing flavonols and the pH of the flower is raised is still uncertain. Since there would be variation in expression of a 3',5'-hydroxylase gene and also in suppression of a gene required for maintenance of low vacuolar pH upon appropriate gene introductions, a range of violet to blue shades is a likely result. The range of color may be the best outcome for commercial purposes, since the producer can select the varieties giving the most desirable colors.

The control of flower color is more complex than simply the control of the anthocyanin biosynthetic pathway, as a number of factors can influence the hue produced by a specific anthocyanin. Nonetheless, the engineering of reduced color confirms our understanding of the role of chalcone synthase and dihydroflavonol reductase in anthocyanin biosynthesis. The proof of our understanding of anthocyanin biosynthesis and control of flower color lies in our ability to predict the outcome of a specific gene introduction. For most desired color modifications this is not yet possible; hence, molecular geneticists attempting to manipulate flower color cannot yet think of themselves as an *engineers*. However, with the ability to introduce genes into a range of ornamental crops already established, and the isolation of additional genes involved in anthocyanin biosynthesis being actively pursued, the biologist's palette is rapidly filling.

ACKNOWLEDGEMENTS

Funding for some of the previously unpublished results reported here was provided by Florigene B.V. Graphics and artwork were produced by Joyce Hayashi and Rob Narbares. I would like to thank Rich Jorgensen for many fruitful discussions, and for introducing me to this research area originally. Many others have contributed to the unpublished work described here, including Karen Ruby, Eunice Ashizawa, Bill Tucker, Alison Morgan, Ebrahim Firoozabady, Karol Robinson, Caroline Napoli, Christine Lemieux, Chaney Nijjar, Joseph Matthew, Brant Girard, and York Moy. Cara Velton and Jay Maddox provided excellent care of the various plants produced throughout this work.

REFERENCES

1. BROUILLARD, R., 1988. Flavonoids and flower colour, In The Flavonoids, J.B. Harborne, editor, Chapman and Hall, London, pp. 539-55.

2   BROUILLARD, R., WIGAND, M.-C., AND CHEMINAT, A., 1990. Loss of colour, a prerequisite to plant pigmentation by flavonoids. Phytochem, 29: 3457-60.

3. TIMBERLAKE, C.F., AND BRIDLE, P., 1975. The anthocyanins, In The Flavonoids, Part 1, Harborne, J.B., Mabry, T.J., and Mabry, H., eds., Academic Press, Inc., New York, pp. 214-66.

4. WIERING, H., AND DE VLAMING, P., 1984. Genetics of pollen and flower color, In Monographs on Theoretical and Applied Genetics 9: Petunia, K.C. Sink, ed. Springer-Verlag, Berling, pp. 49-67.

5. MARTIN, C., CARPENTER, R., COEN, E.S., AND GERATS, T., 1987. The control of floral pigmentation in *Antirrhinum majus*, In Developmental Mutants in Higher Plants, H. Thomas and D. Grierson, eds., Cambridge University Press, Cambridge, pp. 19-52.

6. COE, E.H., NEUFFER, M.G., HOISINGTON, D.A., 1988. The genetics of corn, In Corn and Corn Improvement, G. F. Sprague and J.W. Dudley, eds., ASA, Madison, WI, pp. 81-258.

7. DOONER, H.K., ROBBINS, T.P., AND JORGENSEN, R.A., 1991. Genetic and developmental control of anthocyanin biosynthesis. Annu Rev Genet, 25: 173-99.

8. FORKMANN, G. 1991. The formation of the natural spectrum and its extension by genetic engineering. Plant Breeding, 106: 1-26.

9. GRISEBACH, H., 1980. Recent developments in flavonoid biosynthesis, In Pigments in Plants, Czygan, C.F., ed., Gustav Fischer, Stuttgart, pp. 187-209.

10. HELLER, W., AND FORKMANN, G., 1988. Biosynthesis, In The Flavonoids, J.B. Harborne, ed., Chapman and Hall, London, pp. 399-425.

11. HARBORNE, J.B., 1986. The natural distribution in angiosperms of anthocyanins acylated with aliphatic dicarboxylic acids. Phytochem, 25: 1887-94.

12. YAMAGUCHI, M.-A., TERAHARA, N., AND SHIZIKUISHI, K.-I., 1990. Acetylated anthocyanins in *Zinnia elegans* Acetylated anthocyanins in *Zinnia elegans* flowers. Phytochem, 29: 1269-10.

13. TAKEDA, K., HARBORNE, J.B., AND SELF, R., 1986. Identification

and distribution of malonated anthocyanins in plants of the Compositae. Phytochem, 25, 1337-42.

14. DE VLAMING, P., SCHRAM, A.W., AND WIERING, H., 1983. Genes affecting flower colour and pH of flower limb homogenates in *Petunia hybrida*. Theor Appl Genet, 66: 271-78.

15. REIF, H.J., NEISBACH, U., DEUMLING, B., AND SAEDLER, H., 1985. Cloning and analysis of two genes for chalcone synthase from *Petunia hybrida*. Mol Gen Genet, 199: 208-15.

16. NIESBACH-KLOSGEN, U., BARZEN, E., BERNHARDT, J., ROHDE, W., SCHWARZ-SOMMER, Z., REIF, H.J., WIENAND, U., AND SAEDLER, H., 1987. Chalcone synthase genes in plants: a tool to study evolutionary relationships. J. Mol Evol, 26: 213-25.

17. KOES, R.E., SPELT, C.E., MOL, J.N.M., AND GERATS, A.G.M., 1987. The chalcone synthase multigene family of Petunia hybrida (V30): Sequence homology, chromosomal localization and evolutionary aspects. Plant Mol Biol, 10: 159-69.

18. KOES, R.E., SPELT, C.E., VAN DEN ELZEN, P.J.M., AND MOL, J.N.M., 1989b. Cloning and molecular characterization of the chalcone synthase multigene family of *Petunia hybrida*. Gene, 81: 245-57.

19. FRANKEN, P. NIESBACH-KLOSGEN, U., WEYDEMANN, U., MARECHAL-DROUARD, L., SAEDLER, H., AND WIENAND, U., 1991. The duplicated chalcone synthase genes *C2* and *Whp* (*white pollen*) of *Zea mays* are independently regulated; evidence for translational control of *Whp* expression by the anthocyanin intensifying gene *in*. EMBO J., 10: 2605-12.

20. MARTIN, C., CARPENTER, R., SOMMER, H., SAEDLER, H, AND COEN, R., 1985. Molecular analysis of instability in flower pigmentation of *Antirrhinum majus*, following isolation of the *pallida* locus by transposon tagging. EMBO J, 4: 1625-30.

21. O'REILLY, C., SHEPHERD, N., PEREIRA, A., SCHWARZ-- SOMMER, Z., BERTRAM, I., ROBERTSON, D.S., PETERSON, P.A., SAEDLER, H., 1985. Molecular cloning of the A1 locus of *Zea mays* using the transposable elements En and Mu1. EMBO J, 4: 877-82.

22. BELD, M., MARTIN, C., HUITS, H., STUITJE, A.R., AND GERATS, A.G.M., 1989. Flavonoid synthesis in *Petunia hybrida*: Partial characterization of dihydroflavonol-4-reductase genes. Plant Mol Biol, 13: 491-502.

23. ELKIND, Y., EDWARDS, R., MAVANDAD, M., HEDRICK, S.A., RIBAK, O., DIXON, R.A., and LAMB, C.J. 1990. Abnormal plant development and down-regulation of phenylpropanoid biosynthesis in transgenic tobacco containing a heterologous phenylalanine ammonia-lyase gene. Proc Natl Acad Sci, USA, 87: 9057-61.

24. ECKER, J.R., AND DAVIS, R.W., 1986. Inhibition of gene expression in plant cells by expression of antisense RNA. Proc Natl Acad Sci, USA, 83: 5372-76.

25. SANDLER, S.J., STAYTON, M., TOWNSEND, J.A., RALSTON, M.L., AND BEDBROOK, J.R., 1988. Inhibition of gene expression in transformed plants by antisense RNA. Plant Mol Biol, 11: 301-10.

26. KOES, R.E., SPELT, C.E., AND MOL, J.N.M., 1989a. The chalcone synthase multigene family of *Petunia hybrida* (V30): Differential, light-regulated expression during flower development and UV light induction. Plant Mol Biol, 12, 213-25.

27. VAN DER KROL, A.R., LENTING, P.E., VEENSTRA, J., VAN DER MEER, I.M., KOES, R., GERATS, G.M., MOL, J.N.M., AND STUITJE, A.R., 1988. An anti-sense chalcone synthase gene in transgenic plants inhibits flower pigmentation. Nature, 333: 866-69.

28. VAN DER KROL, A.R., MUR, L.A., DE LANGE, P., GERATS, A.G.M., MOL, J.N.M., AND STUITJE, A.R., 1990b. Antisense chalcone synthase genes in petunia: visualization of variable transgene expression. Mol Gen Genet, 220: 204-12.

29. NAPOLI, C., LEMIEUX, C., AND JORGENSEN, R. 1990. Introduction of a chimeric chalcone synthase gene into petunia results in reversible co-suppression of homologous genes *in trans*. The Plant Cell, 2: 279-89.

30. VAN DER KROL, A.R., MUR, L.A., BELD, M., MOL, J.N.M., AND STUITJE, A.R. 1990a. Flavonoid genes in petunia: addition of a limited number of gene copies may lead to a suppression of gene expression. The Plant Cell, 2: 291-99.

31. TAYLOR, L.P., AND JORGENSEN, R., 1992. Conditional male fertility in chalcone synthase mutants of petunia and maize. J Hered, 83: 11-17.

32. MO, Y., NAGEL, C., AND TAYLOR, L.P., 1992. Biochemical complementation of chalcone synthase mutants defines a role for flavonols in functional pollen. Proc Natl Acad Sci, USA, 89: 7213-17.

33. THE INTERNATIONAL FLORICULTURE QUARTERLY REPORT, April 1992, 3: 56-62.

34. ELOMAA, P., HONKANEN, J., PUSKA, R., SEPPANEN, P., HELARIUTTA, Y., MEHTO, M., KOTILAINEN, M., NEVALAINEN, L, AND TEERI, T.H., 1993. Agrobacterium-mediated transfer of antisense chalcone synthase cDNA to *Gerbera hybrida* inhibits flower pigmentation. Bio/Tech, 11: 508-11.

35. NORIEGA, C., AND SONDAHL, M.R., 1991. Somatic embryogenesis in hybrid tea roses. Bio/Technol, 9: 991-93.

36. FORKMANN, G., AND RUHNAU, B., 1987. Distinct substrate specificity of dihydroflavonol-4-reductase from flowers of *Petunia hybrida*. Z. Naturforsch, 42c: 1146-48.

37. MEYER, P. HEIDMANN, I. FORKMANN, G., AND SAEDLER, H., 1987. A new petunia flower colour generated by transformation of a mutant with maize gene. Nature, 330: 677-78.

38. LINN, F., HEIDMANN, I., SAEDLER, H., AND MEYER, P., 1990. Epigenetic changes in the expression of the maize A1 gene in *Petunia hybrida*: role of numbers of integrated gene copies and state of methylation. Mol Gen Genet, 222: 329-36.

39. MEYER, P., LINN, F., HEIDMANN, I., AND SAEDLER, H. 1990. Engineering of a new flower color variety of petunia, Plant Gene Transfer, Alan R. Liss, Inc., pp. 319-26.

40. MEYER, P., LINN, F., HEIDMANN, I., MEYER, H.Z.A., NIEDENHOF, I., AND SAEDLER, H. 1992. Endogenous and environmental factors influence 35S promoter methylation of a maize A1 gene construct in transgenic petunia and its colour phenotype. Mol Gen Genet, 231: 345-52.

41. JOHN, M.C., AND AMASINO, R.M., 1989. Extensive changes in T-DNA methylation patterns accompany activation of a silent T-DNA *ipt* gene in *Agrobacterium tumefaciens*-transformed plant cells. Mol and Cell Biol, 9: 4298-4303.

42. MATZKE, M.A., PRIMING, M., TRNOVSKY, J., AND MATZKE, A.J.M., 1989. Reversible methylation and inactivation of marker genes in sequentially transformed tobacco plants. EMBO J, 8: 643-49.

43. RIGGS, C.D., AND CHRISPEELS, M.J., 1990. The expression of phytohemagglutinin genes in *Phaseolus vulgaris* is assocatied with organ-specific DNA methylation patterns. Plant Mol Biol, 14: 629-32.

44. KIPLING, RUDYARD, 1890. Blue Roses, published in Gunga Din and Other Favorite Poems, S. Appelbaum, ed., Dover Publications, 1990, p. 61.

45. EUGSTER, C.H., AND MARKI-FISCHER, 1991. The chemistry of rose pigments. Angew Chem Int Ed Engl, 30: 654-72.

46. O'KEEFE, D.P., AND LETO, K.J. 1989. Cytochrome P450 from the mesocarp of avocado (*Persea americana*). Plant Physiol, 89: 1141-49.

47. BOZAK, K.R., YU, H., SIREVAG, R., AND CHRISTOFFERSEN, R.E. 1990. Sequence analysis of ripening related to cytochrome P450 cDNAs from avocado fruit. Proc Natl Acad Sci, USA, 87: 3904-3908.

48. CHUCK, G., ROBBINS, T., NIJJAR, C., RALSTON, E., COURTNEY-GUTTERSON, N., AND DOONER, H.K., 1993. Tagging and cloning of a petunia flower color gene with the maize transposable element *Activator*. The Plant Cell, 5: 371-78.

49. WALBOT, V., 1992. Strategies for mutagenesis and gene cloning using transposon tagging and T-DNA insertional mutagenesis. Annu Rev Plant Physiol Plant Mol Biol, 43: 49-82.

50. SMITH, C.J.S., WATSON, C.F., BIRD, C.R., RAY, J., SCHUCH, W., AND GRIERSON, D. 1990. Expression of a truncated tomato polygalacturonase gene inhibits expression of the endogenous gene in transgenic plants. Mol Gen Genet, 224: 477-81.

51. GORING, D.R., THOMSON, L., AND ROTHSTEIN, S.J. 1991. Transformation of a partial nopaline synthase gene into tobacco suppresses the expression of a resident wild-type gene. Proc Natl Acad Sci, USA, 88: 1770-74.

52. GOTTLOB-McHUGH, S.G., SANGWAN, R.S., BLAKELEY, S.D., VANLERBERGHE, G.C., KO, K., TURPIN, D.H., PLAXTON, W.C., MIKI, B.L., AND DENNIS, D.T. 1992. Normal growth of transgenic tobacco plants in the absence of cytosolic pyruvate kinase. Plant Physiol, 100: 820-25.

53. ZHANG, H., SCHEIRER, D.C., FOWLE, W.H., AND GOODMAN, H.M., 1992. Expression of antisense or sense RNA of an ankyrin repeat-containing gene blocks chloroplast differentiation in *Arabidopsis*. The Plant Cell, 4: 1575-88.

Chapter Five

ENGINEERING ALTERED GLUCOSINOLATE BIOSYNTHESIS BY TWO ALTERNATIVE STRATEGIES

Ragai K. Ibrahim,[*] Supa Chavadej[#], and Vincenzo De Luca[#]

[*]Plant Biochemistry Laboratory
Department of Biology, Concordia University
Montréal, Canada H3G 1M8

[#]Institut de recherche en biologie végétale
Université de Montréal, Montréal, Canada H1X 2B2

INTRODUCTION

The genetic diversity and metabolic variation that exist in Nature provide a spectrum of organisms with various capabilities for substrate assimilation and product synthesis. Both plants and microorganisms have evolved unique biosynthetic pathways for the production of a rich diversity of secondary metabolites that are utilized as medicinal chemicals, antimicrobial compounds and/or signalling molecules that are essential for the plant's survival and its environmental fitness.

Whereas conventional plant breeding has been used successfully for the selection of desirable agronomic traits, it requires long periods of experimentation (ca. 9-12 years) before a genetically stable, new variety is produced. In addition, breeding programs are usually hampered by limitations inherent in the distribution of plant genomes, as well as the genetic and reproductive characteristics of the plants themselves.[1]

The discovery, in the early 1970's, of DNA restriction enzymes and DNA ligases created a new field of research involving recombinant DNA technology. In addition, the recent advances in plant cell and tissue culture techniques, and the genetic transformation of plants using *Agrobacterium*-mediated or direct gene transfer, made possible the incorporation of foreign genes into plants, thus providing new genotypes with desirable characteristics. Genetic engineering of plants avoids the limitations of self-incompatibility and the genetic consequences of inbreeding and/or outbreeding that are encountered when conventional breeding techniques are used to produce new cultivars. In addition to the genetic modification of secondary metabolite synthesis, which is the topic of this volume, other possibilities exist for the direct manipulation of a range of developmental processes: the quality of seed and fruit, flower shape and colour, pollen production, improvement of photosynthetic efficiency of C3 plants, increasing the range of species that can fix nitrogen, improved resistance of plants to disease, pests, herbicides and other environmental stresses, and the use of plants as 'green bioreactors' for the production of value-added chemicals,[2] including mammalian proteins,[3] and biodegradable thermoplastics.[4]

GENETIC MANIPULATION OF PLANT METABOLISM

The different strategies used for the genetic manipulation of plants[1,5] and microorganisms[6,7] have recently been reviewed. Progress in this technology largely depends on the (a) ability to obtain genes encoding desirable traits, which is limited by the low abundance of the target enzymes; (b) proper alignment of the expressed gene with its endogenous substrate and the compartmentation/transport of its product; (c) successful transformation and reproducible regeneration of transgenic plants; (d) proper site of gene integration combined with the level and pattern of gene expression; and (e) the arduous task of performing detailed analysis of transgenic plants and evaluation of their altered metabolic pathways. Nonetheless, an abundance of reports has been forthcoming on the successful expression of foreign genes in crop plants with the aim of enhancing food quality,[8-17] plant resistance to pathogens[18-22] and insect herbivores,[23-25] and conferring tolerance to herbicides.[26-28] This contrasts with the paucity of reports on transformed plants with altered secondary metabolite synthesis[29-37] (Table 1). This lack of progress can be attributed to the slow improvements achieved in the cloning of genes involved in secondary metabolite synthesis and to the poor understanding of the interface between intermediary and secondary metabolism. This is further complicated by the compartmentation of secondary metabolite pathways.[38]

The strategies currently employed in the genetic manipulation of metabolic pathways are summarized below.

Strategies Aimed at Increasing Secondary Metabolite Synthesis

The biosynthetic potential of plant tissue and cell cultures has not been fully realized due to the lack of understanding of the control mechanisms

Table 1. Some cloned genes coding for the enzymes of secondary metabolite synthesis.

Enzyme	Target metabolite[a]
Anthocyanidin 5'-hydroxylase (putative)	Delphinidin[29]
Chalcone synthase	Anthocyanins[30]
Dihydrokaempferol reductase	Pelargonidin[31]
HMG CoA reductase	Steroids[32]
Hyoscyamine β-6-hydroxylase	Tropane alkaloids[33]
Ornithine decarboxylase	Nicotine[34]
Ornithine decarboxylase	Monoterpenoid indole alkaloids[35]
Stilbene synthase	Stilbenes[36]
Tryptophan decarboxylase	Indole alkaloids, auxins[37]

[a]References

involved in metabolite synthesis and accumulation. Molecular strategies may be directed towards (a) the overproduction of simple precursors of secondary metabolites, e.g. amino acids in alkaloid biosynthesis;[34] (b) increasing the concentration of rate limiting enzymes in a specific pathway;[33] (c) creating a new branch for a preexisting pathway;[31] (d) down-regulation of existing reactions using antisense methods (see below); (e) manipulation of regulatory genes whose products may bind DNA and function as transcriptional activators or repressors; and (f) selection of regulatory mutants with increased tissue-independent expression. These strategies have recently been discussed and reviewed by Yamada and Hashimoto.[39]

Antisense RNA Technology

The transformation of plants with DNA sequences complementary to a specific mRNA has been used to inhibit expression of the target gene. It is believed that translation of the mRNA is prevented by the binding of the antisense RNA to the target species via base pairing. This strategy has been successfully used to control the ripening of tomato fruits by the inhibition of either polygalacturonase[11] or 1-aminocyclopropane-1-carboxylate (ACC) synthase,[12] the latter being involved in the biosynthesis of the plant hormone ethylene. In both cases, the decrease in the amount of enzyme activity correlated with the steady state levels of antisense RNA. In contrast, the inhibition of flower pigmentation in transgenic petunia expressing antisense chalcone synthase[30] did not correlate with the antisense RNA levels, which may have been due to incomplete inhibition of the target gene.[40] It is notable, however, that chalcone synthase-deficient mutants of petunia,[41] and transgenic maize that express antisense chalcone synthase,[42] lack the capability to synthesize flavonols which seem to be required for pollen germination and male gametophyte development.[43,44] In general, antisense RNA technology can be exploited to down-regulate, if not eliminate, the formation of undesirable metabolites by inhibiting some of the target enzymes that are involved in their biosynthesis. This approach may be utilized with cyanogenic glycosides by expressing antisense glycosyltransferases;[45] glucosinolates, using an antisense desulfoglucosinolate sulfotransferase[46,47] or a UDP-glucose thiohydroximate glucosyltransferase;[48] in the modification of lignin composition using an antisense phenylpropanoid O-methyltransferase,[49] and the inhibition of sinapine synthesis in crucifers expressing antisense sinapoylcholine synthase.[50] A more challenging example would be the down-regulation of caffeine synthesis and the production of 'naturally decaffeinated' coffee beans! This may be achieved by expressing antisense N-methyltransferases which catalyze the stepwise methylation of xanthine to caffeine.

Modification of Existing Metabolic Pathways

The introduction of foreign genes into plants could also create artificial metabolic sinks that disrupt existing biosynthetic pathways,[51] or the latter may be redirected/channeled to a different branch pathway. Alternatively, the gene product (an enzyme) may redirect an essential precursor of a particular metabolite; compete with another pathway for a limited supply of an essential common precursor; or simply produce novel metabolites.[31] Another strategy involves the overexpression of a rate limiting enzyme for a specific step of a target pathway in order to increase the amount of a desirable metabolite.[33] Numerous attempts to apply these strategies have been made, and two approaches aimed at reducing glucosinolates in *Brassica* form the basis of this Chapter. It should be noted, however, that regardless of the strategy used it is important to determine the impact that the altered metabolic reactions may have on the pool size of intermediates supplying contiguous pathways; the effects on the plant's physiological and developmental processes; and/or the unpredictable potential toxicity of the accumulated products.

GLUCOSINOLATES IN *BRASSICA*

Brassica napus (rapeseed) belongs to the Brassicaceae family, and constitutes an important source of food, condiments and vegetable seed oil. A characteristic feature of the members of this and a few related families is their ability to synthesize mustard oil glucosides (glucosinolates), as well as two other metabolites, erucic acid (a C-22 monounsaturated fatty acid) and sinapine (sinapoylcholine). Over the past 30 years, Canadian scientists have succeeded in breeding new rapeseed cultivars, known as 'canola',[52] through crosses made with the Polish cultivar 'Bronowski', to produce 'double-low' cultivars whose seeds contain no erucic acid and are low in allylglucosinolate content. Canola seed yields a high quality table oil which is low in saturated fat (6%) and high in unsaturated fatty acids (58%), traits that have made it a standard among vegetable oils. This motivated the oilseed industry to spend an estimated $35 million on research in 1992 to further improve canola breeding and genetics.[53] However, the presence of even low amounts of glucosinolates in the protein-rich seed meal reduces its palatability and limits its value as animal feed, except when used in small quantities. Apart from imparting an unpleasant taint to animal products (meat, poultry, milk, eggs), glucosinolates increase the incidence of thyroid disorders in livestock, resulting in reduced weight gain. Furthermore, recent studies on mammalian systems indicate that indolylglucosinolates cause increased activity of a range of oxidative enzymes involved in both activation and inactivation of xenobiotics, as well as the alteration of chemically induced carcinogenesis.[54] Another problem lies in the fact that canola leaves continue to synthesize glucosinolates which are feeding stimulants of lepidopteran larvae. In addition, glucobrassicin, the major indolylglucosinolate of *Brassica spp.*, has recently been reported to be the oviposition stimulant of *Pieris rapae* which causes large crop losses.[55] A brief review of glucosinolate biosynthesis will assist in demonstrating the strategies proposed to further reduce their accumulation in canola.

Glucosinolate   Biosynthesis

Glucosinolates have the general structure shown in Figure 1, where R is derived from the amino acids methionine (allyl-), phenylalanine (benzyl-) or tryptophan (indolyl-). The reactions involved in their formation have been outlined from radiotracer studies[56] and a pathway for their biosynthesis has recently been proposed[57] based on analogies with cyanogenic glycosides. Briefly, the amino acid precursor is first *N*-hydroxylated, then decarboxylated to give rise to the corresponding aldoxime. Transfer of the -SH group from cysteine or methionine to the aldoxime, followed by *S*-glucosylation give rise to the corresponding thiohydroximate and desulfoglucosinolate, respectively. Finally, a sulfotransferase catalyzes the transfer of the sulfate group of 3'-phosphoadenosine 5'-phosphosulfate (PAPS) to give rise to the glucosinolate (Fig. 2). The reactions involved in the chain extention of methionine and the significance of aminotransferases in the biosynthesis of alkenyl glucosinolates have recently been reviewed.[58]

PROPOSED STRATEGIES TO REDUCE GLUCOSINOLATES IN *BRASSICA*

Rather than using antisense RNA technology in order to block glucosinolate biosynthesis at the level of thiohydroximate glucosylation[48] or the sulfation of desulfoglucosinolate,[46] we designed two strategies aimed at modifying the glucosinolate pathway.

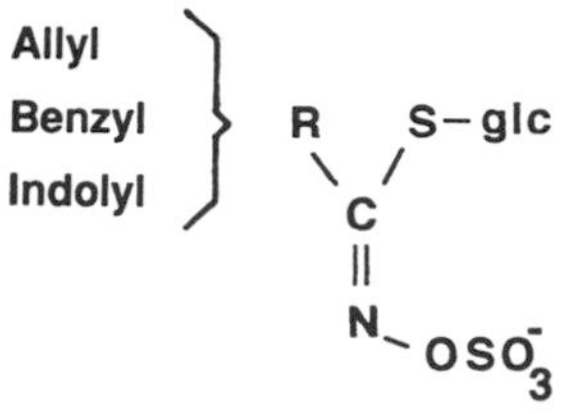

Figure 1: General structure of glucosinolates. R = allyl, benzyl or indolyl residues derived from methionine, phenylalanine and tryptophan, respectively.

Creation of a New Pathway that Competes for the Sulfate Donor PAPS

One strategy involves the introduction of the flavonol 3-sulfotransferase (F3ST, EC 2.8.2.-) gene into *Brassica* with the aim of creating novel, innocuous metabolites, especially flavonol 3-sulfates. This strategy is based on the assumption that the foreign F3ST enzyme will compete with the native desulfoglucosinolate sulfotransferase (DGST) for the limited endogenous pool of 3'-phosphoadenosine 5'-phosphosulfate (PAPS). Since the affinity of the former

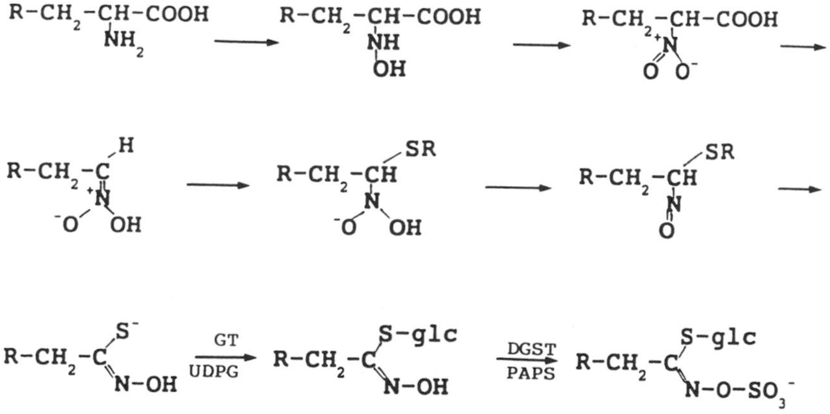

Figure 2: A proposed pathway for the biosynthesis of glucosinolates (R, as in Fig. 1).

enzyme for its substrate and cosubstrate is 1-2 orders of magnitude higher than those of the DGST,[59] it is expected that the sulfation of the endogenous flavonols in canola (kaempferol & isorhamnetin) should predominate over that of the desulfoglucosinolates (Fig. 3).

Flaveria spp. (Asteraceae) accumulate several flavonol mono- to tetrasulfates whose formation is catalyzed by a family of substrate-specific, position-oriented sulfotransferases.[60] The F3ST was purified to electrophoretic homogeneity from young shoot tips of *F. chloraefolia* and used to produce polyclonal antibodies as well as partial amino acid sequencing.[59] The latter probes were used for the isolation and characterization of a cDNA clone coding for the F3ST.[61] The entire coding sequence of F3ST was cloned in the expression vector pBI101 (minus GUS) under the control of CaMV 35S promoter and NPT II as a selctable marker. The resulting plasmid was transferred from *E. coli* to *Agrobacterium tumefaciens* strain LBA4404 by the triparental mating procedure.[62] Explants from canola (glucosinolate positive) as well as potato and tobacco (both glucosinolate negative) were infected with *A. tumefaciens* containing the transformation plasmid, and kanamycin-resistant plants were regenerated from transformed shoots.[63]

In a first transformation experiment, a total of 32, 21 and 13 kanamycin-resistant plants were regenerated from canola, potato and tobacco, respectively (Table 2). Analysis of F3ST gene expression showed that 5, 12 and 9 transgenic canola, potato and tobacco, respectively, exhibited detectable levels of enzyme activity, using quercetin as substrate. However, the overall enzyme activity was relatively low when compared with that of the source plant, *F. chloraefolia* (Table 2). Untransformed controls, as expected, did not exhibit any F3ST activity. Quercetin 3-sulfate was produced by all transgenic plants,although it constituted a minor reaction product in transgenic canola. Unidentified products were also formed, especially in canola (Fig. 4), suggesting the presence of endogenous substrates in the protein preparations. The natural occurrence in canola of desulfoglucosinolate DGST[46-48] renders the analysis of this tissue more complex. The low levels of quercetin 3-sulfate formed in transgenic canola as compared with the unidentified products (Fig. 4) suggests that the specific activity of F3ST was overestimated in these transformants. Furthermore, Western blot analysis revealed that even transgenic plants exhibiting the highest F3ST activity did not accumulate enough enzyme protein to detect immunologically (results not shown).

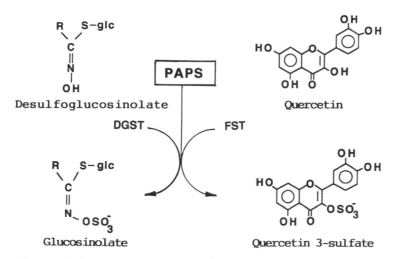

Figure 3: Creation of a new pathway that competes for the sulfate donor PAPS. FST, flavonol sulfotransferase; DGST, desulfoglucosinolate sulfotransferase; PAPS, 3'- phosphoadenosine 5'-phosphosulfate.

Table 2. Expression of F3ST gene in transgenic plants.

Parameter	Canola	Potato	Tobacco
No. of transgenic plants:			
- Analyzed	32	21	13
- With ST activity	5	12	9
Maximum ST activity (pkat/g protein)[a]	2.6[b]	7.2	6.4

[a]Mean ST activity of the positive control (*F. chloraefolia*) is ca. 24 pkat/g protein. Untransformed (control) plants had no ST activity.
[b]Unidentified products (Fig. 4) account for >95% of radioactivity.

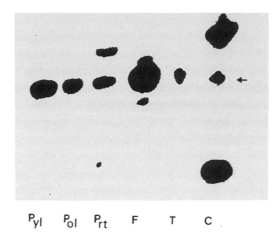

P_{yl} P_{ol} P_{rt} F   T   C

Figure 4: Photograph of an autoradiogram of F3ST reaction products with quercetin as substrate, after TLC in BAW. P, potato (yl, young leaf; ol, old leaf; rt, root); F, *Flaveria*; T, tobacco; C, canola. Arrow indicates position of quercetin 3-sulfate; compounds with higher and lower R_f values are unidentified.

In order to determine if the low level of F3ST activity in transformed canola was due to low gene expression or to post-transcriptional control, the transformants of all these species were analyzed for the presence of F3ST mRNA. Hybridization of F3ST cDNA with RNA from transformants exhibiting highest F3ST enzyme activity indicated that the expression levels were highest in potato, followed by tobacco, whereas no hybridization was observed in canola (Fig. 5). However, the relatively high expression levels in the two former species contrasts with the undetectable level of their F3ST protein on Western blots. This discrepancy may have been due to protein degradation or to inefficient translation. In order to increase the amount of F3ST protein, the pF3ST coding sequence will have to be subcloned downstream of a translational enhancer, such as the 5'-untranslated leader of tobacco mosaic virus RNA.[64]

Whereas the chimeric F3ST gene was expressed in two glucosinolate-negative species, potato and tobacco, no expression could be detected in the target species, canola. This may have been due to the lack of proper coordination of the expressed gene with access to the endogenous substrate, especially if the pathway for flavonoid synthesis is associated with the endoplasmic reticulum membranes, or is otherwise highly compartmentalized in this species.[38] It is

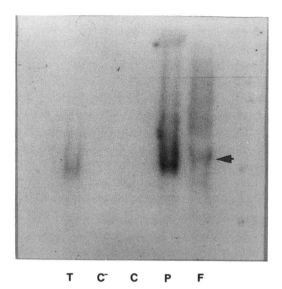

T   C⁻   C   P   F

Figure 5: Northern blot analysis of 20 µg RNA from tobacco
(T), canola (C) and potato (P). Non-transformed canola (C⁻) and
Flaveria (F) served as negative and positive controls,
respectively. Arrow indicates 1.2 kb.

also possible that the lack of expression reflects a high F3ST mRNA instability
in canola, in addition to the low efficiency of transformation and regeneration.

Although there is no available information on the gene encoding DGST
or its sequence homology with that of the F3ST, the expression of the latter
gene in canola may have been inhibited by the endogenous DGST gene by the so
called 'cosuppression' mechanism, also known as sense suppression, a
phenomenon not yet understood (see Chapter by Courtney-Gutterson). The fact
that the pF3ST shares sequence similarities of 29% and 31% with the
hydroxysteroid ST cDNA clone of rat liver and estrogen ST cDNA clone of
bovine placenta, respectively,[61] suggests that it may share an even higher
similarity with the DGST gene. Cosuppression has been observed with the
introduction of an additional copy of the chalcone synthase gene into petunia
with the aim of increasing the intensity of flower color.[65] Instead, the resulting
flowers exhibited either colorless sectors or lacked color completely. This seems
to suggest that the presence of additional homologous chalcone synthase gene

may lead to inactivation of the endogenous gene. A similar mechanism may account for the lack of expression of the F3ST gene in canola, but this phenomenon may also provide an alternative strategy to antisense methods for controlling plant gene expression.

Work in progress is aimed at the construction of a more efficient F3ST chimeric gene in order to optimize both transcription and translation. Alternatively, direct gene transfer and stable transformation with the F3ST gene.may be achieved using microprojectile bombardment of canola shoot tips with *F. chloraefolia* DNA.[66] Further work with both potato and tobacco will investigate the formation of sulfated flavonoids, as well as evaluating the impact of the introduction of this new gene on the sulfate metabolism of transformed plants.

Redirection of the Biosynthetic Precursor Tryptophan

The multiple sinks which simultaneously compete for carbon, nitrogen and other resources in plants can be manipulated by redirecting metabolite flow through desired branch points. A recent strategy used in our laboratory has involved the introduction of the tryptophan decarboxylase (TDC, EC 4.1.1.28) gene in order to create an artificial metabolic sink which redirects tryptophan away from indolylglucosinolate formation and into tryptamine synthesis (Fig. 6).

The enzyme TDC catalyses the decarboxylation of *L*-tryptophan to tryptamine, and it has been purified to homogeneity from the medicinal plant, *Catharanthus roseus*.[67,68] In this plant, TDC participates in the biosynthesis of two commercially important antineoplastic agents, the monoterpenoid indole

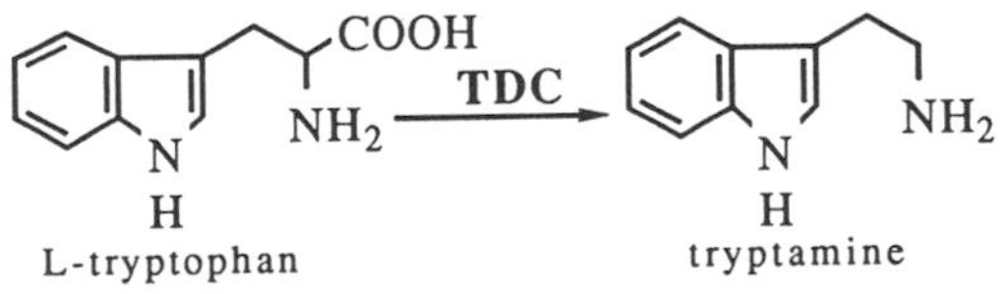

Figure 6: TDC-catalyzed formation of tryptamine.

alkaloids vinblastine and vincristine.[69] The enzyme is characteristic of the plant aromatic decarboxylases that possess high substrate specificity. TDC will accept 5-hydroxy-*L*-tryptophan in addition to *L*-tryptophan, but not *L*-phenylalanine, *L*-tyrosine or *L*-dihydroxyphenylalanine, as substrates.[67] The TDC of *C. roseus*[69] and the substrate-specific tyrosine decarboxylase of *Petroselinum crispum*[71] have recently been cloned and sequenced. The amino acid sequences of both decarboxylases show a high degree of similarity to each other, and to animal aromatic amino acid decarboxylases.[70,71]

EXPRESSION of TRYPTOPHAN DECARBOXYLASE (TDC)

TDC in Canola, Potato and Tobacco

A cDNA clone encoding TDC[70] was inserted, under transcriptional control of CaMV 35S promoter and NPT II as a selectable marker, into the binary Ti plasmid pBI121 after deletion of the GUS gene. The construct was mobilized into the disarmed *A. tumefaciens* strain LBA4404 by the triparental mating procedure,[62] and used to transform canola (*Brassica napus* cv. Westar), tobacco[37] and potato[63] plants. Kanamycin-resistant plants were regenerated, and tested for the presence of TDC mRNA, TDC activity, and tryptamine accumulation. In addition to 40 transgenic tobacco plants,[37] a total of 7 canola and 13 potato plants were generated (Table 3), all of which expressed TDC activity and accumulated tryptamine. The efficiency of production of transgenic plants was variable, with canola being the most difficult. Proof of transformation was obtained by hybridization of the TDC cDNA clone to total RNA extracted from transgenic plants (Fig. 7). Total RNA extracted from 7 transgenic canola (Fig. 7, lane 1C-H), 13 transgenic potato (Fig. 7, lanes 2B-H and 3A-F) and 4 representative transgenic tobacco (Fig. 7, lane 4B-F) was hybridized to a probe derived from the TDC cDNA clone. In contrast, RNA from non-transformed control canola (Fig. 7, lane 1A), potato (Fig. 7, lane 2A) and tobacco (Fig. 7, lane 4A) plants did not retain any of the probe, consistent with the presence of TDC mRNA only in the transgenic plants. A visual comparison of the relative abundance of TDC mRNA expressed in transgenic plants of these 3 species shows that tobacco plants contain the highest levels of TDC mRNA,

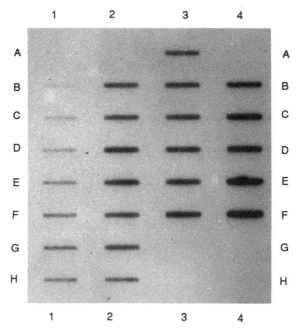

Figure 7: Comparison of relative TDC mRNA levels in wild-type (A) and in different CaMV transformed lines (B-H). Autoradiogram of an RNA slot blot probed with ^{32}P-labeled, 1.6 kb EcoR1 TDC cDNA fragment. Total RNA (30 μg) from young leaves of each plant was applied in each slot. 1, canola; 2 & 3, potato; 4, tobacco.

Table 3. Expression of TDC activity in transgenic plants[a]

Parameter	Canola	Potato	Tobacco
No. of transgenic plants:			
- Analyzed	66	22	50
- With TDC activity	7	13	40
Maximum TDC activity[b]	5.4	17.2	57.6

[a]TDC activity of young leaves of *C. roseus* is 124 pmol tryptophan converted to tryptamine/μg protein/h
[b]pmol tryptophan converted to tryptamine/μg protein/h

followed by potato and canola in decreasing order (Fig. 7). In contrast, TDC mRNA levels in *C. roseus* are not detectable under these experimental conditions,[72] although *C. roseus* TDC specific activities are at least 2- to 6-fold higher than those of any of the transgenic plants produced (Table 3). These results suggest that the level of expression of TDC in transgenic plants is not the only criterion that must be met for increasing TDC activity, but that other undetermined species-specific factors must also be taken into account.

TDC Expression and Accumulation of Tryptamine

The levels of TDC activity and tryptamine content of young leaves of wild-type and CaMV 35S-transformed canola, potato and tobacco plants were compared (Table 4).[73] Transformed tobacco plants accumulate at least 12- and

Table 4. Analysis of transgenic canola plants expressing the TDC gene

Plant	Leaf TDC activity (μmol/gFW)[a]	Leaf trypt- amine content (μmol/gFW)	Leaf gluco- brassicin (nmol/gFW)	Seed indolyl- glucosinolate (μmol/gFW)[b]
Control Plants				
Wild type	1.0	0	103	5.89
GUS-I[c]	0.9	0	121	6.22
TDC Transgenic Plants				
ST-004	1.8	14	38	2.88
ST-115	4.2	75	22	1.09
ST-062	5.8	111	20	0.18

[a]The youngest fully expanded leaves (gFW, 1 gram fresh weight) were used to determine TDC activity, tryptamine and glucosinolate contents. Each measurement (mean ±SD) represents 3 separate plants from each control, and each transgenic line.
[b]Canola seed samples (50 seeds, ca. 100 mg) were extracted and analyzed according to A. Quinsac *et al.* J. Assoc. Off. Anal. Chem. 74, 932, 1991.
[c]Canola plants were transformed with the GUS gene in order to show that the altered glucosinolate profiles obtained with TDC-transformed plants are not due to an artifact of transformation.

50-fold more tryptamine (>1 mg/g fresh weight) than transformed potato and canola plants, respectively. However, the TDC activity required to produce such high levels of tryptamine was only 3- and 10-fold higher than in potato and canola, respectively. When transgenic canola line ST-062 is compared with potato line M9D-05 and tobacco T150-3, it appears that these transgenic lines exhibit almost the same TDC specific activity. However, their tryptamine content amounts to 19, 44 and 201 μg/g fresh weight for canola, potato and tobacco, respectively (Table 5). These results suggest that the accumulation of tryptamine in different transgenic species may depend on the size of, or the access to, the tryptophan pool which may vary from one species to another. In addition, other factors that control the expression of TDC mRNA and TDC activity may contribute to the unpredictable results so often encountered in metabolic engineering experiments of this nature.

Redirection of Tryptophan to Tryptamine Results in the Reduction of Indolylglucosinolate Levels in Canola

Apart from their use in practical applications, transgenic plants that produce tryptamine by redirecting tryptophan away from other pathways can be used to answer other interesting metabolic questions. Will active tryptamine production affect tryptophan pools, and thus affect other processes requiring the participation of the shikimate pathway? Will this pressure on the shikimate pathway also affect other secondary metabolic pathways, such as the accumulation of nicotine in tobacco, steroidal alkaloids in potato and indolylglucosinolates in canola (Fig. 8).

In the case of canola (*B. napus* cv. Westar), this strategy created an artificial metabolic sink as a result of the redirection of tryptophan away from indolylglucosinolates and into tryptamine synthesis (Table 5). This cultivar normally accumulates both allyl- and indolylglucosinolates which are derived from methionine and tryptophan, respectively. Transgenic plants, selected on kanamycin-containing media, were allowed to flower and set seed, and kanamycin-resistant seedlings (T_2 generation) were used in subsequent experiments. Several independent transgenic canola lines were evaluated for TDC enzyme activity, and for the accumulation of the immediate decarboxylation product, tryptamine (Table 5). The results from a few selected lines show a direct correlation among the relative levels of TDC mRNA (Fig. 7, lane 1), the specific activity of TDC found within each transgenic line and the amount of accumulated tryptamine (Table 5).[74] Wild-type control plants, and those

Table 5. Comparison of TDC activity and tryptamine content in young leaves of wild-type and CaMV 35-transformed plants.

Plant	TDC activity[a]	Tryptamine (µg/g FW)
Canola		
Wild-type	0.9	-
ST-004	1.9	trace
ST-005	2.2	10
ST-045	3.6	15
ST-067	4.1	15
ST-053	4.3	18
ST-029	4.3	15
ST-062	5.4	19
Potato		
Wild-type	1.1	-
M9D-17	3.3	35
M9D-02	3.5	22
M9D-05	5.2	44
M9D-23	6.3	45
M9D-11	8.5	73
M9D-24	8.7	49
M9D-12	9.3	54
M9D-13	10.1	38
M9D-10	12.7	85
M9D-08	15.4	43
M9D-16	16.0	58
M9D-25	16.9	95
M9D-21	17.2	80
Tobacco		
Wild-type	1.2	4
T-150-3	6.2	201
T-105-1	17.8	510
T-157-2	28.9	604
T-162-3	50.1	850
T-201-1	57.6	1040

[a]pmol tryptophan converted to tryptamine/µg protein/h

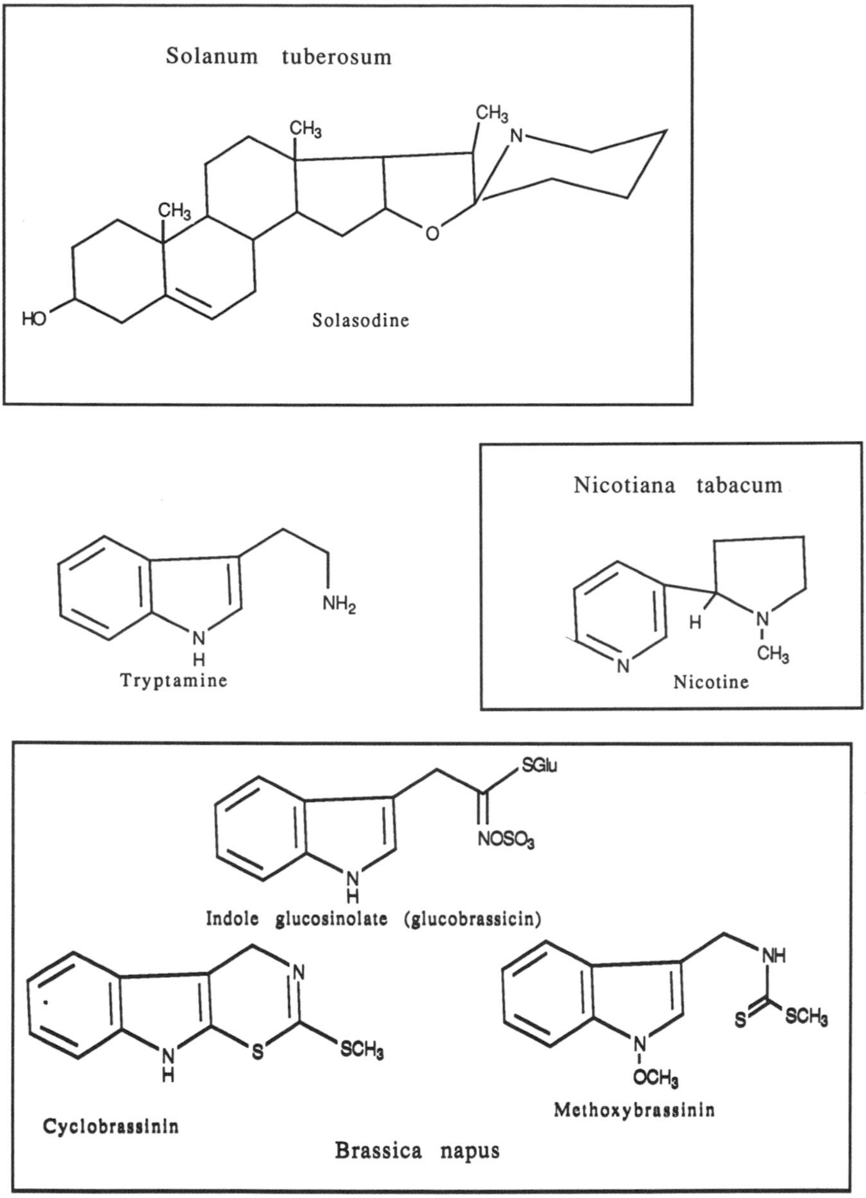

Figure 8: Some characteristic secondary metabolites found in potato, tobacco and canola.

transformed with the GUS gene, exhibited background levels of TDC activity and did not accumulate tryptamine. Unlike the controls, TDC-transformed plants accumulated between 14 and 111 nmol in lines ST-004 and ST-062, respectively. Analysis of glucosinolates in the leaves of transgenic canola plants showed that indolylglucosinolate levels were reduced by up to 80% (Table 4), whereas those of allylglucosinolates remained essentially unchanged (data not shown). A typical HPLC elution profile of glucosinolates shows a drastic reduction of leaf glucobrassicin in transgenic tryptamine-producing leaves (Fig. 9B), as compared to the non-transformed control leaves (Fig. 9A). In contrast, the allylglucosinolate peaks remain unchanged.[75]

Mature seeds produced from each transgenic canola line exhibited much lower levels of indolylglucosinolates than those of control plants. Seeds of transgenic lines ST-004 and ST-062, which exhibited the lowest and highest TDC activities respectively, accumulated 50% and 3% of the levels of indolylglucosinolates present in either controls, wild-type or GUS-transformed plants (Table 5). In contrast, the allylglucosinolate levels in seeds from all transgenic lines remained essentially unaltered (data not shown). These results clearly indicate the feasibility of redirecting tryptophan away from indolylglucosinolate biosynthesis by the heterologous expression of TDC activity in canola.

The TDC Gene in Tobacco Does not Affect the Endogenous Auxin Levels

While several pathways have been proposed for the biosynthesis of auxins in plants,[76] it has been suggested that one pathway involves the TDC-mediated synthesis of tryptamine. To investigate the possible effects of elevated tryptamine accumulation on auxin synthesis, transgenic tobacco plants exhibiting the highest TDC activity and highest tryptamine content were analyzed for IAA levels.[37] Although the tryptamine pool was >260-fold higher in transformed plants than in controls, the IAA levels in both types of plants were comparable (ca. 300 pmol/g fresh weight). This result is consistent with the normal appearance and the lack of any morphological abnormalities observed among plants accumulating high levels of tryptamine.[37] The normal morphological appearance of potato and canola plants that were transformed with the TDC gene also demonstrates that plants can accomodate additional biosynthetic pathways without deleterious effects. Furthermore, the lack of adverse effects of elevated tryptamine levels on auxin production suggests that tryptamine is not an auxin precursor in tobacco. An alternative explanation is

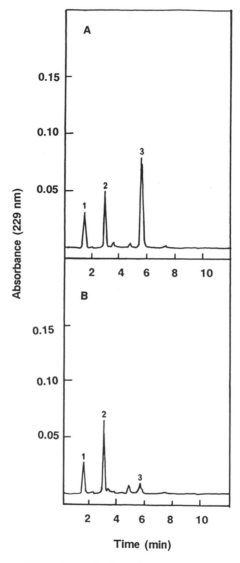

Figure 9: HPLC profile of glucosinolates of non-transformed control leaves (A), and of a typical transgenic canola expressing TDC activity (B). Peaks 1 & 2 are allylglucosinolates derived from methionine; peak 3 is glucobrassicin derived from tryptophan.

that the tryptamine synthesized in these transgenic plants may be sequestered in a compartment (e.g. in the vacuole) where it becomes unavailable for auxin synthesis.

CONCLUSION AND PERSPECTIVES FOR FUTURE WORK

The recent progress in our ability to isolate the genes responsible for the biosynthesis of plant secondary metabolites will have a major impact on the production of agriculturally and medicinally useful plants. Equally, rapid advances in our ability to genetically engineer previously recalcitrant crop and medicinal plants is amplifying the number of target pathways being modified. This technology will also permit the creation of entirely new biosynthetic pathways in different plant species, where the new products might confer a selective advantage such as disease or pest resistance. The isolation of genes for entire biosynthetic pathways is rapidly becoming a reality, and it will soon be possible to transfer these pathways from plants to other organisms better suited for growth in industrial fermenters. This should facilitate the commercial production of an increasing number of natural products that are presently available from plants.

The two examples of metabolic engineering reported in this brief review demonstrate that whereas in some cases spectacular results are obtained, in others, a 'fine-tuning' of a given approach is required to achieve the desired goal. However, detailed analyses of the targeted biochemical pathways and their transgenic counterparts should serve to develop more rational approaches to metabolic engineering of plants.

ACKNOWLEDGEMENTS

The work cited from the authors' laboratories was supported by grants from the Natural Sciences and Engineering Research Council of Canada, and the Department of Higher Education, Government of Québec. We wish to thank Dr. Normand Brisson and Dr. Luc Varin for their valuable contribution to various aspects of this work, and Jacynthe Séguin for her expert assistance.

REFERENCES

1. LINDSEY, K. 1992. Genetic manipulation of crop plants. J. Biotechnol. 26: 1-28.
2. SAITO, K., YAMAZAKI, M., MURAKOSHI, I. 1992. Transgenic medicinal plants: *Agrobacterium*-mediated foreign gene transfer and production of secondary metabolites. J. Nat. Prod. 55: 149-162.
3. HIATT, A., CAFFERKEY, R., BOWDISH, K. 1989. Production of antibodies in transgenic plants. Nature 342: 76-78.
4. POIRIER, Y., DENNIS, D.E., SOMERVILLE, C. 1992. Polyhydroxybutyrate, a biodegradable thermoplastic, produced in transgenic plants. Science 256: 520-523.
5. LAMB, C.J., RYALS, J.A., WARD, E.R., DIXON, R.A. 1992. Emerging strategies for enhancing crop resistance to microbial pathogens. Bio/Technology 10: 1436-1445.
6. DEMAIN L. 1981. Industrial microbiology. Science 214: 987-995.
7. BAILEY, J.E. 1991. Towards a science of metabolic engineering. Science 252: 1668-1675.
8. STARK, D.M., TIMMERMAN, K.P., BARRY, G.F., PREISS, J., KISHORE, G.M. 1992. Regulation of the amount of starch by ADP-glucose phosphorylase. Science 258: 287-292.
9. KLEE, H., HAYFORD, M.B., KRETZMER, K.A., BARRY, G.F., KISHORE, G.M. 1991. Control of ethylene synthesis by expression of a bacterial enzyme in transgenic tomato plants. Plant Cell 3: 1187-1193.
10. SMITH, C.J., WATSON, C.F., RAY, J., BIRD, C.R., MORRIS, P.C., SCHUCH, W., GRIERSON, D. 1988. Antisense RNA inhibition of polygalacturonase gene expression in transgenic tomatoes. Nature 334: 724-726.
11. HAMILTON, A.J., LYCETT, G.W., GRIERSON, D. 1990. Antisense gene that inhibits synthesis of the plant hormone ethylene in transgenic plants. Nature 346: 284-287.
12. OLLER, P.W., MIN-WONG, L., TAYLOR, L.P., PIKE, D.A., THEOLOGIS, A. 1991. Reversible inhibition of tomato fruit senescence by antisense RNA. Science 254: 437-439.
13. WORRELL, A.C., BRUNEAU, J.M., SUMMERFELT, K., BOERSIG, M., VOELKER, T.A. 1991. Expression of a maize sucrose phosphate synthase in tomato alters carbohydrate partioning. Plant Cell 3: 1121-1130.

14. VOELKER, T.A., WORRELL, A.C., ANDERSON, L., BLEIBAUM, J., FAN, C., HAWKIN, D.J., RADKE, S.E., MAELOR-DAVIS, H. 1992. Fatty acid biosynthesis redirected to medium chains in transgenic oilseed plant. Science 257: 72-74.

15. ARONDEL, V., LEMIEUX, B., HWANG, I., GIBSON, S., GOODMAN, H.M., SOMERVILLE, C. 1992. Map-based cloning of a gene controlling *Omega*-3 fatty acid desaturation in *Arabidopsis*. Science 258: 1353-1355.

16. ALLENBACH, S., SIMPSON, R.B. 1990. Manipulation of methionine-rich protein genes in plant seeds. Trends Biotechnol. 8: 156-162.

17. FLAVELL, R.B., GOLDSBOROUGH, A.P., ROBERT, L.S., SCHNICK, D., THOMPSON, R.D. 1989. Genetic variation in wheat HMW glutin subunits and the molecular basis of bread-making quality. Bio/Technology 7: 1281-1285.

18. BROGLIE, K., CHET, I., HOLLIDAY, M., CRESSMAN, R., BRIDDLE, P., KNOWLTON, C., MAUVAIS, C.J., BROGLIE, R. 1991. Transgenic plants with enhanced resistance to the fungal pathogen *Rhizoctinia solani*. Science 254: 1194-1197.

19. NEALE, A.D., WAHLEITHNER, J.A., LUND, M., BONNETT, H.G., KELLY, A., MEEKS-WAGNER, D.R., PEACOCK, W.L., DENNIS, F.S. 1990. Chitinase, _-1,3-glucanase, osmotin and extensin are expressed in tobacco explants during flower formation. Plant Cell 2: 673-684.

20. HAIN, R., REIF, H.J., KRAUSE, E., LANGEBARTELS, R., KINDL, H., VORNAM, B., WISE, W., SCHMELZER, E., SCHREIER, P.H., ST_CKER, R.H., STENZEL, K. 1993. Disease resistance results from foreign phytoalexin expression in a novel plant. Nature 361: 153-156.

21. LOESCH-FRIES, L.S., MERLO, D., ZINNEN, T., BURHOP, L., HILL, K., KRAHN, K., JARVIA, N., NELSON, S., HALK, E. 1988. Expression of alfalfa mosaic virus RNA4 in transgenic plants confers virus resistance. EMBO J. 6: 1845-1851.

22. POWELL-ABEL, P.A., NELSON, R.S., HOFFMAN, DE-B, ROGERS, S.G., FRALEY, R.T., BEACHY, R.N. 1986. Delay of disease development in transgenic plants that express the tobacco mosaic virus coat protein. Science 232: 738-743.

23. HILDER, V.A., GATEHOUSE, A.M., SHEERMAN, S.E., BAKER, R.F., BOULTER, D. 1987. A novel mechanism of insect resistance engineered into tobacco. Nature 330: 160-163.

24. FISCHOFF, D.A., BOWDISH, K.S., PERLAK, F.J., MARRONE, P.G., MCCORMICK, S.M., NIEDERMEYER, J.G., DEAN, D.A., KUSANO-KRETZMERK, K., MEYER, E.J., ROCHESTER, D.E., ROGERS, S.G., FRALEY, R.T. 1987. Insect tolerant transgenic tomato plants. Bio/Technology 5: 807-813.

25. PERLACK, F.J., DEATON, R.W., ARMSTRONG, T.A., FUCHS, R.F., SIMS, S.R., GREENPLATE, J.T., FISCHOFF, D.A. 1990. Insect resistant cotton plants. Bio/Technology 8: 939-943.

26. SHAH, D.M., HORSCH, R.B., KLEE, H.J., KISHORE, G.M., WINTER, J.A., TUMER, N.E., HIRONAKE, C.M., SANDERS, P.R., GASSER, C.S., AYKENT, S., SIEGEL, N.R., ROGERS, S.G., FRALEY, R.T. 1986. Engineering herbicide tolerance in transgenic plants. Science 233: 478-481.

27. CHEUNG, A.Y., BOGORAD, l., VAN MONTAGU, M., SCHELL, J. 1988. Relocating a gene for herbicide tolerance: a chloroplast gene is converted into a nuclear gene. Proc. Natl. Acad. Sci. USA 85: 391-395.

28. COMAI, L., FACCIOTTI, D., HIATT, W.R., THOMPSON, D., ROSE, R.E. STALKER, D.M. 1985. Expression in plants of a mutant *aro*A gene from *Salmonella typhimurium* confers tolerance to glyphosate. Nature 317: 741-744.

29. ROBERTS, P. 1991. Bouquets to Australia for the first blue rose. Fin. Rev., August 11, p. 28.

30. VAN DER KROL, A.R., LENTING, P.E., VEENSTRA, J., VAN DER MEER, I.M., KOES, R.E., GERATS, A.G.M., MOL, J.N.M., STUITJE, A.R. 1988. An antisense chalcone synthase gene in transgenic plants. Nature 333: 866-869.

31. MEYER, P., HEIDMANN, I., FORKMANN, G., SAEDLER, H. 1987. A new petunia flower colour. Nature 330: 677-678.

32. CHAPPELL, J., PROULX, J., WOLF, F., CUELLAR, R.E., SAUNDERS, C. 1991. Is hydroxymethylglutaryl CoA reductase a rate limiting step for isoprenoid metabolism? Plant Physiol. 96: S127.

33. MATSUDA, M., OKABE, S., HASHIMOTO, T., YAMADA, Y. 1991. Molecular cloning of hyoscyamine β6-hydroxylase, a 2-oxoglutarate-dependent dioxygenase,from cultured roots of *Hyoscyamus niger*. J. Biol. Chem. 266: 9460-9469.

34. HAMILL, J.D., ROBBINS, R.J., PARR, A.J., EVANS, D.M., FURZE, J.M., RHODES, E.J.C. 1990. Effects of overexpressing yeast

ornithine decarboxylase gene on nicotine levels in hairy roots of *Nicotiana* rustica. Plant Mol. Biol. 15: 27-38.

35. KUTCHAN, T.M. 1989. Expression of enzymatically active cloned strictosidine synthase from higher plant *Rauwulfia serpentina* in *E. coli*. FEBS Lett. 257: 127-130.

36. HAIN, R., BIESELER, B., KINDL, H. SCHÖDER, H., STOCKER, R. 1990. Expression of stilbene synthase gene in *Nicotiana tabacum* results in synthesis of the phytoalexin resveratrol. Plant Mol. Biol. 15: 325-335.

37. SONGSTAD, D.D., DE LUCA, V., BRISSON, N., KURZ, W.G.W., NESSLER, C. 1990. High levels of tryptamine accumulation in transgenic tobacco expressing tryptophan decarboxylase gene. Plant Physiol. 94: 1410-1413.

38. LUCKNER, D., DIETRICH, B., LERBS, W. 1980. Cellular compartmentation and channeling of secondary metabolism in microorganisms and higher plants. In: Progress in Phytochemistry, Vol. 6 (R. Reinhold, J.B. Harborne, T. Swain, eds.), Pergamon Press, Oxford, pp. 103-142.

39. YAMADA, Y., HASHIMOTO, T. 1990. Possibilities for improving yields of secondary metabolites in plant cell cultures. In: Progress in Plant Cellular and Molecular Biology (H.J.J. Nijkamp, L.H.W. Van der Plass, J. Van Aartrijk, eds.), Kluwer, Dordrecht, Netherlands, pp. 781-790.

40. HIATT, W.R., KRAMER, M., SHEEHY, R.E. 1989. The application of antisense RNA technology to plants. In: Genetic Engineering, Vol. 11 (J.K. Setlow, ed.), Plenum, New York, pp. 49-63.

41. TAYLOR, L.P., JORGENSEN, R. 1992. Conditional male fertility in chalcone synthase-deficient petunia. J. Hered. 83: 11-17.

42. VAN DER MEER, I.M., STAM, M.E., VAN TUNEN, A.J., MOL, J.N.M., STUITJE, A.R. 1992. Antisense inhibition of flavonoid biosynthesis in petunia results in male sterility. Plant Cell 4: 253-262.

43. MO, Y., NAGEL, C., TAYLOR, L.P. 1992. Biochemical complementation of chalcone synthase defines a role for flavonols in functional pollen. Proc. Natl. Acad. Sci. USA 89: 7213-7217.

44. YLSTRA, B., TOURAEV, A., MORENO, R.M., STR_GER, E., VAN TUNEN, A.J., VICENTE, O., MOL, J.N.M., HEBERLE-BORS, E. 1992. Flavonols stimulate development, germination and tube growth of tobacco pollen. Plant Physiol. 100: 902-907.

45. CONN, E.E. 1991. The metabolism of a natural product: Lessons learned from cyanogenic glycosides. Planta Med. 57:S1-9.

46. GLENDENING, T.M., POULTON, J.E. 1990. Partial purification and characterization of a 3'-phosphoadenosine 5'-phosphosulfate:desulfoglucosinolate sulfotransferase from cress (*Lepidium sativum*). Plant Physiol. 94: 811-818.

47. JAIN, J.C., GROOTWASSINK, J.W.D., KOLENOVSKY, A.D., UNDERHILL, E.W. 1990. Purification and properties of 3'-phosphoadenosine 5'-phosphosulfate: desulfoglucosinolate sulfotransferase from *Brassica juncea*. Phytochemistry 29: 1425-1428.

48. JAIN, J.C., GROOTWASSINK, J.W.D., REED, D.W., UNDERHILL, D.W. 1990. Persistent copurification of enzymes catalyzing the sequential glucosylation and sulfation in glucosinolate biosynthesis. J. Plant Physiol. 136: 356-361.

49. GOWRI, G., BUGOS, R.C., CAMPBELL, W.H., MAXWELL, C.A., DIXON, R.A. 1991. Stress responses in alfalfa (*Medicago sativa* L.). X. Molecular cloning and expression of S-adenosyl-L-methionine: caffeic acid 3-O-methyltransferase, a key enzyme of lignin biosynthesis. Plant Physiol. 97: 7-14.

50. STRACK, D., KNOGGE, W., DAHLBENDER, B. 1983. Enzymatic synthesis of sinapine from 1-O-sinapoyl-β-D-glucose and choline by a cell-free system from developing seeds of red radish (*Raphanus sativum*). Z. Naturforsch. 38c: 21-27.

51. LEFEBVRE, D.D. 1990. Expression of mammalian metallothionein suppresses glucosinolate synthesis in *Brassica campestris*. Plant Physiol. 93: 522-524.

52. Plant Biotechnology Institute. 1992. From Rapeseed to Canola: The billion dollar success story. National Research Council of Canada, Saskatoon, Canada, pp. 1-79.

53. KIDD, G. 1993. Is pursuing improved canola an unctuous aim? Bio/Technology 11: 448-449.

54. MCDANIEL, R., MCLEAN, A.E.M. 1988. Chemical and biological properties of indole glucosinolates (glucobrassicins): a review. Fd. Chem. Toxic. 26: 59-70.

55. RENWICK, J.A., RADKE, C.D., SACHDEV-GUPTA, K., STÄDLER, E. 1992. Leaf surface chemicals stimulating oviposition by *Pieris rapae* (Lepidoptera) on cabbage. Chemoecology 3: 33-38.

56. UNDERHILL, E.W. 1980. Glucosinolates. In: Encyclopedia of Plant Physiology, Vol. 8 (E. Bell, B. Charlwood, eds.) Springer-Verlag, New York, pp. 493-511.

57. POULTON, J.E., MOLLER, B.L. 1993. Glucosinolates. In: Methods in Plant Biochemistry, Vol. 9 (J.P. Lea, ed.), Academic Press, London, pp. 209-237.

58. CHAPPLE, C.C.S., ELLIS, B.E. 1992. Secondary metabolite profiles of crucifer seeds: Biogenesis, role and prospects for directed modification. In: Biosynthesis and Molecular Regulation of Amino Acids in Plants (B.K. Singh, H.E. Flores, J.C. Shannon, eds.), American Society of Plant Physiologists, Rockville, MD, pp. 239-248.

59. VARIN, L., IBRAHIM, R.K. 1992. Novel flavonol 3-sulfotransferase: Purification, kinetic properties and partial amino acid sequence. J. Biol. Chem. 267: 1856-1863.

60. VARIN, L. 1992. Flavonoid sulfation: Phytochemistry, enzymology and molecular biology. Rec. Adv. Phytochem. 26: 233-254.

61. VARIN, L., DE LUCA, V., IBRAHIM, R.K., BRISSON, N. 1992. Molecular characterization of two plant flavonol sulfotransferases. Proc. Natl. Acad. Sci. USA 89: 1286-1290.

62. BEVAN, M. 1984. *Agrobacterium* vectors for plant transformation. Nucleic Acids Res. 12: 8711-8722.

63. HORSCH, R.B., FRY, J.E., HOFFMAN, N.L., EICHHOLTZ, D., ROGERS, S.G., FRALEY, R.T. 1985. A simple and general method for transferring genes into plants. Science 227: 1229-1231.

64. GALLI, D.R., SLEAT, D.E., WATTS, J.W., TURNER, P.C., WILSON, T.M.A. 1987. The 5'-leader sequence of tobacco mosaic virus RNA enhances the expression of foreign gene transcripts in vitro and in vivo. Nucleic Acids Res. 15: 3257-3263.

65. NAPOLI, C., LEMIEUX, C., JORGENSEN, R. 1990. Introduction of a chimeric chalcone synthase gene into petunia results in reversible cosuppression of homologous genes *in trans*. Plant Cell 2: 279-289.

66. KLEIN, T.M., GOFF, S.A., ROTH, B.A., FROMM, M.E. 1990. Applications of the particle gun in plant biology. In: Progress in Plant Cellular and Molecular Biology (H.J.J. Nijkamp, L.H.W. Van der Plas, J. Van Aartrijk, eds.), Kluwer, Dordrecht, Netherlands, pp. 56-66.

67. NOÉ, W., MOLLENSCHOTT, C., BERLIN, J. 1984. Tryptophan decarboxylase from *Catharanthus roseus* cell suspension cultures: Purification, molecular and kinetic data of the homogeneous protein. Plant Mol. Biol. 3: 281-288.

68. ALVAREZ-FERNANDEZ, J., OWEN, T.G., KURZ, W.G.W., DE LUCA, V. 1989. Immunological detection and quantitation of tryptophan decarboxylase in developing *Catharanthus roseus* seedlings. Plant Physiol. 91: 79-84.

69. DE LUCA, V., KURZ, W.G.W. 1988. Monoterpene indole alkaloids (*Catharanthus roseus*). In: Cell Culture and Somatic Cell Genetics of Plants, Vol. 5 (F. Constabel, I.K. Vasil, eds.) Academic Press, New York, pp. 386-401.

70. DE LUCA, V., MARINEAU, C., BRISSON, N. 1989. Molecular cloning and analysis of cDNA encoding a plant tryptophan decarboxylase: Comparison with animal Dopa decarboxylases. Proc. Natl. Acad. Sci. USA 86: 2582-2586.

71. KAWALLECK, P., KELLER, H., HAHLBROCK, K., SCHEEL, D., SOMSSICH, I.E. 1993. A pathogen-responsive gene of parsley encodes tyrosine decarboxylase. J. Biol. Chem. 268: 2189-2194.

72. ROEWER, I.A., CLOUTIER, N., NESSLER, C.L., DE LUCA, V. 1992. Transient induction of tryptophan decarboxylase and strictosidine synthase in cell suspension cultures of *Catharanthus roseus*. Plant Cell Rep. 11: 86-89.

73. CHAVADEJ, S., BRISSON, N., DE LUCA, V. 1993. Expression of tryptophan decarboxylase gene in transformed plants. (submitted).

74. CHAVADEJ, S., BRISSON, N., MCNEIL, J.N., DE LUCA, V. 1993. The redirection of tryptophan in transgenic canola changes the mustard oil glycoside profile (submitted).

75. CHAVADEJ, S., BRISSON, N., DE LUCA, V. 1993. Changes in glucosinolate patterns of tryptophan decarboxylase-transformed canola plants (*Brassica napus*) (submitted).

76. REINECKE, D.M., BANDURSKI, R.S. 1990. Auxin biosynthesis and metabolism. In: Plant Hormones and Their Role in Plant Growth and Development (P.J. Davies, ed.), Kluwer, Dordrecht, Netherlands, pp.24-42.

Chapter Six

GENETIC MANIPULATION OF LIGNIN AND PHENYLPROPANOID COMPOUNDS INVOLVED IN INTERACTIONS WITH MICROORGANISMS

Richard A. Dixon[1], Carl A. Maxwell[2], Weiting Ni[1],
Abraham Oommen[1], and Nancy L. Paiva[1]

[1]Plant Biology Division, Samuel Roberts Noble Foundation,
P. O. Box 2180, Ardmore, OK 73402

[2]DuPont Agricultural Products, Stine Haskell Research Center
P. O. Box 30, Elkton Road
Newark, DE 19714

INTRODUCTION

Increasing knowledge of the biochemistry of plant secondary product synthesis, and the cloning of biosynthetic pathway genes, has opened up the possibility of engineering novel pathways or reducing unwanted metabolites by genetic engineering strategies. Such approaches should lead to significant improvements in agronomic performance and post-harvest processing. At present, the necessary knowledge base is most advanced in the area of phenylpropanoid derivatives and certain alkaloids. Further advances in our understanding of terpenoid biochemistry, and our ability to genetically transform cereals and large seeded legumes, are required to underpin fuller exploitation of genetic manipulation of secondary metabolism for plant improvement.

One main focus of our laboratory is the biochemistry and molecular biology of the antimicrobial isoflavonoid phytoalexin pathway, and related areas of phenylpropanoid metabolism, in the forage legume alfalfa. We here describe potential applications of the information we have obtained from this sytem for the genetic manipulation of lignin, nodulation gene inducers and phytoalexins.

THE MULTIPLE FUNCTIONS OF PHENYLPROPANOID-DERIVED METABOLITES IN LEGUMES

In most plants, the phenylpropanoid pathway supplies precursors for the structural polymer lignin and for flavonoid flower pigments and UV protectants. In legumes, flavonoid derivatives also play crucial roles in interactions with microorganisms. Thus, symbiotic nitrogen fixing *Rhizobia* and *Bradyrhizobia* recognize flavones, flavanones, chalcones or isoflavones released to the rhizosphere as signals for the activation of their nodulation genes, whereas isoflavonoids (isoflavones, isoflavans and pterocarpans) are the major structural

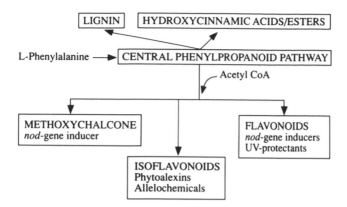

Fig.1. Biosynthetic inter-relationships between phenylpropanoid derived secondary metabolites in a legume such as alfalfa.

class of phytoalexins in legumes. It is also possible that isoflavonoids may be involved in allelopathy and defense against herbivores and pathogenic nematodes in alfalfa.

The biosynthetic inter-relationships between the various classes of functionally distinct phenylpropanoid compounds in legumes are summarized in Fig. 1. It is obvious from this scheme that genetic manipulation of reactions common to the synthesis of more than one product will likely impact on the formation of these different products. Less obvious, but nevertheless likely, is the possibility of cross-talk between specific branch pathways through endogenous mechanisms of flux-sensing. Thus, plants altered in their capacity to form specific phenylpropanoid-derived compounds must be tested for predicted or unexpected phenotypic effects involving related pathways.

BIOCHEMISTRY AND MOLECULAR BIOLOGY OF PHENYLPROPANOID AND FLAVONOID SYNTHESIS IN ALFALFA

The isolation of genes encoding enzymes of the central phenylpropanoid pathway and the lignin, flavonoid and isoflavonoid branch pathways is the starting point for the genetic manipulation of these pathways, both quantitatively (e.g. by over-expression or antisense strategies) or qualitatively (e.g. by introducing genes from species with structurally different end-products). Our understanding of the enzymology and molecular biology of these pathways in alfalfa was the subject of a recent review,[1] the salient features from which are summarized in Table 1. Since reviewing this area, three further genes for enzymes of the pathway have been cloned, encoding: cinnamic acid 4-hydroxylase (CA4H),[2] a cytochrome P450 catalyzing the second step in the central phenylpropanoid pathway; acetyl CoA carboxylase (B. Shorrosh and R.A. Dixon, unpublished results), the enzyme which converts acetyl CoA to malonyl CoA, one of the two substrates of chalcone synthase; and chalcone O-methyltransferase. The identity of the CA4H clone was confirmed by expression studies in yeast.[2] Very little is known about plant cytochrome P450s at the molecular level; knowledge of the sequence of CA4H provides clues for the design of oligonucleotide primers for polymerase chain reaction amplification of other plant P450 sequences. A specific target now is the isoflavone 2'-hydroxylase P450, which has been suggested to be the rate limiting step in pterocarpan phytoalexin synthesis in chickpea.[3] Increasing acetyl CoA carboxylase levels may also increase the flux into the flavonoid branch pathway under conditions of induced flavonoid synthesis.

It will be noted that all the enzymes listed in Table 1, except for the lignin-specific caffeic acid O-methyltransferase, are induced on exposure of alfalfa cell cultures to fungal elicitor macromolecules, and that this induction results from increased transcript levels. Understanding how the activation of defense gene promoters is effected and coordinated is a major area of our program which is discussed briefly at the end of this article. More detailed reviews of this topic have appeared.[4,5]

GENETIC MANIPULATION OF REDUCED LIGNIN LEVELS

Reasons for Modifying Lignin Content and Composition

Lignin is a major structural polymer of secondarily thickened plant vascular tissue and fibers. It imparts mechanical strength to stems and trunks, and hydrophobicity to conducting vessels. Localized accumulation of lignin around pathogen infection sites is an important mechanism of disease resistance, particularly in cereals.[6] Lignin is a major waste product of the paper industry, its removal by harsh chemical treatments being both expensive and environmentally unfriendly. Furthermore, lignification strongly limits the biodegradability of

Table 1. Properties of phenylpropanoid/flavonoid biosynthetic enzymes and expression of their genes in alfalfa.

Enzyme	Subunit Mr (kDa)	No. of subunits	No. of genes[a]	Elicitor induction of: Enzyme[b]	Transcripts[c]	Tissue-specific expression of transcripts[d]
L-Phenylalanine ammonia-lyase	79	4	~6	++	++	R, N, S, M, L, P, B, F
Cinnamic acid 4-hydroxylase	58	1	~2	+	++	ND
4-Coumarate: CoA ligase	ND[e]	1	ND	+	++	ND
Chalcone synthase	43	2	>7	++	++	R, N, S, M, L, P, B, F
Chalcone isomerase	28	1	ND	+	++	R, N, S, M, L, P, B, F
Isoflavone synthase	ND	ND	ND	+	ND	ND
Isoflavone O-methyltransferase	41	1	ND	+++	ND	ND
Isoflavone 2'-hydroxylase	ND	1?	ND	++	ND	ND
Isoflavone reductase	37.5	1	1	++	++	R, N, S, L, P
Pterocarpan synthase	ND	1	ND	+	ND	ND
Chalcone 2'-O-methyltransferase	43	1	~2	+	+	R
Caffeic acid 3-O-methyltransferase	41	1	~2	-	-	R, N, S, M, L, P, B, F
Acetyl CoA carboxylase	>200	ND	>1	ND	++	ND

[a]Based on complexity of Southern hybridization pattern (not copy no.)

[b] - = not induced, + = <10 fold, ++ = 10-100 fold, +++ = >100 fold

[c] - = not induced, + = <25-fold, ++ = >25-fold

[d]R = roots, N = root nodules, S = stems, M = apical meristems, L = leaves, P = petioles, B = buds, F = flowers. The organ in which the major expression is observed is underlined.

[e]ND = not determined

plant cell wall material, and lignin levels have been shown to be negatively correlated with forage digestibility.[7,8] Sound economic reasons therefore exist to justify attempts at altering lignin composition or levels by genetic engineering. Ideally, engineered changes in lignin should increase extractability of other cell wall components without compromising the plant's rigidity, water conductance or disease resistance.

Target Enzymes

As the topic of genetic manipulation of lignin is covered in more detail elsewhere in this volume (Chapter 12), only brief comments on possible ways of manipulating lignin levels and/or composition will be made here. As lignin is a polymer of phenylpropane units, initial flux into the lignin biosynthetic pathway is determined by the activity of L-phenylalanine ammonia-lyase (PAL). This enzyme is encoded by multigene families in legumes,[1] but it is not known whether one or more gene family members are specifically involved in lignification. PAL activity may, however, be rate limiting for lignification. Thus, transgenic tobacco plants in which PAL expression has been reduced through a co-suppression effect associated with introduction of a foreign PAL

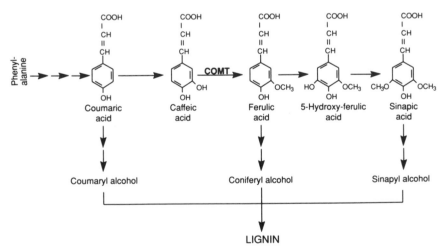

Fig. 2. Biosynthesis of lignin monomers. COMT = caffeic acid 3-0-methyltransferase.

gene from bean, exhibit strongly reduced lignin levels and wilt more readily than wild type plants.[9] The extent of lignin reduction parallels the level of PAL co-suppression (N.P. Bate, R.A. Dixon, C.J. Lamb and W. Ni, unpublished results). However, since reduction of PAL activity will undoubtedly result in pleiotropic effects, this is not a recommended strategy for practical applications unless it should prove possible to target antisense RNA or ribozymes to specific PAL transcripts associated with lignification. The different PAL genes in some plants (e.g. bean) show widely divergent nucleotide sequences in their first exons,[10] which may facilitate such an approach.

Two enzymes specific for monolignol biosynthesis, cinnamyl alcohol dehydrogenase and caffeic acid 3-O-methyltransferase (COMT), are attractive targets for down-regulation by antisense RNA, as are the peroxidases or laccases involved in polymerization of the monolignols (see Chapter 11). COMT from alfalfa is a bi-specific enzyme, methylating both caffeic acid and 5-hydroxy ferulic acid (Fig. 2). Reducing COMT activity should therefore equally reduce production of coniferyl alcohol and sinapyl alcohol, the major lignin building blocks in dicotyledonous angiosperms, and would be predicted to result in an overall decrease in lignin quantity. However, if such a manipulation resulted in an increased incorporation of coumaryl residues into lignin, the lignin may become less biodegradable due to increased inter- and intra-molecular cross-linking. The recent cloning of ferulate 5-hydroxylase from *Arabidopsis* (C. Chapple, personal communication) provides a means of manipulating the coniferyl: sinapyl alcohol ratios to alter lignin cross-linking.

Reduction of Lignin by Antisense Expression of Caffeic Acid O-Methyltransferase

We recently reported the cloning of the bi-specific COMT from alfalfa.[11] COMT sequences appear to be highly conserved in the plant kingdom; the alfalfa and aspen sequences are 86% identical at the amino acid level,[11] and the not full length tobacco lignin-specific COMT is 71% identical to the alfalfa COMT.[12] We have introduced a 0.43 kb portion of the alfalfa COMT coding sequence, in the antisense orientation, into tobacco using *Agrobacterium*-mediated transformation. This resulted in reduced COMT activity and thioglycolate-extractable lignin levels in a number of independent transformants.[13] The reduced lignin levels segregated with the transgene in the R1 population. The reduction in lignin could be strikingly demonstrated by phloroglucinol staining of horizontal sections through the top six internodes of antisense and control plants. In antisense plants, virtually no red staining was observed in the upper internodes; however, by the sixth internode, staining was near control levels.[13] These observations could be explained by the expression pattern of the cauliflower mosaic virus 35S promoter used to drive the antisense construct, by higher endogenous levels of COMT expression in the older vascular tissues, or by the expression in the lower internodes of COMT isoforms with lower sequence identity to the antisense transgene. The diameters of the vessels in the antisense plants appeared greater than in controls, presumably as a result of the less heavily lignified walls allowing increased expansion.

Alkaline nitrobenzene oxidation analysis of thioglycolate extracted lignin from antisense and control plants revealed no major differences in the relative proportions of coumaryl, coniferyl and sinapyl monomers.[13] Thus, down-regulation of COMT appears to be a useful strategy for manipulating lignin levels. Furthermore, these data suggest that lignin composition is not determined simply by the relative availabilities of the three monolignols.

Preliminary results suggest that the COMT antisense strategy can also be used to reduce lignin levels in alfalfa stems, and studies are now in progress to determine the effects of such manipulations on forage digestibility. A small reduction in lignin levels may have proportionally larger effects on digestibility.[7] A major concern now is to determine the extent to which lignin levels can be reduced without adverse effects on the plant's response to water stress and pathogen attack.

STRATEGIES FOR GENETIC MANIPULATION OF NODULATION GENE INDUCERS

Biosynthetic Inter-relationships between Alfalfa *nod* Gene Inducers

The initiation of a symbiotic interaction between bacteria of the genus *Rhizobium* or *Bradyrhizobium* and roots of leguminous plants is initiated by the activation of the nodulation (*nod*) genes of the bacteria. These genes encode enzymes required for the synthesis of specific nodulation factors recently shown to be acylated and sulfated lipo-oligosaccharides.[14] The nodulation factors induce root hair curling, infection, and cortical cell divisions in the plant roots[14,15] as initial stages in the formation of nitrogen-fixing nodules.

Bacterial *nod* genes are activated by host plant components released into the rhizosphere from seeds and roots. These *nod* gene inducers are flavonoid-derived compounds. *R. meliloti*, the alfalfa symbiont, is responsive to luteolin and chrysoeriol (two 5-hydroxy-substituted flavones released from alfalfa seeds)[16] and to 4',7-dihydroxyflavone, 4',7-dihydroxyflavanone, 4,4'-dihydroxy-2'-methoxychalcone and formononetin malonyl glucoside (5-deoxy derivatives released from alfalfa roots).[17,18] The biosynthetic origins and relationships between these compounds are shown in Fig. 3. Isoflavonoids such as daidzein and formononetin inhibit *nod* gene activation in some *Rhizobium* species,[19] but act as the major *nod* gene inducers for the soybean symbiont *Bradyrhizobium japonicum*.[20] In addition to *nod* gene induction, 5-hydroxyflavonoids released from alfalfa seeds can enhance the growth rate of *R. meliloti*, suggesting that alfalfa can influence the composition of the rhizosphere microbe population.[21]

The A-ring 5-deoxy substitution pattern of the root-released *nod* gene inducers arises from the activity of chalcone reductase, which reduces the polyketide formed from condensation of 4-coumaroyl CoA and malonyl CoA prior to its release from chalcone synthase.[22] This enzyme therefore defines the branch point between seed and root released *nod* gene inducers, and represents a potential target for manipulation of the relative levels of these compounds. Two other key reactions are catalyzed by *O*-methyltransferases; the B-ring methylation of luteolin to yield chrysoeriol and the A-ring methylation of the 2',4,4'-trihydroxychalcone. *O*-methylation of the trihydroxyhalcone prevents isomerization to the corresponding flavanone by chalcone isomerase, thereby theoretically limiting the formation of 4',7-dihydroxyflavanone and -flavone; this potential competition would not, however, affect the synthesis of luteolin and chrysoeriol. 4,4'-Dihydroxy-2'-methoxychalcone is the most potent of the *nod* gene inducers released from alfalfa roots, inducing *nodABC-lacZ* fusions at a concentration of around 1nM, some 10-fold lower than required for activation by luteolin.[17] Furthermore, methoxychalcone activates *nodABC* transcription through interaction with the product of the regulatory *nodD1* and *nodD2* genes, whereas luteolin only activates transcription via *nodD1*.[23]

Limitations to Nodulation Efficiency

The relative importance to the establishment of the *Rhizobium*/legume symbiosis of the *nod* gene inducers released from seeds to those released from roots is not clear. It is, however, believed that flavonoid *nod* gene inducers can, under some circumstances, be limiting for root nodulation and symbiotic nitrogen fixation.[24,25] Thus, it has been shown that a line of alfalfa selected through two generations for increased nitrogen fixation and growth contained 60% higher concentrations of *nod* gene inducers in its roots than the parent line. Conversely, the parent line established more nodules and fixed more nitrogen on addition of 10 µM luteolin to the rhizosphere. It would therefore seem feasible to

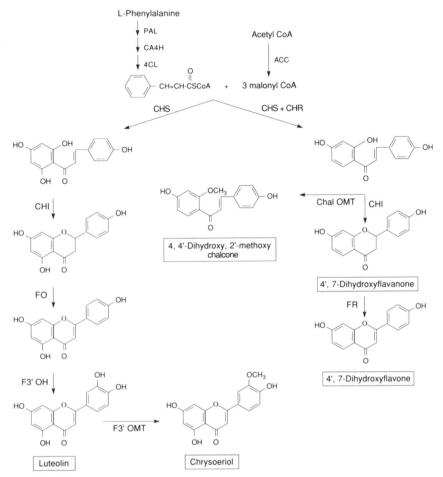

Fig. 3. Biosynthesis of nodulation gene inducers in alfalfa. Luteolin and chrysoeriol are released from seeds, whereas 4'-7-dihydroxyflavone, 4'7-dihydroxyflavanone and 4,4'-dihydroxy, 2'-methoxychalcone are released from roots. Enzymes are: PAL, phenylalanine ammonia-lyase; CA4H, cinnamic acid 4-hydroxylase; 4CL, 4-coumarate: CoA ligase; ACC, acetyl CoA carboxylase; CHS, chalcone synthase; CHR, chalcone reductase; CHI, chalcone isomerase; ChalOMT, chalcone 2'-*O*-methyltransferase; FO, flavanone oxidase; F3'OH, flavone 3'-hydroxylase, F3' OMT, flavone 3'-O-methyltransferase.

be able to genetically manipulate nodulation efficiency by increasing production of one or more of the *nod* gene inducers shown in Fig. 3. It is unlikely that CHI is limiting for production of either the 5-hydroxy or 5-deoxy series of compounds. Potential targets for genetic manipulation would therefore be flavanone oxidase, flavone 3'-hydroxylase (an α-ketoglutarate-dependent dioxygenase), flavone 3'-*O*-methyltransferase or chalcone *O*-methyltransferase (chalcone OMT). Of these enzymes, we have chosen to investigate the chalcone OMT, as it produces the most potent of the *nod* gene inducers of alfalfa in a single enzymatic step.

Cloning and Expression of Alfalfa Chalcone OMT

We recently purified to near homogeneity an OMT from alfalfa cell suspension cultures which catalyzes the 2'-*O*-methylation of 2',4,4'-trihydroxychalcone[26]. This enzyme exhibits very strict substrate specificity, and is inactive against all chalcones tested other than the trihydroxychalcone and its 4-methoxy derivative. Thus, chalcone OMT does not act on the chalcone substrate of the 5-hydroxy flavonoid derivatives released from alfalfa seeds, but its activity could theoretically block cyclization of 2',4,4'-trihydroxychalcone, and therefore production of the 5-deoxy flavone/flavanone root-released *nod* gene inducers (Fig. 3) and the antimicrobial isoflavonoids of alfalfa. Chalcone OMT is restricted to roots in alfalfa seedlings, where its activity increases from 48 h post-germination concomitant with the release of 4-4'-dihydroxy, 2'-methoxy-chalcone (Fig. 4). Treatment of roots with $CuCl_2$ inhibits appearance of chalcone OMT activity; $CuCl_2$ is an abiotic elicitor that induces isoflavone OMT activity in roots, but has no effect on caffeic acid OMT activity. Chalcone OMT is a monomer of Mr 45 kDa, and exhibits a K_M value of 2.2 μM for its chalcone substrate (approximately 4-fold lower than the K_M of alfalfa CHI).[26]

Using oligonucleotide probes based on N-terminal protein sequence information, we have now cloned the chalcone OMT and confirmed its activity following expression in *E. coli* .[27] Chalcone OMT genes are found in all the legumes we tested (alfalfa, *Medicago truncatula*, pea, and soybean), but were not found in tobacco, parsley, wheat or *Arabidopsis*. *In situ* hybridization has shown that chalcone OMT transcripts are localized to epidermal and cortical cells starting a few mm behind the root tip. Chalcone isomerase transcripts are also found in these cells, suggesting the need for a regulatory mechanism to control flux into the competing isomerization and O-methylation reactions.

Experiments are now in progress to manipulate the levels of methoxychalcone in transgenic alfalfa by overexpression and antisense inhibition

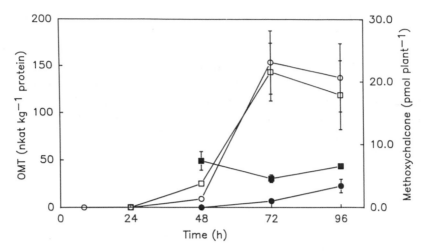

Fig. 4. Release of 4,4'-dihydroxy-2'-methoxychalcone from control (O) or CuCl$_2$-treated (●) roots of alfalfa seedlings, and corresponding activities of chalcone OMT in extracts from control (□) and CuCl$_2$-treated (■) seedlings immediately after collection of root exudates. Time = hours post-germination. Data from Ref. 26.

using the cloned chalcone OMT sequence. Increasing chalcone OMT activity should not affect production of luteolin or chrysoeriol, but may decrease the levels of 4',7-dihydroxy flavanone/flavone. In addition to providing a possible means of increasing nodulation efficiency, these studies should increase our knowledge of the regulation of *nod* gene inducer synthesis and release.

STRATEGIES FOR GENETIC MANIPULATION OF PHYTOALEXINS

Phytoalexins are low molecular weight, antimicrobial secondary metabolites, produced by plants mainly in response to attack by pathogens. Certain structural classes of phytoalexins are associated with certain plant families, such as phenolic phytoalexins in legumes, terpenoids in solanaceous plants, and polyacetylenes in the Asteraceae. In legumes, most of the identified phytoalexins are isoflavonoids or pterocarpans, although stilbenes and conjugated acetylenes have been found in some species. Often a mixture of biosynthetically-related phytoalexins is produced by the plant. In alfalfa, the major phytoalexin produced in our model cell culture system and infected plants is medicarpin, although two fungitoxic medicarpin derivatives, vestitol and sativan, are also known to accumulate in infected leaves.

Successful pathogens have evolved several mechanisms for overcoming a host's phytoalexin arsenal. One mechanism is to avoid inducing, or to actively suppress, the synthesis of phytoalexins by the host. For example, in resistant interactions between certain soybean cultivars and specific races of *Phytophthora megasperma* f.sp. *glycinea*, the host rapidly synthesizes large amounts of the phytoalexin glyceollin and the spread of the pathogen is contained, whereas in susceptible interactions the host synthesizes very little phytoalexin despite the spread of the pathogen throughout the host tissue.[28] Another mechanism involves degradation of the host's phytoalexin(s), either by very general enzymes such as peroxidases or by highly specialized detoxification pathways, some of which are induced by the phytoalexins themselves.[29] A third mechanism is insensitivity to phytoalexins, either due to lack of uptake or lack of the necessary sensitive metabolic target.

The above mechanisms employed by pathogens for escaping the host's phytoalexins suggest various strategies for the genetic manipulation of phytoalexin biosynthesis to overcome these mechanisms. These can be classed into two general categories. The first involves changing the regulation and/or timing of phytoalexin synthesis, by making production constitutive, for example. The second is based on modification of the structure of the phytoalexin(s) produced, either by modifying slightly the native phytoalexins, or by introducing new biosynthetic pathways into the plant. Specific strategies for these types of manipulations are outlined below, emphasizing examples which are directly applicable to alfalfa. Potential structural modifications to isoflavonoid phytoalexins are summarized in Fig. 5.

Structure-Activity Relationships: Potential Modifications to Existing Phytoalexin Pathways

Several series of structurally-related isoflavonoids have been isolated or synthesized and tested in bioassays against plant pathogens and other microorganisms. A large increase in antimicrobial activity has been correlated with the prenylation of certain isoflavonoids; the addition of 5-carbon side chains to genistein (isoflavone), dalbergioidin (isoflavanone), and demethylmedicarpin (pterocarpan) to form respectively wighteone, kievitone, and phaseollidin greatly

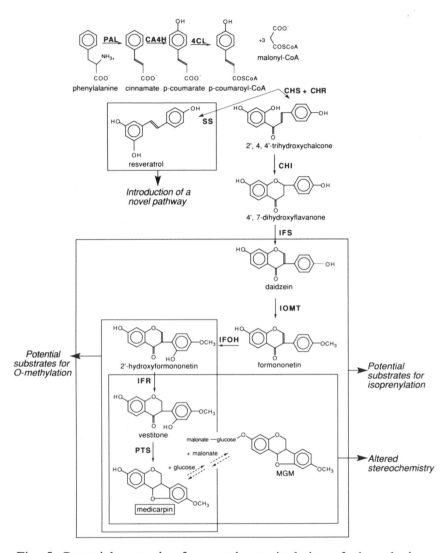

Fig. 5. Potential strategies for genetic manipulation of phytoalexin structures in alfalfa (see text). Enzymes are: PAL, phenylalanine ammonia-lyase; CA4H, cinnamic acid 4-hydroxylase; 4CL, 4-coumarate: CoA ligase, CHS, chalcone synthase; CHR, chalcone reductase; CHI, chalcone isomerase; IFS, isoflavone synthase; IOMT, isoflavone *O*-methyltransferase; IFOH, isoflavone 2'-hydroxylase; IFR, isoflavone reductase; PTS, pterocarpan synthase; SS, stilbene synthase. MGM is medicarpin malonyl glucoside.

increased the antifungal activity.[30-32] Prenylated isoflavonoids are common in certain legume species such as lupin (Lupinus) and green bean (*Phaseolus*) but have not been reported in alfalfa. Prenyltransferase activity capable of prenylating medicarpin has been detected in bean cultures[33] and isoflavonoid-prenylating activities have been assayed in wounded lupins.[34] Since suitable substrates for these enzymes already exist in alfalfa as intermediates in the phytoalexin pathway (Fig. 5), cloning of prenyltransferases from either source and expression in alfalfa is likely to result in production of new, more fungitoxic compounds (Fig. 6).

It is thought that prenylation increases fungitoxicity by increasing the lipophilicity or membrane permeability of a compound. The antifungal activity of isoflavonoids may also be increased by methylation of free hydroxyl groups, possibly by increasing lipophilicity or by protecting hydroxyl groups from oxidative detoxification reactions. Alfalfa methyltransferases acting on a wide range of isoflavonoids and pterocarpans have been partially characterized;[35] if cloned and over-expressed in alfalfa they may beneficially alter the phytoalexin profile (Fig. 5). An O-methyltransferase from pea which carries out the final step in pisatin biosynthesis has been purified.[36] This enzyme is highly active towards 6a-hydroxylated pterocarpans, but inactive towards medicarpin itself. Co-expression of the pea O-methyltransferase with a plant or fungal 6a–hydroxylase would result in the production of homopisatin (also known as variabilin) (Fig. 6). Methylation of the C3 hydroxyl of medicarpin would block one common detoxification route used by some alfalfa pathogens, while addition of the 6a-hydroxyl may help maintain sufficient solubility (some studies indicate that one free hydroxyl is required for bioactivity). The relative toxicities of medicarpin and homopisatin are currently being compared in bioassays using alfalfa pathogens in order to determine the potential utility of such modifications.

Most pterocarpan phytoalexins that have been analyzed possess the 6aR,11aR stereochemistry at the bridgehead carbons in the center of the molecule, although some contain the opposite 6aS,11aS configuration. The RS and SR isomers are theoretically forbidden due to steric considerations.[37] Alfalfa produces exclusively the 6aR,11aR or (-) isomer of medicarpin, whereas a few other legumes produce pterocarpans of the opposite stereochemistry, or mixtures of the two forms (Fig. 6). While there are no differences in the lipophilicity and other general chemical properties of the RR and SS pairs of pterocarpan isomers, the stereochemistry can have drastic effects on the antifungal activity against certain pathogens. In bioassays with several fungal pathogens of pea and red clover, two legumes which produce the phytoalexin 6aR,11aR or (-) maackiain, the (+) isomer of maackiain was significantly more toxic than the natural host

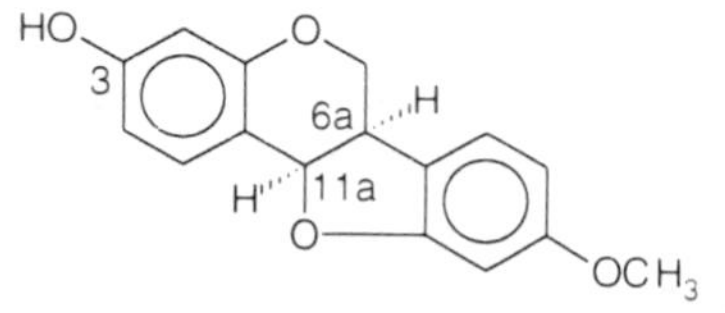

(-) or (6aR,11aR)-medicarpin

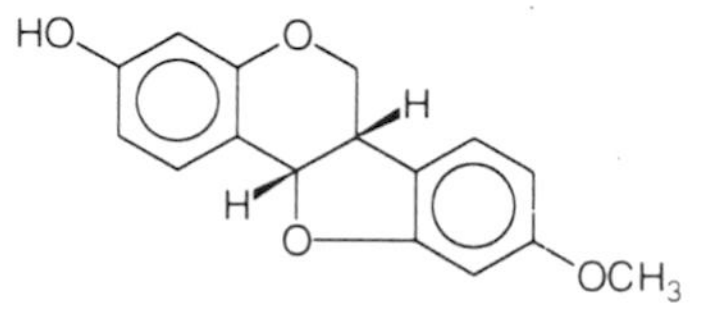

(+) or (6aS,11aS)-medicarpin

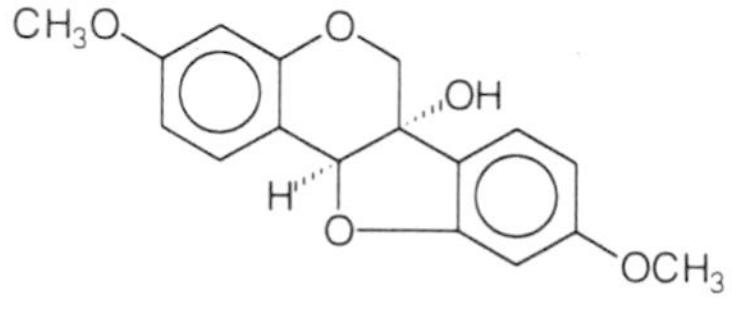

(-)- homopisatin

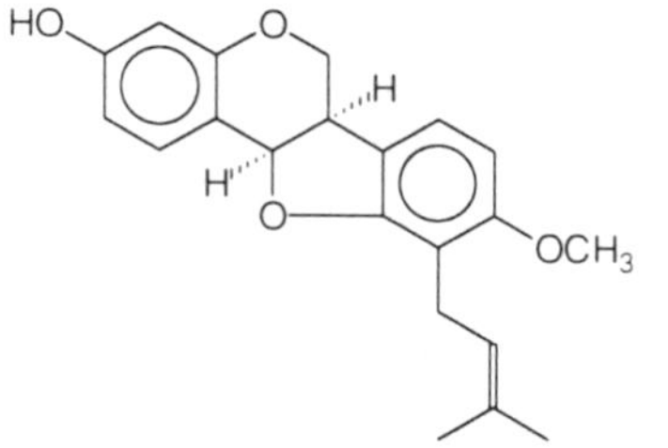

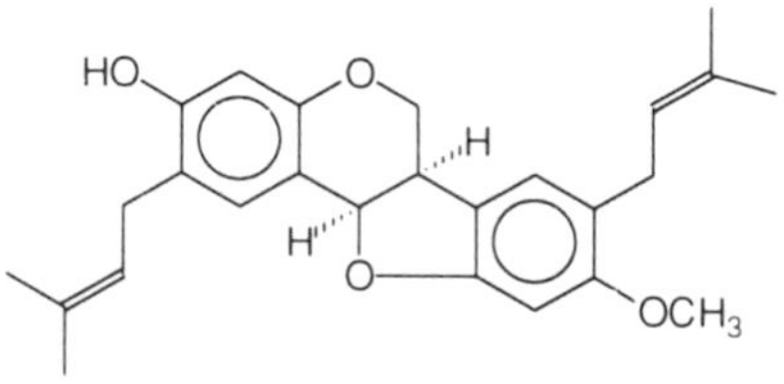

possible prenylated forms of medicarpin

Fig. 6. Structures of medicarpin stereoisomers and medicarpin derivatives which could be produced in alfalfa by genetic manipulation strategies.

isomer.[38] As an explanation of such phenomena, it has been suggested that pathogens have evolved very effective but very specialized machinery to destroy their natural host's phytoalexins. In a previous study, one isolate of a red clover pathogen could rapidly degrade (-) maackiain but was unable to metabolize (+) maackiain.[39]

Because many alfalfa pathogens show increased sensitivity to the (+) isomer of maackiain,[38] and because maackiain and medicarpin are structurally very similar, it is possible that genetic manipulation of alfalfa to produce (+) medicarpin might increase its resistance to certain pathogens. The relative

toxicities of (-) and (+) medicarpin against several alfalfa pathogens are now being compared to determine whether or not the trends observed with (-) and (+) maackiain are maintained. We are also investigating the enzymological differences between (-) and (+) pterocarpan-forming legumes, to determine which genes would need to be transferred to alter the stereochemistry of the phytoalexins produced. Isoflavone reductase (IFR) in alfalfa cell extracts or expressed in *E. coli* from an alfalfa IFR cDNA clone produces 3R-vestitone, creating the first of two chiral centers in (-) or 6aR,11aR medicarpin (Ref. 40 and N. L. Paiva, unpublished results). Surprisingly, only (-) IFR activity has been detected in crude extracts of pea and peanut, legumes which produce (+) pisatin and (+) medicarpin, respectively (N. L. Paiva, Y. Sun and G. Hrazdina, unpublished results); it was originally predicted that these legumes would contain (+) IFR enzymes which would produce 3S-isoflavanones. Various explanations for this finding are being explored, including the possible presence of an isoflavanone or pterocarpan epimerase.

Engineering New Phytoalexin Pathways

A general trend often noted in phytoalexin research over the years has been that successful pathogens of a specific host plant are insensitive to the phytoalexins of that host, whereas non-pathogens are sensitive to the same phytoalexins.[38,41] A corollary is that a pathogen, although sensitive to the phytoalexins of its natural hosts, might be insensitive to phytoalexins from plants on which it is not a pathogen. Genetic engineering techniques now allow for the introduction of new phytoalexin pathways from totally unrelated species into valuable crop plants. The first successful application of this strategy involved the production of a stilbene phytoalexin (common to grapevine, peanut, and pine species) in tobacco (a species known to produce terpenoid but not phenolic phytoalexins).[42] A fragment of grapevine genomic DNA containing two stilbene synthase genes was transferred into tobacco plants, resulting in inducible accumulation of the stilbene resveratrol (Fig. 5). Stilbene synthase utilizes the same substrates as chalcone synthase, malonyl CoA and 4-coumaroyl CoA, compounds which are found in all flavonoid-producing plants. These transgenic plants showed increased resistance to infection by *Botrytis cinerea*, thereby representing the first example of the direct manipulation of secondary metabolites to increase disease resistance.

The above strategy should also be successful if applied in alfalfa, possibly even more so. Like tobacco, alfalfa does not accumulate stilbene phytoalexins. It may, however, accumulate much higher levels of stilbenes than

tobacco if transformed with a stilbene synthase gene, as all of the genes encoding enzymes for the conversion of phenylalanine to 4-coumaroyl CoA, plus acetyl CoA carboxylase which synthesizes malonyl CoA, are known to be elicitor-inducible in alfalfa (Ref. 1 and see above). This suggests that the pools of the stilbene precursors may be much higher during induced defense in alfalfa than in tobacco.

Stilbene synthase is a rare example of a single enzyme which can carry out the production of a fungitoxic compound from common biosynthetic intermediates. The introduction of most other phytoalexin pathways into new hosts would require the transfer of several genes, but many phytoalexin-specific genes have not yet been cloned. For example, if one wished to produce medicarpin in tobacco, genes encoding all the enzymes of the pathway in Figure 5 beyond the CHI stage would have to be transferred.

Engineering Modifications to the Regulation of Existing Phytoalexin Pathways

Increased resistance to certain pathogens might be obtained by changing the timing of phytoalexin production or how much phytoalexin is produced. To ward off fungi which are sensitive to a particular set of phytoalexins but which fail to induce, or which may actually suppress, phytoalexin biosynthesis, the production of phytoalexins could be made constitutive. Two possible strategies can be envisioned. One is to clone each of the genes encoding the necessary enzymes in a pathway, and transform each back into the plant under the control of a constitutive promoter such as the CaMV 35S promoter. For a multistep pathway such as that leading from phenylalanine to medicarpin, the procedure is conceivable but would require the manipulation of more than ten genes. A possible alternative is to change the amounts of specific transcription factors present in the cells in such a way that the transcription of phytoalexin biosynthetic enzyme genes becomes constitutive. This could be achieved by either lowering the levels of suppressing factors or increasing the level of activating factors. Currently, there is some evidence that defense genes may have regulatory factor-binding sites in common, and therefore modifying the expression of a single transcription factor may affect the regulation of several genes in the pathway. Promoters from inducible PAL and CHS genes from bean share a number of highly conserved cis-elements,[43] and work is in progress to purify and clone the transcription factors which bind to these regions.[44] This would then make possible the manipulation of the levels of these factors by sense over-expression or antisense technology.

In certain instances, the amount of phytoalexin in a particular cultivar may be low due to a rate-limiting step in the pathway. A cultivar of chickpea susceptible to the fungus *Ascochyta* was found to produce much less medicarpin after inoculation than a resistant cultivar.[45] The activities of early pathway enzymes were quite similar between the two cultivars, but the levels of isoflavone 2'- and 3'-hydroxylase activities in the susceptible cultivar were much lower, leading the authors to conclude that these enzymes are the rate-limiting steps in the accumulation of medicarpin and maackiain in this cultivar. Increasing the level of these enzymes by introducing more copies of their genes may therefore improve the resistance of this cultivar. It is apparent, however, that increasing one activity will quickly result in a new rate limiting step; in the above study, PTS and IFR were also lower in the susceptible cultivar and could become rate-limiting if IFOH activity were increased.

One major concern regarding the constitutive accumulation of phytoalexins is the potential adverse effects on the plant and possibly the consumer. With many phytoalexins, it has been shown that the major accumulation is in a narrow zone around the infection site, whereas healthy tissues are free of these compounds. In alfalfa and other legumes which produce medicarpin and maackiain, these compounds accumulate constitutively at low levels in the roots, as malonylated glucoside conjugates (Fig. 5). Many phytoalexins are potentially toxic to the host and it is generally thought that this localized production, conjugation and sequestration in the vacuole are all possible mechanisms to protect the host from its defensive weapons. In the report of engineered stilbene production in tobacco,[42] no mention of adverse effects on the host were described, but this work used a pathogen-inducible promoter and only low levels of stilbenes were produced.

New phytoalexins may adversely affect the plant's interaction with symbionts. Most plants can form mycorrhizal associations with beneficial fungi, and legumes form nitrogen-fixing nodules in association with rhizobia, as discussed above. Plants must therefore have evolved mechanisms to avoid killing their symbionts with phytoalexins. There is evidence that the pterocarpan phytoalexin glyceollin plays a role in the soybean/*Bradyrhizobium japonicum* symbiosis. *Bradyrhizobium* strains which effectively nodulate soybean are sensitive to glyceollin but contain an inducible tolerance mechanism for protection against the host phytoalexin.[46] Ineffective (non-fixing) nodules contain much higher levels of glyceollin, suggesting that the plant is actively trying to remove the non-functional bacteria.[47] Introduction of a new phytoalexin may therefore inadvertently prevent a desirable symbiont from becoming established. Even if engineered phytoalexins are not toxic to the plant, they may have anti-quality effects, depending on the end use of the plant and on which parts accumulate the compounds. In two classic cases, such problems were seen in traditional breeding programs which selected for increased resistance to pests in celery and forage clovers.[41] The celery, while producing well in the field, contained high levels of antifungal psoralens which caused photoactivated blistering in workers who handled the produce. The clover, while performing well as a forage crop in Australia, contained high levels of estrogenic isoflavones and coumestrol, and caused infertility problems in sheep.

Phytoalexin-specific Gene Promoters

To avoid many of the concerns about constitutive production of non-native phytoalexins described above, the use of pathogen-inducible promoters would be beneficial. In theory, a defense-related promoter from the host would be ideal, such that the new phytoalexin-producing or -modifying genes are only expressed when and where the host would normally produce phytoalexins. There are several examples of foreign pathogen-inducible genes retaining their inducibility in transgenic tobacco, although the level of expression may be reduced below that in the species of origin. The peanut stilbene synthase promoter drives low levels of stilbene synthase expression,[42] while the bean chitinase,[48] bean CHS[49] and rice chitinase[50] promoters all drive localized, inducible expression of β-glucuronidase (GUS) in tobacco, demonstrating that there must be some conservation of the signal tranduction pathways for defense gene activation in different plants. However, the developmental regulation of defense response gene promoters may vary considerably depending on the species in which the transgene is inserted[51] (see below).

We have recently focussed attention on the regulation of the *IFR* gene from alfalfa. Expression of this gene (as indicated by mRNA levels) is closely correlated with IFR enzyme activity and phytoalexin accumulation under a number of conditions. Low expression is observed in roots where low levels of a medicarpin conjugate accumulate; no expression is observed in healthy, medicarpin-free leaves, while rapid, strong induction is observed in elicited cell cultures and pathogen-infected leaves (N. L. Paiva, unpublished results). We are interested in determining the *cis*-elements and *trans*-factors involved in the developmental and pathogen-inducible regulation of this gene, and their relations to the regulatory factors for pathway genes (e.g. *PAL, CHS*) which are also involved in non-defense-related processes such as the production of lignin, pigments, *nod*-gene inducers, and UV protectants. We have constructed *IFR* promoter-GUS fusions in binary vectors and introduced these into transgenic alfalfa and tobacco plants. In alfalfa, as little as 435 nucleotides of the promoter confer correct developmental and pathogen-inducible expression in whole plants and plant cell cultures, based on previous metabolite and mRNA analysis[40] (A. Oommen, R.A. Dixon, and N. L. Paiva, unpublished results). A longer promoter fragment (765 bp) confers a similar pattern of expression, but it is not yet known if the longer fragment confers higher expression. Such a phytoalexin-specific promoter might have useful practical applications in manipulating the phytoalexin profile in alfalfa. Potential medicarpin-modifying enzymes such as the prenyltransferases and *O*-methyltransferases described above, could be expressed under the control of a promoter such as the phytoalexin-specific *IFR* promoter to guarantee expression in the same cells and at the same time as the endogenous pathway genes. We plan on comparing the use of the inducible *IFR* promoter and constitutive CaMV 35S promoter for the phytoalexin engineering strategies outlined above. We are also using the longer fragment of the *IFR* promoter to drive the expression of anti-sense *IFR* constructs in an attempt to block phytoalexin biosynthesis in alfalfa.

   In tobacco, a plant which does not accumulate isoflavonoids or pterocarpans, alfalfa *IFR* promoter-GUS fusions have revealed a developmental expression pattern very different from that in alfalfa. The longer construct produced a very similar pattern of constitutive, cortex-specific root expression to that found in alfalfa roots, but also gave very high levels of expression in the above-ground plant parts, particularly xylem-associated cells, flower parts and seeds, tissues where the gene is not normally expressed in alfalfa. The shorter construct was almost completely inactive, except possibly in certain flower parts. The longer construct appears to be very weakly induced in transgenic tobacco cell cultures. Thus, while providing a useful tool for genetically engineered phytoalexin manipulations in alfalfa, the *IFR* promoter serves as a clear example of how an engineered defense gene might be incorrectly expressed in a foreign transgenic host.

REFERENCES

1.   DIXON, R.A., CHOUDHARY, A.D., DALKIN, K., EDWARDS, R., FAHRENDORF, T., GOWRI, G., HARRISON, M.J., LAMB, C.J., LOAKE, G.J., MAXWELL, C.A., ORR, J., PAIVA, N.L. 1992. Molecular biology of stress-induced phenylpropanoid and isoflavonoid biosynthesis in alfalfa. In: Phenolic Metabolism in Plants, (H.A. Stafford, R.K. Ibrahim, eds), Plenum Press, New York, pp 91-138.

2.   FAHRENDORF, T., DIXON, R.A. 1993. Stress responses in alfalfa
      (*Medicago sativa* L.). XVIII. Molecular cloning and expression of the
      elicitor-inducible cinnamic acid 4-hydroxylase cytochrome P450.
      Arch. Biochem. Biophys.305:509-515.

3.   GUNIA, W., HINDERER, W., WITTKAMPF, U., BARZ, W. 1991.
      Elicitor induction of cytochrome P450 monooxygenases in cell
      suspension cultures of chickpea (*Cicer arietinum* L.) and their
      involvement in pterocarpan phytoalexin biosynthesis. Z. Naturforsch.
      46C: 58-66.

4.   DIXON, R.A., BHATTACHARYYA, M.K., HARRISON, M.J.,
      FAKTOR, O., LAMB, C.J., LOAKE, G.J., NI, W., OOMMEN, A.,
      PAIVA, N.L., STERMER, B., YU, L.M. 1993. Transcriptional
      regulation of phytoalexin biosynthetic genes. In: Advances in
      Molecular Genetics of Plant-Microbe Interactions, (E.W. Nester,
      D.P.S. Verma, eds), Kluwer Academic Publishers, Dordrecht, pp
      497-509.

5.   LINDSAY, W.P., LAMB, C.J., DIXON, R.A. 1993. Microbial
      recognition and activation of plant defense systems. Trends in
      Microbiol. 1:181-187.

6.   WALTER, M.H. 1992. Regulation of lignification in defense. In: Plant
      Gene Research. Genes Involved in Plant Defense (T. Boller, F.
      Meins, eds.), Springer, Wien, pp 327-353.

7.   CASLER, M.D. 1987. *In vitro* digestibility of dry matter and cell wall
      constituents of smooth bromegrass forage. Crop Sci. 27: 931-934.

8.   ALBRECHT, K.A., WEDIN, W.F., BUXTON, D.R. 1987. Cell wall-
      composition and digestibility of alfalfa stems and leaves. Crop Sci.
      27:735-741.

9.   ELKIND, Y., EDWARDS, R., MAVANDAD, M., HEDRICK, S.A.,
      RIBAK, O., DIXON, R.A., LAMB, C.J. 1990. Abnormal plant
      development and down regulation of phenylpropanoid biosynthesis in
      transgenic tobacco containing a heterologous phenylalanine
      ammonia-lyase gene. Proc. Natl. Acad. Sci. USA 87: 9057-9061.

10.   CRAMER, C.L., EDWARDS, K., DRON, M., LIANG, X., DILDINE,
      S.L., BOLWELL, G.P., DIXON, R.A., LAMB, C.J., SCHUCH,
      W. 1989. Phenylalanine ammonia-lyase gene organization and
      structure. Plant Mol. Biol. 12: 367-383.

11.   GOWRI, G., BUGOS, R.C., CAMPBELL, W.H., MAXWELL, C.A.,
      DIXON, R.A. 1991. Stress responses in alfalfa (*Medicago sativa* L.).
      X. Molecular cloning and expression of S-adenosyl-L-methionine:

caffeic acid 3-O-methyltransferase, a key enzyme of lignin biosynthesis. Plant Physiol. 97: 7-14.

12. JAECK, E., DUMAS, B., GEOFFROY, P., FAVET, N., INZE, D., VAN MONTAGU, M., FRITIG, B., LEGRAND, M. 1992. Regulation of enzymes involved in lignin biosynthesis: induction of O-methyltransferase mRNAs during the hypersensitive reaction of tobacco to tobacco mosaic virus. Mol. Plant Microbe. Interact. 5: 294-300.

13. NI, W., PAIVA, N.L., DIXON, R.A. 1993. Reduced lignin in transgenic plants containing an engineered caffeic acid O-methyltransferase antisense gene. Transgenic Pres., in press.

14. LEROUGE, P., ROCHE, P., FAUCHER, C., MAILLET, F., TRUCHET, G., PROME´, J.C., DEBNARIE´, J. 1990. Symbiotic host-specificity of Rhizobium meliloti is determined by a sulphated and acylated glucosamine oligosaccharide signal. Nature 344: 781-784.

15. TRUCHET, G., ROCHE, P., LEROUGE, P., VASSE, J., CAMUT, S., DE BILLY, F., PROME´, J.C., DENARIE´, J. 1991. Sulphated lipo-oligosaccharide signals of Rhizobium meliloti elicit root nodule organogenesis in alfalfa. Nature 351: 670-673.

16. HARTWIG, U.A., MAXWELL, C.A., JOSEPH, C.M., PHILLIPS, D.A. 1990. Chrysoeriol and luteolin released from alfalfa seeds induce nod genes in Rhizobium meliloti. Plant Physiol., 92: 116-122.

17. MAXWELL, C.A., HARTWIG, U.A., JOSEPH, C.M., PHILLIPS, D.A. 1989. A chalcone and two related flavonoids released from alfalfa roots induce nod genes in Rhizobium meliloti. Plant Physiol. 91: 842-847.

18. DAKORA, F., JOSEPH, C.M., PHILLIPS, D.A. 1993. Alfalfa (Medicago sativa L.) root exudates contain isoflavonoids in the presence of Rhizobium meliloti. Plant Physiol. 101: 819-824.

19. FIRMIN, J.L., WILSON, K.E., ROSSEN, L., JOHNSTON, A.W.B. 1986. Flavonoid activation of nodulation genes in Rhizobium reversed by other compounds present in plants. Nature 324: 90-92.

20. KOSSLAK, R.M., BOOKLAND, R., BARKEI, J., PAAREN, H.E., APPELBAUM, E.R. 1987. Induction of Bradyrhizobium japonicum common nod genes by isoflavones isolated from Glycine max. Proc. Natl. Acad. Sci. USA 84: 7428-7432.

21. HARTWIG, U.A., JOSEPH, C.M., PHILLIPS, D.A. 1991. Flavonoids released naturally from alfalfa seeds enhance growth rate of

Rhizobium meliloti. Plant Physiol. 95: 797-803.

22. WELLE, R., SCHRODER, G., SCHILTZ, E., GRISEBACH, H., SCHRODER, J. 1991. Induced plant responses to pathogen attack. Analysis and heterologous expression of the key enzyme in the biosynthesis of phytoalexins in soybean (*Glycine max* L. Merr. cv. Harosoy 63). Eur. J. Biochem. 196: 423-430.

23. HARTWIG, U.A., MAXWELL, C.A., JOSEPH, C.M., PHILLIPS, D.A. 1990. Effects of alfalfa *nod* gene-inducing flavonoids on *nodABC* transcription in *Rhizobium meliloti* strains containing different *nodD* genes. J. Bacteriol. 172: 2769-2773.

24. KAPULNIK, Y., JOSEPH, C.M., PHILLIPS, D.A. 1987. Flavone limitations to root nodulation and symbiotic nitrogen fixation. Plant Physiol. 84: 1193-1196.

25. JAIN, V., GARG, N., NAINAWATEE, H.S. 1990. Naringenin enhanced efficiency of *Rhizobium meliloti*-alfalfa symbiosis. World J. Microbiol. 6: 434-436.

26. MAXWELL, C.A., EDWARDS, R., DIXON, R.A. 1992. Identification, purification, and characterization of S-adenosyl-L-methionine: isoliquiritigenin 2'-O-methyltransferase from alfalfa (*Medicago sativa* L.). Arch. Biochem. Biophys. 293: 158-166.

27. MAXWELL, C.A., HARRISON, M.J., DIXON, R.A. 1993. Molecular characterization and expression of alfalfa isoliquiritigenin 2'-O-methyltransferase, an enzyme specifically involved in the biosynthesis of an inducer of *Rhizobium meliloti* nodulation genes. Plant J., in press.

28. BHATTACHARYYA, M.K., WARD, E.W.B. 1986. Resistance, susceptibility, and accumulation of glyceollins I-III in soybean organs inoculated with *Phytophthora megasperma* f.sp. *glycinea*. Physiol. Mol. Plant Pathol. 29: 227-237.

29. VAN ETTEN, H.D., MATTHEWS, D.E., MATTHEWS, P.S. 1989. Phytoalexin detoxification: Importance for pathogenicity and practical implications. Annu. Rev. Phytopathol. 27: 143-164.

30. INGHAM, J.L., KEEN, N.T., HYMOWITZ, T. 1977. A new isoflavone phytoalexin from fungus-inoculated stems of *Glycine wightii*. Phytochemistry 16: 1943-1946.

31. WOODWARD, M.D. 1979. Phaseoluteone and other 5-hydroxyisoflavonoids from *Phaseolus vulgaris*. Phytochemistry 18: 363-365.

32. ADESANYA, S.A., O'NEILL, M.J., ROBERTS, M.F. 1986. Structure-

related fungitoxicity of isoflavonoids. Physiol. Mol. Plant Pathol. 29: 95-103.

33. BIGGS, D.R.M., WELLE, R., VISSER, F.R., GRISEBACH, H. 1987. Dimethylallylpyrophosphate: 3,9-dihydroxypterocarpan 10-dimethylallyl transferase from *Phaseolus vulgaris*. FEBS Lett. 220: 223-226.

34. ATTUCCI, S., GULICK, P., IBRAHIM, R.K. 1993. Differential hybridization strategy for cloning plant isoflavone prenyltransferases. (Abstr) Plant Physiol. Supplement 102: 114.

35. EDWARDS, R., DIXON, R.A. 1991. Isoflavone O-methyltransferase activities in elicitor-treated cell suspension cultures of *Medicago sativa*. Phytochemistry 30: 2597-2606.

36. PREISIG, C.L., MATTHEWS, D.E., VAN ETTEN, H.D. 1991. Purification and characterization of S-adenosyl-L-methionine: 6a-hydroxymaackiain 3-O-methyltransferase from *Pisum sativum*. Plant Physiol. 91: 559-566

37. DEWICK, P. 1988. Isoflavonoids. In: The Flavonoids: Advances in Research Since 1980, (J.B. Harborne, ed), Chapman and Hall, New York, p 166.

38. DELSERONE, L.M., MATTHEWS, D.E., VAN ETTEN, H.D. 1992. Differential toxicity of enantiomers of maackiain and pisatin to phytopathogenic fungi. Phytochemistry 31: 3813-3819.

39. VAN ETTEN, H.D., MATTHEWS, P.S., MERCER, E.H. 1983. (+)-Maackiain and (+) medicarpin as phytoalexins in *Sophora japonica* and identification of the (-) isomers by biotransformation. Phytochemistry 22: 2291-2295.

40. PAIVA, N.L., EDWARDS, R., SUN, Y., HRAZDINA, G., DIXON, R.A. 1991. Molecular cloning and expression of alfalfa isoflavone reductase, a key enzyme of isoflavonoid phytoalexin biosynthesis. Plant Mol. Biol. 17: 653-667.

41. SMITH, D.A. 1982. Toxicity of phytoalexins. In: Phytoalexins, (J.A. Bailey, J.W. Mansfield, eds), John Wiley and Sons, New York, pp 218-252.

42. HAIN, R., REIF, H.-J., KRAUSE, E., LANGEBARTELS, R., KINDL, H., VORNAM, B., WIESE, W., SCHMELZER, E., SCHREIER, P.H., STOCKER, R.H., STENZEL, K. 1993. Disease resistance results from foreign phytoalexin expression in a novel plant. Nature 361: 153-156.

43. DIXON, R.A., LAMB, C.J. 1990. Regulation of secondary metabolism

at the biochemical and genetic levels. In: Secondary Products from Plant Tissue Culture, (B.V. Charlwood and M.J.C. Rhodes eds), Clarendon Press, Oxford, pp 103-118.

44.   YU, L.M., LAMB, C.J., DIXON, R.A. 1993. Purification and biochemical characterization of proteins which bind to the H-box *cis*-element implicated in transcriptional activation of plant defense genes. Plant J. 3:805-816.

45.   DANIEL, S., TIEMANN, K., WITTKAMPF, U., BLESS, W., HINDERER, W., BARZ, W. 1990. Elicitor-induced metabolic changes in cell cultures of chickpea (*Cicer arietinum* L.) cultivars resistant and susceptible to *Ascochyta rabiei*. Planta 182: 270-278.

46.   PARNISKE, M., AHLBORN, B., WERNER, D. 1991. Isoflavonoid-inducible resistance to the phytoalexin glyceollin in soybean Rhizobia. J. Bacteriol. 173: 3432-3439.

47.   PARNISKE, M.C., FISCHER, H.-M., HENNECKE, H., WERNER, D. 1991. Accumulation of the phytoalexin glyceollin I in soybean nodules infected by a *Bradyrhizobium japonicum nifA* mutant. Z. Naturforsch. 46C: 318-320.

48.   ROBY, D., BROGLIE, K., CRESSMAN, R., BIDDLE, P., CHET, I., BROGLIE, R. 1990. Activation of a bean chitinase promoter in transgenic tobacco plants by phytopathogenic fungi. Plant Cell 2: 999-1007.

49.   STERMER, B.A., SCHMID, J., LAMB, C.J., DIXON, R.A. 1990. Infection and stress activation of a bean chalcone synthase promoter in transgenic tobacco. Mol. Plant-Microbe Interact. 3: 381-388.

50.   ZHU, Q., DOERNER, P.W., LAMB, C.J. 1992. Developmental and stress regulation of a rice chitinase gene promoter in transgenic tobacco. Plant J. 3: 203-212.

51.   BENFEY, P.N., CHUA, N.-H. 1989. Regulated genes in transgenic plants. Science 244: 174-181.

Chapter Seven

THE GENETIC ORIGINS OF BIOSYNTHESIS AND LIGHT-RESPONSIVE CONTROL OF THE CHEMICAL UV SCREEN OF LAND PLANTS

Richard Jorgensen

Department of Environmental Horticulture
University of California
Davis, California 95616-8587

INTRODUCTION

Most land plants possess the capacity to protect themselves from UV light, and do so by producing pigments that absorb efficiently in the UV-A and UV-B regions of the spectrum while allowing transmission of nearly all

Genetic Engineering of Plant Secondary Metabolism,
Edited by B.E. Ellis *et al.*, Plenum Press, New York, 1994

photosynthetically useful wavelengths.[1,2] These UV-absorbing pigments are mainly phenylpropanoids and flavonoids. This chapter summarizes current understanding of the mechanism of UV protection in higher land plants, evaluates the information available from lower land plants and their green-algal relatives, and then considers the possible evolutionary origins of this use of chemical filters for selectively screening UV light from solar radiation. It is proposed that photocontrol over the biosynthesis of UV-absorbing phenylpropanoids and flavonoids may have evolved in concert with the evolution of the high biosynthetic activity necessary for UV protection. The toxicity of phenylpropanoids and flavonoids has been postulated to have been a barrier to the evolution of an effective chemical UV screen, and that some means for sequestering these compounds and/or for controlling their synthesis probably evolved prior to, or in concert with, the evolution of high rates of biosynthesis. The original photoreceptor and signal transduction system is speculated to have been based on photoisomerization of a phenylpropanoid ester and a pre-existing product feedback mechanism for controlling phenylpropanoid biosynthesis. Understanding the original mechanism for photocontrol of the chemical UV screen of land plants could be valuable for understanding the adaptability of extant land plants to rising levels of solar UV-B radiation and may suggest genetic strategies for engineering improved UV tolerance in crop plants.

THE CHEMICAL UV SCREEN OF HIGHER LAND PLANTS

The hypothesis that flavonoids have a UV-protective function in higher land plants is supported by the following observations about these compounds: (1) they absorb efficiently in the UV, (2) they are localized preferentially in the epidermis, (3) they accumulate rapidly following UV irradiation, and (4) the most efficient wavelength for induction (290 - 300 nm) is also the most damaging.[2,3]

Flavones and flavonols absorb strongly in both the UV-B and the UV-A regions of the spectrum and are primarily responsible among flavonoids for UV protection in plants. Among the various phenylpropanoids, the substituted cinnamic acids, cinnamyl aldehydes, and cinnamyl alcohols are all potential UV protectants. These compounds, because they possess an exocyclic double bond that is conjugated with an aromatic ring, absorb UV light about 20-fold more effectively than do phenylalanine and tyrosine. They also have absorbance maxima better placed in the important UV-B region. Anthocyanins are often mentioned erroneously as UV protectants, perhaps because their synthesis is responsive to UV light in some systems; however, they absorb poorly in the UV-A and UV-B and actually play little role in UV protection.[4,5]

Higher land plants typically sequester phenylpropanoids in vacuoles, often as glycosides, and in cell walls in various forms.[6] Cinnamic acids are found often as esters in the walls of epidermal cells, including in the cuticle. Cinnamyl alcohols, of course, are found in various cell wall polymers, including lignin, suberin, and wound lignins, but probably not for use in UV absorption. Flavonoids are sequestered in vacuoles, cell walls, and cuticles in a variety of bound forms, and even as free precipitate on plant surfaces.[5] Appropriate localization is likely to be important for effective UV protection and determined by this need. However, interpretation is not straightforward since patterns of localization and sequestration are also determined by the toxicity of these compounds and the variety of functions they possess such as disease resistance.

In higher land plants, the synthesis of phenylpropanoids, flavones and flavonols is controlled developmentally, can be induced by stress and disease, and is responsive to three photoreceptor systems. One photoreceptor is phytochrome which mainly responds to red and far red light, but also to blue and UV to some extent. Another is the so-called blue light photoreceptor (sometimes "cryptochrome"), which is sensitive to blue and UV-A light and thought to be based on a flavin. The third is the UV-B photoreceptor which is relevant mainly to flavonoid biosynthesis and which is largely uncharacterized. Each of the three photoreceptors can exert control over flavonoids at the level of gene transcription. Among higher plants, a great diversity of different modes of flavonoid photocontrol occur (reviewed by Beggs et al.[4]). In parsley roots, for example, neither blue light nor red light can induce flavonoid synthesis, but each can modulate the effect of UV-B light, which is required for synthesis. This is in contrast to parsley leaves, in which blue and red light each induce flavonoid production (via the blue light receptor and phytochrome), though much less effectively than does UV-B. Complicating the interpretation of UV-treatment experiments is the fact that high levels of UV-B (and UV-C) light also induce stress responses, which in turn induce flavonoid synthesis. Since these have an action spectrum at lower levels of UV-B that is distinct from the induction of flavonoid synthesis, it is unlikely that these responses relate to a specific UV photoreceptor, but rather are simply a reaction to the damage caused by the absorption of UV light.[2] This does not rule out the existence of a UV-B photoreceptor controlling flavonoid biosynthesis, but illustrates the experimental difficulties in studying it. The advantages presented by investigation of the immediate ancestors of land plants will be considered in this chapter.

Measuring Contributions to the UV Screen

The relative importance of flavonoids as compared to phenylpropanoids in UV protection is somewhat uncertain and may vary among plant species. Recently, this question has been addressed with two new approaches, one optical and the other genetic. Measurement of the depth of penetration of UV-B and visible light into plants with fiber optic microprobes is a new methodology for making possible the direct measurement of UV-B absorption by cell layers.[7,8] Using this technology, the UV-screening properties of conifer needles have been investigated during foliar development at varying depths in the epidermis and mesophyll.[9] It was shown that attenuation of UV light by the epidermis was not correlated with the presence of soluble compounds (mostly flavonoids), but with insoluble phenolics, possibly ferulic acid and other phenylpropanoids co-polymerized with cutin and lignin in the epidermal cell wall (which is exceptionally thick and lignified in conifers).

The other new approach to this problem is the use of mutants diminished in flavonoid and/or phenylpropanoid production. Li et al[10] have shown that *Arabidopsis* chalcone synthase mutants with reduced flavonoid levels are hypersensitive to UV-B light, but also that mutants reduced in both flavonoids and phenylpropanoids are even more sensitive, implying that both flavonoids and phenylpropanoids play roles in UV screening. Further work with additional mutants seems likely to provide much more detailed information than is now available on the particular compounds important for UV protection. Reverse genetic approaches can also be expected to yield useful information on the role of specific phenylpropanoids and flavonoids in UV protection biology. These approaches make it possible to use a molecularly cloned gene to create a "dominant negative mutation" by interfering witht the expression of the homologous, endogenous gene in a plant. This can be done by antisense suppression, sense suppression, and various other means.[11] These approaches can be applied to phenylpropanoid and flavonoid genes in any plant species in which whole, fertile transgenic plants can be produced, such as *Arabodopsis* and *Petunia*. Many flavonoid genes have been molecularly cloned (for review, see Dooner *et al.*),[12] and dominant negative mutations can be produced. It will be particularly interesting to assess the result of elimination of flavone and flavonol synthases when these gene sequences also become available. Combined with efforts to localize these compounds and to determine their chemical state (i.e., unbound, esterified, glycosylated, etc.), genetic approaches promise to improve understanding of the UV screening capabilities of higher plants.

THE ROLE OF UV PROTECTION IN THE ORIGIN OF LAND PLANTS

The immediate ancestors of the first land plants are believed to have been green algae inhabiting the shallow littoral zone in bodies of fresh water. Water absorbs UV-B light inefficiently (it takes a several meter column of water to reduce it to non-damaging levels), and so the effects of solar UV radiation on aquatic organisms in shallow water are substantial.[13] Thus, it is presumed that the shallow-dwelling aquatic ancestors of land plants would have benefitted from UV-protecting pigmentation, and probably required it. This, taken together with the near ubiquity of flavonoids and phenylpropanoids in major land plant groups, has led to the suggestion that the capabilities of modern land plants for producing UV-absorbing pigments might have arisen first in such freshwater algae.[5,14]

To illustrate further the likely benefits to shallow-water-dwelling algae of possessing (and controlling the amounts of) UV-absorbing pigments, the example of UV-protectants produced by some corals and anemones is valuable. These corals and anemones are each comprised of symbiotic animal and algal partners that are protected from solar UV because the organism produces UV-absorbing mycosporine-like amino acids (MAAs). The MAAs appear in response to long-term UV exposure, and the concentration of MAAs is correlated with depth of the organism in the sea and with photosynthetic sensitivity of the algal partner to UV light, implying that MAAs play a natural role in UV protection.[15] Another example is that of scytonemin, a UV-absorbing pigment of many cyanobacteria living in high light environments. Scytonemin accumulates in extracellular sheaths in response to high light, absorbs most of the incident UV-A and much of the UV-B, and provides resistance to UV-A-caused photoinhibition.[16]

These examples demonstrate the value of a light-responsive chemical UV screen to organisms living in environments subject to high levels of solar radiation. The benefits of such a screen to land plants are clear, given that nearly all are obligate phototrophs and occupy terrestrial habitats, often in full sunlight. As indicated earlier, phenylpropanoids and flavonoids have been implicated as the primary protectants. Given the potential for rising levels of solar UV-B in the near future, detailed understanding of the regulation of land plants' chemical UV screen may be of great importance. It is through its regulatory system that plants can most quickly adapt to increasing levels of UV-B. It is the perspective of this chapter that the capacity for response to high UV-B levels in extant land plants has its basis in the evolution of UV protection in the algal ancestors of land plants. How can the capabilities and potential of the genetic systems that regulate the chemical UV screen of land plants be better understood?

THE EVOLUTION OF UV PROTECTION

Charophycean Algae

The extinct algal ancestors of land plants are believed to be related to present-day members of the green-algal class Charophyceae. The Charophyceae share with land plants a broad array of features distinguishing them from all other green algae. In fact, the most advanced charophyceans, though clearly algae, can be described in many respects as extremely primitive plants. Many authorities consider them to be closer to bryophytes, especially to hornworts, than to other algae (e.g., ref. 17). The most plant-like of the Charophyceae is *Coleochaete*, which retains its zygotes on the maternal thallus, a likely step towards the retention and nurturing of the zygote within the tissue of the gametophyte, a process defining the embryophytes (reviewed by Graham).[18] *Coleochaete* grows in high-light, littoral-zone, freshwater habitats as an epiphyte or epilith. The other charophyceans to be addressed here belong to the order Charales which includes the common and ecologically important genera *Chara* and *Nitella*, known collectively as charophytes. Charophytes are common in shallow freshwater habitats with high light levels and low nutrient content, but many species can adapt to very low light levels through a plant-like photomorphogenic response involving phytochrome. They require a sandy or muddy substrate to which they attach themselves via rhizoids, and they grow by apical cell division resembling meristematic growth in plants (reviewed by Grant).[19]

Unfortunately, nothing is known of the capacity for these algae to protect themselves from UV radiation, and only limited information exists on even the presence of flavonoids and phenylpropanoids in charophyceans. Although flavones, a principal class of plant UV protectants, have been reported to be present at low levels in the charophyte *Nitella hookeri*, Markham suggests that this result should be accepted only with reservation.[20] No comprehensive survey for flavonoids in the Charophyceae has been reported, and there has been no published attempt to detect flavonoids in the most plant-like of these algae, *Coleochaete*.

With respect to phenylpropanoids, both *Coleochaete* and *Staurastrum* (a more primitive charophycean) have been found to be able to produce substantial amounts of a lignin-like polymer in their cell walls, probably for the purpose of microbial defense.[21,22] This implies a capability to synthesize hydroxycinnamyl alcohols and to localize them outside the plasma membrane, though this remains to be demonstrated. If confirmed, this would suggest that charophycean algae have the biochemistry for basic phenylpropanoid synthesis, including p-coumaroyl Coenzyme A, which is not only required for hydroxycinnamyl alcohol biosynthesis, but is also the phenylpropanoid substrate for chalcone synthase, the first step in flavonoid biosynthesis. Clearly, much work needs to be done before we have a definitive picture of phenylpropanoid and flavonoid biosynthesis in charophycean algae.

Lower Land Plants

If it were the case that UV protection originated in the green-algal ancestors of land plants, one might expect to find evidence that all lower groups of land plants (e.g., bryophytes and lower tracheophytes such as ferns, fern allies and horsetails) possess this protection. The information available comes from chemical taxonomic studies, which at best are only indirect regarding UV protection biology. All land plant groups, including all three subclasses of bryophytes, produce cinnamyl alcohols in one form or another.[23,24] All groups of tracheophytes produce flavonoids, but flavonoids have been detected only in some of the tested species of bryophytes. This might seem to suggest that flavonoid biosynthesis evolved sometime during the evolution of bryophytes. However, it has been argued that flavonoid biosynthesis has been lost, or progressively reduced, in many lineages of bryophytes[20] and that its functions have been replaced by other biochemical pathways now highly developed in these lineages, e.g., terpenoids for defense purposes in liverworts,[25] or that compounds were simply at too low a concentration to be detected. [26] Thus, it is possible that flavonoid biosynthesis evolved prior to the origin of all extant land plant groups; this possibility needs further investigation.

Evolutionary Diversification of Gene Functions

Another factor that confounds interpretation of the evolutionary history of flavonoids and interferes with attempts to trace it back to its origins is the evolutionary diversification of the chalcone synthase enzyme. Higher plant stilbene synthases are now known to be homologous to chalcone synthases.[27] Thus, the ancestral function of the gene is an open question.[28] Bibenzyls, especially lunularic acids, are a class of compounds known to be widespread in liverworts.[24] The precursors for bibenzyl synthesis, dihydro-*p*-coumaric acid and acetyl CoA, are similar to the precursors for flavonoid biosynthesis, and so, given a likely common origin of chalcone synthase and stilbene synthase, it is necessary to consider that bibenzyl synthase might also be homologous to chalcone synthase. Thus, in order to understand the genetic origins of UV protection biochemistry in land plants, detailed knowledge of the functional evolution of chalcone synthase, stilbene synthase, and bibenzyl synthase in early land plants and their nearest algal ancestors is needed.

Evolution of Photocontrol over Phenylpropanoid and Flavonoid Biosynthesis

The production of UV-absorbing compounds by an organism, or the mere presence of genes responsible for their synthesis, does not demonstrate that an organism has evolved the capacity to use those compounds for UV protection, i.e., to produce them in sufficient quantities and at appropriate times and locations. Potentially more informative is the mode of control over biosynthesis. Because UV protection is most beneficial in high light environments, it might be expected that biosynthesis of UV-absorbing pigments is responsive to light.[2] There are two reasons for suggesting this. First, because high metabolite concentrations are required for UV protection, light responsiveness would ensure that production occurs only when necessary, preventing wasteful production. Second, because most phenolics are somewhat phytotoxic, photocontrol over their synthesis would limit any deleterious side-effects of their accumulation to circumstances in which they are beneficial[29].

However, the fact that flavonoids and phenylpropanoids have been the object of considerable exploitation during the evolution of higher plants, has resulted in a wide range of functions besides UV protection. Accompanying the many functional roles of these two large families of compounds are a complex web of interacting controls over biosynthesis, including induction by stress, light, development, and disease [5,30] Thus, understanding the relationships between function and control in higher land plants can be exceedingly difficult. Because of this, investigations in lower land plants or higher green algae may be helpful in determining the true role of photocontrol in UV protection.

Prior to the evolution of some means to sequester phenylpropanoids in lower land plants or charophycean algae, these compounds must have been present only at low concentrations because of their toxicity. For this reason it has been proposed that the first evolutionary advances in phenylpropanoid biochemistry may have been driven by another function, e.g., as internal physiological regulators or signal molecules.[28] Later, their potential value as anti-fungal agents could have been exploited through the evolution of some means for vacuolar targetting or transport into the wall, followed by the evolution of higher biosynthetic rates. Whether the levels of phenylpropanoids that are effective for defense would also be adequate to provide *some* UV protection is unknown and needs to be investigated. Presumably though, the production of sufficient levels of phenylpropanoids for *full* UV-B protection would have evolved subsequently to their exploitation for defensive purposes, and then perhaps only by subjecting their expression to control by light in order to link the production of large quantities of pigments to the need for them.

Thus, UV light *inducibility* of UV-absorbing compounds may have been a key requirement for exploitation of these compounds as efficient UV protectants. If so, this would be a key criterion also for assessing whether charophycean algae, and by extrapolation their ancestors in common with the higher land plants, have some capacity for UV protection, and for determining which biosynthetic genes and which biochemical products are likely to play a role in UV protection.

A SPECULATION REGARDING THE ORIGINS OF PHOTOCONTROL OVER UV PROTECTION AND THE NATURE OF THE UV-B PHOTORECEPTOR

Many green algae possess signal transduction systems for both phytochrome and a blue light photoreceptor.[31] It is possible that the phenylpropanoid pathway could have been coupled first to one of these photoreceptor systems, perhaps during charophycean evolution. There seems to be little evidence for a specific UV-B responsive signal transduction system in green algae, so it might seem unlikely that UV-B was the first wavelength to control UV protectants in plant evolution. Such a hypothesis, however, overlooks a more direct route for placing UV protectants under photocontrol, i.e., the possibility that UV-B protection and UV-B responsiveness could have sevolved in concert. This hypothesis goes as follows.

Prior to playing a UV-protective role, phenylpropanoids probably played a role in physiological control and/or microbial defense.[28] Photocontrol of the pathway could have been unnecessary at this early evolutionary stage. However, one mode of control that probably would have existed at this stage is a product feedback mechanism, limiting production when the product was abundant and stimulating production when it was depleted. In higher plants, *trans-*

cinnamic acid application is known to prevent transcription of PAL genes in cultured cells.[32] Based on his studies of this feedback system, Engelsma[33] made the interesting proposal that photoisomerization of *trans*-cinnamic acids to *cis*-cinnamic acids could regulate phenylpropanoid biosynthesis because depletion of the *trans* isomer would demand more production of phenylpropanoids through the product feedback mechanism (assuming that the *cis* isomer is not capable of mediating feedback). Thus, a plant possessing negative feedback control over PAL expression by *trans*-cinnamic acid levels also necessarily possesses UV-responsive control over PAL expression, simply because UV light would deplete *trans*-cinnamic acid and so cause an increase in PAL expression.

The action spectra for UV-B photoreceptor control of flavonoid production in various plants have peaks at 290 - 300 nm.[4,34] Phenylpropanoid esters have the following maxima: 280 - 290 for cinnamoyl, 310 - 315 for coumaroyl, 320 - 325 nm for feruloyl, and 330 - 335 for caffeoyl and sinapoyl esters.[35] This might point to cinnamoyl esters as the best phenylpropanoid candidate for the chromophore of a UV-B photoreceptor system, perhaps with some contribution from coumaroyl esters as well, (although the chemical environment and glycosylation state of the 4-hydroxyl position of hydroxycinnamic acids must also be considered because of their known effects in shifting absorption maxima). Thus, UV-B photocontrol could be a simple outcome of a negative product feedback system that is specific for the *trans* isomer. It makes sense to consider also that it could have been the original (and most primitive) mode of photocontrol over phenylpropanoids in evolution.

Implications of the Hypothesis

What would be the advantage of such a system? For the aquatic ancestors of plants, control of this pathway by the UV-B region of the spectrum could have offered an economical means for preventing high level expression in deep water, where little UV-B would penetrate, and directing high level expression near the surface, where damaging UV-B is most abundant. Thus, the production of substantial amounts of UV-absorbing pigments would occur only to the extent they are needed (i.e., in a depth-dependent manner). Although other photoreceptors might have been able to accomplish this (e.g., deep water is depleted in yellow light), a direct biochemical relationship between the photoreceptor and the UV-absorbing pigment might have provided the most straightforward evolutionary path for establishing production tuned to its need.

What are the implications of this hypothesis for modern plants? If the ancestors of land plants evolved the capacity to induce their chemical UV screen in response to levels of UV-B light through a simple feedback mechanism, we might expect many extant plants to have retained this system, though perhaps in an embellished or suppressed form. If so, it may be possible to increase the responsiveness of the system to UV-B light by one or a few transgenic modications, in agricultural plants at least. It is less clear how this knowledge could be used to adapt the many species of wild plants to higher UV-B levels because transgenic modification of tens of thousands of wild plant species would be impractical. Perhaps though, better understanding of the regulation of biosynthesis will suggest adaptive strategies that cannot yet be foreseen.

Complexity of Control in Higher Land Plants

Subsequent to the evolution of a light-responsive UV screen, flavonoid and phenylpropanoid biosynthesis diversified further to produce deoxyanthocyanins, anthocyanins, and isoflavonoids. Because anthocyanins do not absorb in the UV-B, but in visible wavelengths, they may play a role in photoprotection in those wave lengths. A question that arises is why is there often a different light requirement for anthocyanin synthesis than for flavonoid synthesis.[4] The answer presumably is that there is a value in being able to control the pathways separately, and to express them under different circumstances, because of their distinct functions in light protection. Over time, control over pigment biosynthesis could have become more complex, resulting in the diversity of photocontrol strategies now exhibited by higher plants. The fact that in some systems it is known that UV-B induces anthocyanins while visible light induces other flavonoids is not an argument against the hypothesis that UV-B responsiveness was the original mode of photocontrol over flavonoids. Rather, because anthocyanins evolved prior to the appearance of angiosperms, many opportunities have existed for evolution of increasingly complex photocontrol of the phenylpropanoid pathway and its flavonoid branches during the last 150 million years and before. Thus, the only taxa in which a valid test of the hypothesis can be made are taxa closely related to the one in which the system first evolved.

REFERENCES

1.   CALDWELL, M. M, ROBBERECHT, R. and FLINT..S. D 1983. Internal filters: prospects for UV-acclimation in higher plants. Physiol Plant 58:445-450.

2.   WELLMAN, E. 1983. UV radiation in photomorphogenesis. In: Photo-
      morphogenesis, (W Schropshire, Jr., and H Mohr, eds.), pp. 745-
      756, Springer-Verlag, Berlin.

3.   SCHMELZER, E, JAHNEN, W., HAHLBROCK,. K. 1988. *In situ*
      localization of light-induced chalcone synthase mRNA, chalcone
      synthase, and flavonoid end products in epidermal cells of parsley
      leaves. PNAS 85:2989-2993.

4.   BEGGS, C.J, WELLMAN, E., GRISEBACH, .H. 1986. Photocontrol of
      flavonoid biosynthesis. In: Photomorphogenesis in plants, (R.E
      Kendrick and G.H.M Kronenberg, eds), Nijhoff: Dordrecht. pp. 467-
      499.

5.   STAFFORD, H.A. 1990. Flavonoid metabolism. CRC Press: Boca
      Raton, Fla.

6.   YAMAMOTO, E., BOKELMAN, G.H. LEWIS, N.G. 1989.
      Phenylpropanoid metabolism in cell walls. In: Plant cell wall
      polymers, (N.G. Lewis and M.G. Paice, eds.) Am Chem Soc,
      Washington, DC., pp. 68 - 88.

7.   VOGELMANN, T.C., MARTIN, G. CHEN, G, BUTTRY, D. 1991.
      Fibre optic microprobes and measurement of light microenvironment
      within plant tissues. Adv Bot Res 18:255-295.

8.   VOGELMANN, T. C. 1993. Plant tissue optics. Ann Rev Pl Physiol
      and Pl Mol Biol 44:231-251.

9.   DELUCIA, E.H, DAY, T.A., VOGELMANN, T.C. 1992. Ultraviolet-B
      and visible light penetration into needles of subalpine conifers during
      foliar development. Plant Cell Environ 15:921-929.

10.   LI, J., OU-LEE, T-M., RABA, R., AMUNDSON, R.G, Last, R.L.
      1993. Arabidopsis flavonoid mutants are hypersensitive to UV-B
      irradiation. Plant Cell 5:171-179.

11.   JORGENSEN, R. 1992. Silencing of plant genes by homologous
      transgenes. Ag. Biotech News Inform. 4:265N-273N

12.   DOONER, H.K, ROBBINS, T.R., JORGENSEN, R.A. 1991. Genetic
      and developmental control of anthocyanin biosynthesis. Annu. Rev.
      Genet. 25:173-199.

13.   SMITH, R.C. 1989. Ozone, middle ultraviolet radiation, and the aquatic
      environment. Photochemistry and Photobiology 50:459-468.

14.   LOWRY, B, LEE, D., HEBANT, C. 1980. The origin of land plants: a
      new look at an old problem. Taxon 29:183-197.

15.   SHICK, J.M., LESSER, M.P., STOCHAJ,.W.R. 1991. Ultraviolet
      radiation and photooxidative stress in zooxanthellate anthozoa: the sea

anemone *Phyllodiscus semoni* and the octocoral *Clavularia* sp. Symbiosis 10:145-173.

16.   GARCIA-PICHEL, F., SHERRY, N.D., CASTENHOLZ, R.W. 1992. Evidence for an ultraviolet sunscreen role of the extracellular pigment scytonemin in the terrestrial cyanobacterium *Chlorogloeopsis* sp. Photochem Photobiol 56:17-15.

17.   PICKETT-HEAPS, J.D. 1975. Green algae: Structure, reproduction, and evolution. Sinauer: Sunderland, Mass.

18.   GRAHAM, L.E. 1985. The origin of the life cycle of land plants. Am Sci 73:178-186.

19.   GRANT, M.C. 1990. Phylum Chlorophyta, Class Charophyceae, Order Charales. In: Handbook of Protoctista. (L Margulis et al. eds), Jones and Bartlett:Boston. pp.641-648.

20.   MARKHAM, KR. 1988. Distribution of flavonoids in the lower plants and its evolutionary significance. In: The Flavonoids. (JB Harborne,ed), pp. 427 - 468, Chapman and Hall: London .

21.   DELWICHE, C.F, GRAHAM, L.E, THOMSON, N. 1989. Lignin-like compounds and sporopollenin in *Coleochaete*, an algal model for land plant ancestry. Science 245:399-401.

22.   GUNNISON, D., AND ALEXANDER, M. 1975. Basis for the resistance of several algae to microbial decomposition. Appl Microbiol 29:729-738.

23.   TAKEDA, R., HASEGAWA, J., SINOZAKI, K. 1990. Phenolic compounds from Anthocerotae. In: Bryophytes: their chemistry and chemical taxonomy, (HD Zinsmeister and R Mues,eds.), pp. 201-207, Clarendon, Oxford.

24.   GORHAM, J. 1990. Phenolic compounds, other than flavonoids, from bryophytes. In: Bryophytes: their chemistry and chemical taxonomy, (H.D Zinsmeister and R Mues, eds.), pp. 171-200, Clarendon, Oxford.

25.   MARKHAM, K.R. 1990. Bryophyte flavonoids, their structures, distribution, and evolutionary significance. In: Bryophytes: their chemistry and chemical taxonomy, (H.D Zinsmeister and R. Mues, eds), pp. 143-161, Clarendon, Oxford.

26.   MUES, R. 1990. The significance of flavonoids for the classification of bryophyte taxa at different taxonomic ranks. In: Bryophytes: their chemistry and chemical taxonomy, (H.D Zinsmeister and R Mues, eds), pp. 421-436, Clarendon, Oxford.

27.   SCHRODER J., AND SCHRODER, G. 1990. Stilbene and chalcone

synthases: related enzymes with key functions in plant-specific pathways. Z. Naturforsh. 45c:1-8.

28   STAFFORD, H.A. 1991. Flavonoid evolution: an enzymic approach. Plant Physiol 96:680-685.

29.   RHODES, M.J.C. 1985. The physiological signficance of plant phenolic compounds. Annu. Proceed. Phytochem. Soc. Eur. 25:99-117.

30.   HAHLBROCK, K., SCHEEL, D. 1989. Physiology and molecular biology of phenylpropanoid metabolism. Annu Rev Plant Physiol Plant Mol Biol 40:347-369.

31.   DRING, M.J. 1988. Photocontrol of development in algae. Ann Rev Plant Physiol Plant Mol Biol 39:157-174.

32.   MAVANDAD, M, EDWARDS, R. LIANG, X., LAMB, C.J., DIXON, R.A. 1990. Effects of trans-cinnamic acid on expression of the bean phenylalanine ammonia-lyase gene family. Plant Physiol 94:671-680.

33.   ENGELSMA, G. 1974. On the mechanism of the changes in phenylalanine ammonia-lyase activity induced by ultraviolet and blue light in gherkin hypocotyls. Plant Physiol 54:702-705.

34.   HASHIMOTO, T, SHICHIJO, C., YATSUHASHI, H. 1991. Ultraviolet action spectra for the induction and inhibition of anthocyanin synthesis in broom sorghum seedlings. J. Photochem. Photobiol. B, 11:353-363.

35.   IBRAHIM, R, BARRON, D. 1989. Phenylpropanoids. In: Methods in plant biochemistry, vol. 1, Plant phenolics, (J.B Harborne, ed.) pp. 75-112, Academic Press, London.

Chapter Eight

GENETIC CONTROL OF MONOTERPENE BIOSYNTHESIS IN MINTS (*MENTHA*: LAMIACEAE)*

Rodney Croteau and Jonathan Gershenzon

Institute of Biological Chemistry
Washington State University
Pullman, WA 99164-6340 USA

INTRODUCTION

The genetic basis of plant monoterpene composition has been the subject of detailed investigation, most notably an extensive series of crossing experiments performed on species of mints (*Mentha*) by M.J. Murray and his co-workers. This research, which began in the 1950s, identified a number of genes

---

*This chapter is dedicated to Merritt J. Murray and his many colleagues for their pioneering research in *Mentha* genetics.

Genetic Engineering of Plant Secondary Metabolism,
Edited by B.E. Ellis *et al.*, Plenum Press, New York, 1994

regulating the production of the principal essential oil components in *Mentha*.[1,2] More recently, Croteau and associates have examined the biosynthesis of the major monoterpenes of the commercial *Mentha* species (peppermint, native spearmint and Scotch spearmint). Using both *in vivo* and *in vitro* experimental systems,[3-5] the sequence of intermediates has been established and the relevant enzymes have been isolated and characterized. In the last year, a gene encoding a key enzyme of the pathway has been isolated, sequenced and functionally expressed in a bacterial host.[6]

The availability of substantial amounts of both biochemical and classical genetic data provides an unprecedented opportunity to understand the genetic control of an extended pathway of plant secondary metabolism. Unfortunately, there has been little attempt to integrate these two approaches thus far. In this paper, we assess the classical crossing studies of Murray and collaborators from a modern biochemical perspective and attempt to assign the identified genes to the actual chemical transformations they control. For organizational clarity, each biosynthetic reaction is discussed in the appropriate biogenetic sequence, starting with the reaction of the ubiquitous, acyclic isoprenoid metabolite, geranyl pyrophosphate, the precursor of all regular monoterpenes (Fig. 1). Description of each enzyme-catalyzed step is followed by an account of the genes that appear to be associated with it. By this cohesive approach, we hope to clarify what is known about both the biosynthetic pathway and its genetic underpinnings.

The genus *Mentha* (family Lamiaceae) consists of about 25 species, a roughly equal number of primary and secondary hybrids, and a large number of intraspecific chemical races or chemotypes.[7-9] The taxonomy of the group is complex, particularly in the section *Mentha* which includes the commercial species, due to great morphological variation, frequent hybridization, persistence of hybrids via vegetative propagation, and a long history of human involvement in cultivation and dispersal.[8,10] The commercial cultivars are vegetatively propagated (some types are, in fact, sterile or nearly so, necessitating such clonal propagation) insuring that the essential oil produced by this perennial crop is "true to type". The *Mentha* species of major agronomic interest[11] (Table 1) include peppermint (*M.* x *piperita* L., a sterile, high menthone/menthol selection of *M. aquatica* L. x *M. spicata* L.[10,12]), native spearmint (*M. spicata* L.) and Scotch spearmint (*M.* x *gracilis* Sole, a high carvone selection of *M. arvensis* L. x *M. spicata* L.[13,14]). Pennyroyal (*M. pulegium* L.) and cornmint (*M. arvensis* L. var. *piperascens* Malinv.) are also of some commercial significance and, although few biochemical studies have been carried out with these species,[15] their close chemical and taxonomic relationship to peppermint permits their inclusion here.

The great diversity of essential oil composition in *Mentha* has provided geneticists with the necessary tools for examining the inheritance patterns of the various monoterpene constituents. Genetic studies in the genus have also been

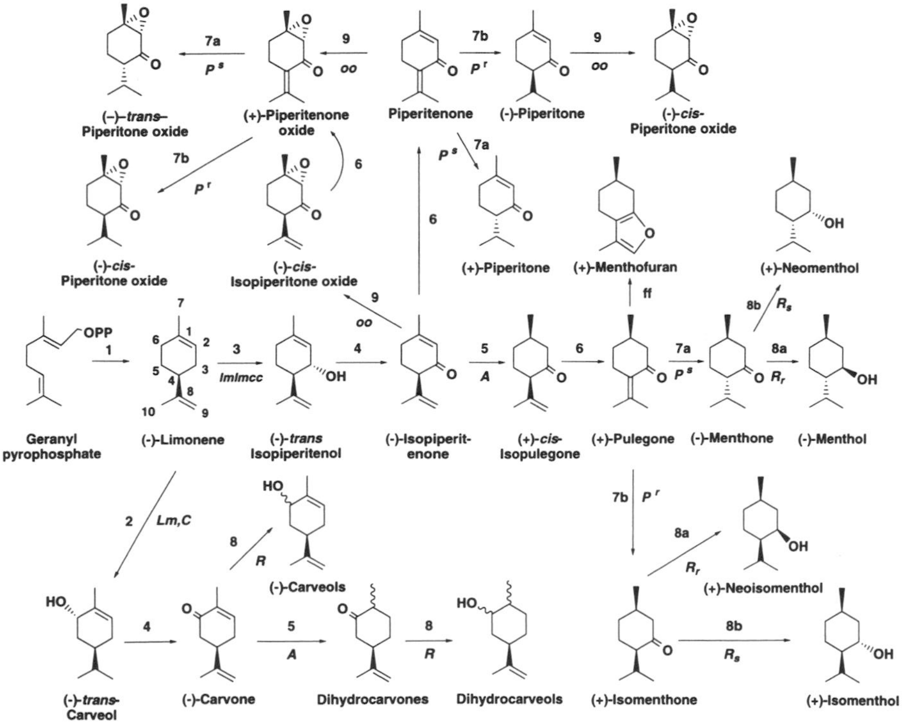

Fig. 1. Pathways of *p*-menthane monoterpene biosynthesis in *Mentha* species. The numbered enzymatic steps and the associated genes (designated by letters) are described in the text and in Table 2.

facilitated by the fact that monoterpene composition is only weakly influenced by changing environmental conditions, and that developmental variation, although significant in some species (e.g., peppermint[16]), is predictable and

Table 1. Representative composition of the essential oil (% of total) of commercial *Mentha* species. Only major components (> 1%) are tabulated, and the data are from Lawrence[2] or our own studies[a].[15,23]

Compound	*M. spicata* Spearmint	*M. x gracilis* Scotch spear.	*M. x piperita* Peppermint	*M. pulegium* Pennyroyal	*M. arvensis* Cornmint
Pinenes and other olefins	4.9	2.5	2.8	1.0	2.0
Limonene	12.1	15.6	1.2	tr	1.5
1,8-Cineole	2.6	1.8	1.9	tr	tr
Dihyrocarvones	1.6	1.1			
Carveols & dihydrocarveols (and esters)	4.6	2.5			
Sabinene hydrates	1.4	tr	1.2		
Carvone	67.8	68.6			
Menthone	tr	1.0	23.1	16.3	4.4
Isomenthone	tr	tr	3.4	4.3	2.4
Menthol	tr	tr	46.7		75.8
Other menthol isomers		tr	2.4		2.1
Menthyl acetate			2.7	tr	6.4
Other menthyl acetates			2.3		tr
Pulegone		tr	3.4	tr	tr
Isopulegones			tr	68.1	
Menthofuran			2.5	2.2	
Piperitone		tr	tr	1.1	tr
Isopiperitenone			tr	tr	tr
Piperitenone			tr	2.4	tr
Percent of Total	95.0	93.1	93.6	95.4	94.6

[a]tr: indicates trace; no entry indicates that the compound was not detected.

Table 2. Monoterpene biosynthetic enzymes of *Mentha* species and their genetic associations.

Enzyme number	Name	Genetic association
1	(-)-limonene synthase	no gene assigned; expressed in *ii* genotype; limonene accumulates in *iiLmcc* genotype
2	(-)-limonene-6-hydroxylase	controlled by *C* gene; expressed in *LmC* genotype
3	(-)-limonene-3-hydroxylase	controlled by *Lm* gene; expressed in *lmlmcc* genotype
4	*p*-menthenol dehydrogenase	no gene identified; appears to be present in all *Mentha* sp. examined
5	*p*-menthadienone-$\Delta^{1,2}$-reductase	controlled by *A* gene; yields (3*R*)-stereochemistry
6	*p*-menthenone isomerase	no gene identified; appears to be widespread in *Mentha* sp.
7	*p*-menthenone-$\Delta^{4,8}$-reductase	controlled by *P* gene; *P^r* and *P^s* alleles described for (4*R*)-reduction (enzyme 7b) and (4*S*)-reduction (enzyme 7a); *P^r* mostly dominant over *P^s*
8	*p*-menthanone-3-keto-reductase	controlled by *R* gene; probably two *R*-type genes exist, one (*R$_r$*) for the production of (3*R*)-alcohols (enzyme 8a) and one (*R$_s$*) for the production of (3*S*)-alcohols (enzyme 8b)
9	(-)-limonene-3-hydroxylase/ 3-keto-(1*S*,2*S*)-epoxidase	probably controlled by *O* gene; expressed in *oo* genotype
10	*p*-menthanol acetyltransferase	controlled by *E* gene
11	*p*-menthanol glucosyltransferase	no gene identified

thus easily accounted for.[1] In addition, the taxonomic complexities of the genus have not significantly impeded genetic analyses. Similar genes are present naturally in a wide range of species (e.g., the *P* gene is found in *M. arvensis*, *M. spicata*, *M. x piperita*, *M. crispa*, *M. pulegium* and *M. gattefossei*[17]), and the extensive variation in ploidy level,[10] which radically alters gene dosage levels, apparently has no major influence on the segregation and expression of genes controlling monoterpene composition.[18-20] Whereas compositional diversity provides the grist of the geneticist's mill, the progress of biochemical and molecular genetic investigations of monoterpene formation has been aided by the availability of clonal populations of fixed oil composition, since plant material from many individuals must be combined in this work while avoiding the confounding influence of variation.

BIOCHEMISTRY AND GENETICS OF MONOTERPENE METABOLISM IN *MENTHA*: A CORRELATION

The monoterpene constituents of the essential oils of peppermint and spearmint are distinguished by the position of oxygenation on the *p*-menthane ring.[2] Peppermint (*M. x piperita*) and related species (*M. aquatica*, *M. arvensis*, *M. pulegium*) produce almost exclusively monoterpenes bearing an oxygen function at C3, such as pulegone, menthone and menthol, whereas spearmint species (*M. spicata*, *M. x gracilis* and *M. crispa*) produce almost exclusively monoterpenes bearing an oxygen function at C6, typified by carvone (Table 1 and Fig. 1; the numbering system employed here is based on limonene, the parent *p*-menthane monoterpene of *Mentha*). The genetic basis of the C3- and C6-oxygenation patterns was an early target of Murray's research[21] that ultimately led to a lengthy series of hybridization experiments in which strains or varieties that contained a particular compound were crossed with strains lacking that compound, followed by numerous back-crosses between the F_1 hybrids and either parent. Analysis of the progeny of these crosses demonstrated that the occurrence of many C3- and C6-oxygenated monoterpenes is controlled by dominant/recessive alleles that segregate independently in classical Mendelian fashion. However, the *p*-menthane oxygenation pattern itself was established to be under the control of two closely linked loci discussed below, *Lm* and *C*.[1]

Biosynthetic investigations have demonstrated that the regiospecificity of oxygenation is established very early in the monoterpene biosynthetic sequence where (-)-(4*S*)-limonene, the first cyclic olefin to arise from the common isoprenoid intermediate geranyl pyrophosphate and the precursor of *both* oxygenated series,[15] is hydroxylated at C3 to yield (-)-*trans*-isopiperitenol (in peppermint-type species) or at C6 to afford (-)-*trans*-carveol (in spearmint-type species).[22] The remaining enzymatic machinery responsible for the subsequent redox transformations of isopiperitenol to menthol (enzymes **4-8a**, Fig. 1) is present in both peppermint and spearmint species, as demonstrated by cell-free assay.[23] However, the products in spearmint species are different due to the absence of the C3-oxygenated precursor. The C6-oxygenated product *trans*-carveol, instead, is oxidized to carvone by the same dehydrogenase responsible for conversion of *trans*-isopiperitenol to isopiperitenone, but carvone is a poor substrate, or not a substrate, for the subsequent steps (**5** to **8**, Fig. 1) involving double bond reduction and migration.[23] The biosynthetic capabilities of peppermint and spearmint, and the gene designations assigned to each enzymatic transformation, are summarized in Figure 1 and described in Table 2. It should again be noted that the term "C6-oxygenation" is based on the numbering of the limonene precursor and thus reflects the biosynthetic origin of these derivatives. In the older literature, this pattern is often referred to a "C2-oxygenation" based on numbering of the product (e.g., carvone). For "C3-oxygenation", the numbering systems are the same, whether based on the limonene precursor or the derived products.

Unambiguous biochemical evidence has revealed the pathway and enzymology of limonene metabolism to be rather straightforward, although for many years considerable confusion surrounded these early metabolic steps. Murray's extensive crossing experiments did not identify any genes that regulate limonene formation, and this, coupled to the complex organization of the genes responsible for the conversion of limonene to other cyclic monoterpenes[1,24] (*vide infra*), obscured the role of this olefin precursor and the origins of the C3- and C6-oxygenation patterns. The failure to demonstrate, in any but trace quantities, oil constituents thought to be key intermediates of the pathways led to several improvised biogenetic proposals[1,2] that also served to further confuse this issue. Thus, α-terpineol and terpinolene (Fig. 2) which were once proposed as intermediates en route to C3-oxygenated compounds are, in fact, produced biosynthetically in only trace amounts, whereas isopiperitenol, a true intermediate, does not accumulate in peppermint as it is rapidly transformed (via isopiperitenone and isopulegone) to pulegone, leaving very small amounts of piperitenone and piperitone (also once suggested as intermediates) as dead-end side products.[15,23] Furthermore, the reluctance to accept double bond migration as a likely biosynthetic step,[1] although well precedented in the ketosteroid isomerase reaction,[25] delayed appreciation of the central role played by the $\Delta^{8,9} \rightarrow \Delta^{4,8}$ isomerization sequence of the pathway. Finally, the inability of

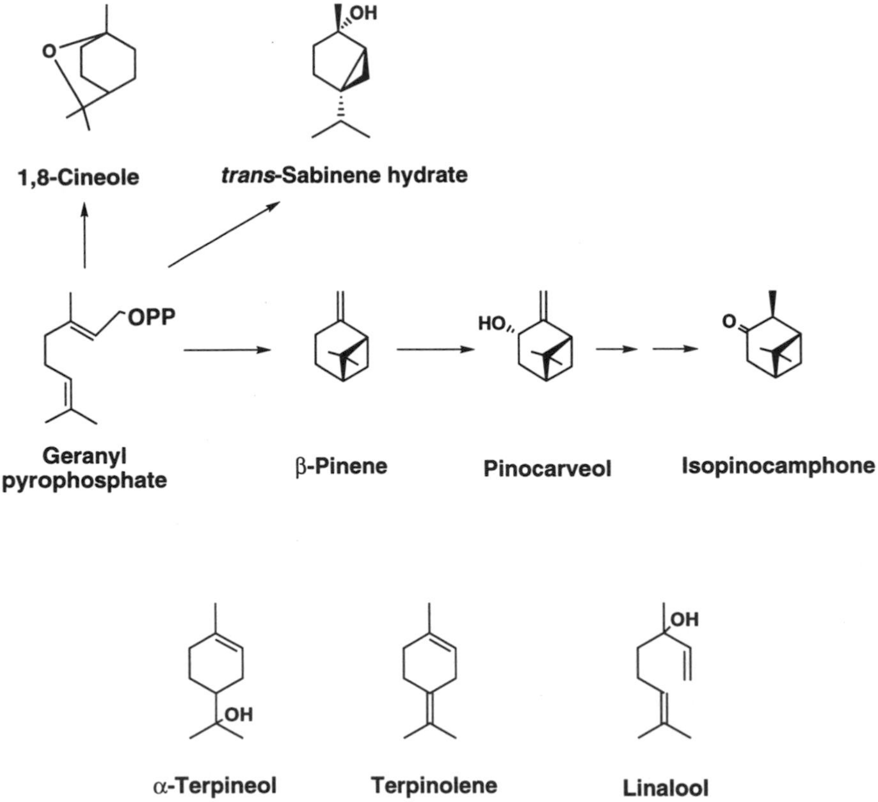

1,8-Cineole   ***trans*-Sabinene hydrate**

Geranyl pyrophosphate   **β-Pinene**   **Pinocarveol**   **Isopinocamphone**

α-Terpineol   **Terpinolene**   **Linalool**

Fig. 2. Alternate fates of geranyl pyrophosphate in *Mentha* species, and the structures of α-terpineol, terpinolene and linalool.

earlier analytical methods to allow assignment of absolute configuration to metabolites available only in small quantities limited stereochemical inferences about possible precursor-product relationships. Given a half-dozen reasonable biogenetic proposals, with neither chemical nor genetic evidence as a firm basis for selecting the correct route, a direct biochemical approach was needed to provide the experimental grounds for the pathways illustrated (Fig. 1).

It is now well established that the monoterpenes of *Mentha* and other members of the Lamiaceae are produced and accumulated in glandular trichomes, highly specialized secretory structures found on the surfaces of leaves, young stems and parts of the inflorescence.[26-28] The pathways and enzymes described here were determined primarily by combination of *in vivo* and *in vitro* studies employing cell clusters isolated from glandular trichomes, or cell-free extracts derived therefrom.[29,30]

Cyclization of Geranyl Pyrophosphate to (-)-Limonene

Geranyl pyrophosphate:(-)-limonene cyclase (enzyme **1**; also called limonene synthase) has been demonstrated in soluble enzyme extracts of all of the commercial *Mentha* species[15,23] and purified to homogeneity from gland cell extracts of peppermint and native spearmint.[31] The enzyme has been characterized in some detail,[31-33] and shown to be a fairly hydrophobic monomer of molecular weight ~56,000, with pH optimum near 6.7, isoelectric point of 4.35, and k_{cat} of about 0.3/s. The enzyme requires only a divalent metal ion as a cofactor (Mn^{2+} preferred), and the K_m value measured for the active geranyl pyrophosphate-metal ion complex is 1.8 μM. Inhibition and substrate protection studies indicate that limonene synthase bears essential histidine and cysteine residues at or near the active site.[32,33] The enzyme produces primarily optically pure (-)-(4S)-limonene (94%) with trace amounts of myrcene, α-pinene and β-pinene as co-products.[32]

Experiments with alternate substrates and substrate analogs have confirmed many elements of the reaction mechanism,[32] including the binding and ionization of the substrate-metal ion complex, the preliminary isomerization of geranyl pyrophosphate to linalyl pyrophosphate (a bound intermediate capable of cyclization), and the participation of a series of carbocation•pyrophosphate anion pairs in the reaction sequence. Based on the electrophilic isomerization-cyclization cascade catalyzed by limonene cyclase, a mechanism-based inactivator (suicide substrate) of the enzyme has been designed.[34] The basic properties of (4S)-limonene synthase from both *M.* x *piperita* and *M. spicata* are identical, and in both properties and mechanism of action they are similar to other terpenoid cyclases of higher plants.[4]

Polyclonal antibodies were generated in rabbits against the SDS-denatured limonene synthase of spearmint, and immunoblotting analysis revealed that these antibodies were very specific for the limonene synthase from all *Mentha* species tested, suggesting that this cyclase protein is very similar, if not identical, among these species.[35] However, no immunological cross-reactivity was observed with limonene synthases from Valencia orange (*Citrus sinensis*, Rutaceae) or wormseed (*Chenopodium ambrosioides*, Chenopodiaceae). Furthermore, the antibody preparation did not detectably cross-react with other monoterpene cyclases from species of the Lamiaceae, Asteraceae, and Apiaceae, or from conifer species, and no cross-reactivity was demonstrated toward several sesquiterpene cyclases of higher plant and microbial origin. Although the antibody preparation was highly selective for denatured limonene synthases from *Mentha*, the antibodies did not recognize the native protein. Nevertheless, specificity for the target enzyme was unambiguously demonstrated when the antibody preparation was shown to cross-react with the cyclase protein expressed in *Escherichia coli* that harbored the corresponding limonene synthase cDNA gene from *M spicata* (*vide infra*). With the specificity of the antibody confirmed, tissue printing and immunogold-cytochemical methods were employed to demonstrate that the limonene synthase was specifically localized in the leucoplasts of the oil gland secretory cells.[36]

For the purpose of isolating the limonene synthase cDNA,[6] internal amino acid sequences of the purified protein from spearmint oil glands were obtained by CNBr cleavage and V8 proteolysis, and were utilized to design three distinct oligonucleotide probes. These probes were subsequently employed to screen a spearmint cDNA library, and four clones were isolated. Three of these cDNA isolates were full-length and were functionally expressed in *E. coli*, yielding a peptide that is immunologically recognized by the polyclonal antibodies raised against the purified limonene synthase from spearmint,[35] and that is catalytically active in generating from geranyl pyrophosphate a product distribution absolutely identical to that of the native enzyme (principally limonene with small amounts of the coproducts $\alpha-$ and β-pinene and myrcene). The longest open reading frame is 1800 nucleotides, and the deduced amino acid sequence contains a putative plastidial transit peptide of approximately 90 amino acids and a mature protein of about 510 residues corresponding to the native enzyme. This evidence is entirely consistent with the localization of the mature 56 kDa cyclase within the gland cell leucoplasts.[36]

Peppermint (*M. x piperita*) is an allohexaploid thought to have originated from a hybridization event between spearmint (*M. spicata*) and *M. aquatica*.[10,12] Thus, it is not surprising that both peppermint and spearmint contain ostensibly the same limonene synthase,[31] as indicated above. RNA blot hybridization of limonene synthase DNA to spearmint and peppermint poly(A)$^+$ RNA verified the presence of homologous sequences in both species and indicated that the limonene cyclase mRNA transcript was about 2,400 nucleotides long, as expected. Several nucleotide differences among the three full-length clones in the 5'-untranslated region suggested the presence of several limonene synthase genes and/or alleles in the tetraploid spearmint genome, and genomic DNA blot hybridization analysis confirmed the presence of a small gene family.

Sequence comparisons with a sesquiterpene cyclase, *epi*-aristolochene synthase from tobacco,[37] and a diterpene cyclase, casbene synthase from castor bean (Mau, C.J.D., West, C.A. unpublished), demonstrated a significant degree of similarity among these three terpenoid cyclases, the first three examples of this large family of catalysts to be described from higher plants. As indicated above, recent evidence has implicated histidine and cysteine residues at the active site of limonene synthase and several other terpenoid cyclases,[32,33] and a search of the aligned sequences of limonene synthase, *epi*-aristolochene synthase, and casbene synthase revealed the presence of four such conserved residues. The limonene cyclase sequence does not resemble any of the published sequences for microbial sesquiterpene cyclases or prenyltransferases from a wide range of organisms, enzymes that, like the monoterpene cyclases, employ allylic pyrophosphate substrates and exploit similar electrophilic reaction mechanisms.[38,39] The sequence (I,L,V)XDDXXD occurs in two of the three homologous domains of most prenyltransferases, and it has been suggested that these aspartate-rich elements function in binding the divalent metal ion complexed pyrophosphate moiety of the prenyl substrates.[40] The terpenoid cyclases would be expected to exhibit similar substrate binding requirements, and most do contain this motif, or a very similar one.[6,38]

Based on the correlation of enzyme activity with glandular trichome development and monoterpene production, limonene synthase would appear to catalyze a rate-controlling step of monoterpene biosynthesis in peppermint.[41] The limonene cyclase cDNA provides a useful tool for examining the developmental regulation of monoterpene biosynthesis in *Mentha* and for determining directly the role of this gene product in the control of essential oil yield.

Alternate Fates of Geranyl Pyrophosphate

All *Mentha* species must produce geranyl pyrophosphate, the precursor of the monoterpenes, since all produce monoterpenes of one type or another. The responsible enzyme, geranyl pyrophosphate synthase, has been purified and characterized,[42] and this chain length-specific prenyltransferase is localized in the glandular trichomes of sage (*Salvia officinalis*, Lamiaceae) where it presumably functions to supply this key precursor.[43] A few rare strains and chemotypes of

Mentha accumulate geraniol, geranyl acetate and related acyclic derivatives to the level of about 30% of the essential oil, but never, apparently, to the exclusion of the more typical cyclohexanoid monoterpene types.[9,44] Such a condition likely reflects a limited capacity for cyclization, with the consequence that unutilized geranyl pyrophosphate is hydrolyzed and the geraniol otherwise metabolized by conjugation or redox transformations.

Linalool (Fig. 2) and linalyl acetate are the main constituents in the oils of hybrids of several *Mentha* taxa, including *M. citrata*[9,45] (now considered to be a variety of *M. aquatica*[10]). From crosses involving *M. citrata*, Murray and Lincoln[19] demonstrated that a dominant allele designated *I* allows the accumulation of linalool and linalyl acetate, and largely prevents the production of the more typical cyclic monoterpenes. The biosynthesis of linalool has not been very thoroughly investigated, although in cell-free systems from lavender (*Lavandula officinalis*) it has been shown to be derived from geranyl pyrophosphate with divalent metal ion as cofactor (Croteau, unpublished results); the activity thus resembles a monoterpene cyclase in its substrate requirement. Since the mechanism of cyclization requires the initial isomerization of geranyl pyrophosphate to a linalyl intermediate,[4] it is tempting to suggest that the *I* allele encodes an enzyme that catalyzes an abortive cyclization.

The dominant *I* allele nearly completely prevents the formation of the cyclic ketones, such as carvone and menthone, probably by inhibiting the formation of limonene itself. Therefore, it stands to reason that the recessive allele *i* must allow the cyclization of geranyl pyrophosphate to limonene, although this point has never been explicitly addressed. Any *Mentha* taxon with high levels of either C3- or C6-oxygenated *p*-menthanes must obviously also produce the precursor of limonene, and thus bear the doubly recessive *ii* combination. However, according to Murray,[1,24,46,47] individuals with high limonene content *per se* occur only when the dominant *Lm* allele (preventing C3-oxygenation) is separated from *both* the dominant *I* allele and the dominant *C* allele (promoting C6-oxygenation) as a result of rare quadrivalent pairing and crossing over between homeologous chromosomes (see below for further discussion of *I*, *Lm* and *C* interactions).

Minor components of peppermint and spearmint oil, including 1,8-cineole, *trans*-sabinene hydrate, β-pinene and isopinocamphone (Fig. 2), achieve prominence in certain *Mentha* taxa,[1,8] and at least in one instance a relevant gene has been identified. Lincoln et al.[48] examined the monoterpene composition of hybrids resulting from crosses between *M. citrata* (linalool type) and *M. aquatica* (C3-oxygenated *p*-menthane type), and reported that if a gene designated I_S is separated from the linked I gene responsible for linalool production and substituted into *M. aquatica* of the recessive *ii* genotype (normally allowing limonene production and metabolism), the resulting hybrids produce high amounts of β-pinene and isopinocamphone, as well as high levels of other atypical monoterpenes. Although the enzymes catalyzing the formation of these normally minor products have not yet been examined in *Mentha* species, 1,8-cineole synthase and pinene synthases have been well-documented in sage (*S. officinalis*, Lamiaceace)[49,50] and sabinene hydrate synthase has been characterized from marjoram (*Majorana hortensis*, Lamiaceae).[51,52] The limonene synthase of *Mentha* discussed above does produce very small amounts of β-pinene.[32] The transformation of β-pinene to isopinocamphone has been described in cell-free extracts of hyssop (*Hyssopus officinalis*, Lamiaceae) and shown to involve three steps: cytochrome P-450-dependent hydroxylation of β-pinene to *trans*-pinocarveol, dehydrogenation of *trans*-pinocarveol to pinocarvone, and reduction of the exocyclic double bond to yield isopinocamphone (Fig. 2).[53] Since neither limonene-3-hydroxylase (peppermint) nor limonene-6-hydroxylase (spearmint) utilize β-pinene as a substrate,[22] the initial hydroxylation of β-pinene in isopinocamphone-accumulating *Mentha* taxa must involve another cytochrome P-450 hydroxylase.

Hydroxylation of (-)-Limonene: 3- vs 6-Hydroxylation

The cytochrome P-450-dependent (-)-limonene hydroxylases (enzymes **2** and **3**) have been examined in microsomal preparations from epidermal gland extracts of peppermint, native spearmint and Scotch spearmint.[22,23] The substrate specificity, regio- and stereochemistry of oxygen insertion, and other characteristics of these oxygenases have been determined, with the very high selectivity for (-)-(4*S*)-limonene as substrate being notable.[22] In Scotch and native spearmint, this O_2/NADPH-requiring reaction gives rise exclusively to (-)-*trans*-carveol. (-)-*trans*-Isopiperitenol cannot be detected as a product, indicating the presence of only (-)-limonene-6-hydroxylase (enzyme **2**).[22,23] Conversely, microsomes obtained from peppermint glands produce only (-)-*trans*-isopiperitenol when incubated with (-)-limonene; (-)-*trans*-carveol is not detected as a product, indicating the exclusive presence of (-)-limonene-3-hydroxylase (enzyme **3**). These results are entirely consistent with the accumulation of C6-oxygenated *p*-menthane derivatives in spearmint and of C3-oxygenated *p*-menthane compounds in peppermint, and they establish the hydroxylases as the key enzymes that determine the type of monoterpenes formed in most *Mentha* species.

The (-)-limonene-6-hydroxylase from spearmint has been purified to homogeneity, and the homogeneous protein employed to obtain amino acid sequence information and to raise polyclonal antibodies in preparation for isolating the corresponding cDNA. (-)-Limonene-6-hydroxylase antiserum recognizes peptides from other *Mentha* species on immunoblots, including a 55 kDa cytochrome P-450 species from peppermint that is presumed to be the (-)-limonene-3-hydroxylase. The antibody preparation also cross-reacts with monoterpene hydroxylases from other members of the Lamiaceae,[54] indicating the presence of common epitopes; however, the specificity of the preparation has not been extensively examined. Limonene-3-hydroxylase and limonene-6-hydroxylase share many gross properties, and the immunological evidence suggests that these two proteins are structurally similar. Nevertheless, these enzymes are readily distinguished by the regiochemistry of the reaction catalyzed and their differential sensitivity to azole-type inhibitors,[22] features which require at least minor differences in active site structure.

Curiously, limonene-6-hydroxylase is apparently less efficient than its counterpart, limonene-3-hydroxylase, since there is always a larger amount of residual limonene in taxa with C6-oxygenated monoterpenes than in those with C3-oxygenated monoterpenes (Table 1). The phenomenon, however, is not readily explained by kinetic differences between the hydroxylases, since the K_m and V_{rel} values for limonene (based on total microsomal protein) are very similar.[22] The observation probably relates to the relative limonene synthesis/hydroxylation balances, or to differential channeling or other compartmentation effects.

Two closely linked diallelic loci are considered to control the metabolic fate of limonene.[1,18,21,47] The dominant *Lm* allele is postulated to prevent the C3-oxidation of limonene to isopiperitenol whereas the recessive *lm* allele is thought to allow this conversion. In contrast, the dominant *C* allele is proposed to stimulate C6-oxidation while its recessive counterpart *c* does not promote this conversion. The two dominant alleles are believed to be tightly linked. This model explains the results of numerous experiments[1,18,24,55,56] and accounts for the facts that C3- and C6-oxidation are mutually exclusive (ignoring trace constituents) and that C6-oxidation is dominant to C3-oxidation. The effects of the various possible allelic combinations can be summarized as follows (assuming an *ii* genotype since the epistatic *I* allele leads to the accumulation of linalool and linalyl acetate rather than cyclic monoterpenes):

LmLmCC, LmLmCc, LmlmCC, LmlmCc -

C6-oxygenated products are formed (commonly observed)

LmLmcc, Lmlmcc -
limonene accumulates (observed infrequently; only when *Lm* is
separated from *C* by crossing over between homeologous chromosomes
in an unusual quadrivalent pairing)
lmlmCC, lmlmCc -
should allow both C3- and C6-oxygenated products to be formed (this is
a combination never before reported)
lmlmcc -
C3-oxygenated products are formed (commonly observed)

These results can be rationalized if *Lm* represents a regulatory gene that
controls the expression of limonene-3-hydroxylase and *C* is a structural or a
regulatory gene that controls C6-hydroxylation. Control of hydroxylase
expression could be mediated by regulatory genes since, although the major *p*-
menthane monoterpenes of the spearmint-type species are all products of C6-
hydroxylation, trace amounts of C3-hydroxylation products also occur in the
essential oil of these taxa, attesting to the presence of low levels of limonene-3-
hydroxylase activity. Or, perhaps, limonene-6-hydroxylase simply has a greater
affinity for substrate *in vivo* than limonene-3-hydroxylase, so that its presence
results predominantly in C6-hydroxylated products. (The K_m values for
limonene measured *in vitro*, however, are not substantially different.[22]) It seems
unlikely that the roles of *Lm* and *C* in *Mentha* can be clarified until the
structures of the hydroxylase genes, and the *Lm* and *C* genes and gene products,
are known and the molecular level interactions between these entities deciphered.
Interestingly, in *Satureja douglasii*, another member of the Lamiaceae possessing
both C3-oxygenated *p*-menthane and C6-oxygenated *p*-menthane chemical races,
the production of C3-oxygenated monoterpenes is dominant to the production of
C6-oxygenated types, in contrast to the situation in *Mentha*.[57]
    Additional findings on the genetic control of monoterpene
hydroxylation in mints have emerged from a biochemical evaluation of a
γ-radiation-induced mutant of Scotch spearmint (*M.* x *gracilis*) that produces a
peppermint-type oil (C3-oxygenation pattern), unlike the spearmint-type oil
(C6-oxygenation) found in the wild type.[23] *In vitro* measurement of all of the
enzymes responsible for the production of both the C3-oxygenated and C6-
oxygenated families of monoterpenes from (-)-limonene indicated that both the
mutant and wild type possessed a virtually identical complement of enzymes,
with the exception of the microsomal, cytochrome P-450-dependent (-)-limonene

hydroxylase; the C6-hydroxylase producing (-)-*trans*-carveol in the wild type had been replaced by a C3-hydroxylase producing (-)-*trans*-isopiperitenol in the mutant. Additionally, the mutant, but not the wild type, could carry out the cytochrome P-450-dependent, stereospecific (1*S*,2*S*)-epoxidation of the α,β-unsaturated bond of the ketones formed via C3-hydroxylation to produce piperitenone oxide, and *cis*- and *trans*-piperitone oxide. All of the evidence was consistent with the same protein (enzyme 9) carrying out both oxygenase activities (i.e., C3-hydroxylation and 3-keto-C1,C2-epoxidation), and there is precedent for such bifunctional P-450 cytochromes in other systems.[58] Therefore, these results suggest that irradiation resulted in either mutation of the C6-hydroxylase structural gene, converting it to a 3-hydroxylase-1,2-epoxidase, or mutation of a regulatory gene (perhaps *Lm*?) to a form which suppresses 6-hydroxylation and activates a nascent 3-hydroxylase-1,2-epoxidase. If the latter possibility is true, the 3-hydroxylase-1,2-epoxidase may also represent the enzyme expressed in the recessive *oo* genotype considered to be responsible for the production of 1,2-oxides.[1,2,59] The coupling of 3-hydroxylase and 1,2-epoxidase activity in a single gene product may explain the origin of a number of hybrids in which monogenic differences result in the production of 1,2-epoxides as well as the anticipated C3-ketones.[60] This bifunctional cytochrome P-450 is probably related to, but distinct from, the C3- or C6-limonene hydroxylases of peppermint and spearmint, respectively, neither of which possess epoxidase activity. The putative hydroxylase/epoxidase may also hydroxylate β-pinene to *trans*-pinocarveol (Fig. 2), but its specificity has not yet been fully examined. The stereochemical origin of the (4*R*)- and (4*S*)-isopropyl substituents of *trans*- and *cis*-piperitone-(1*S*,2*S*)-oxides is described below with reference to *P* gene control of the double bond reductases involved in menthone and isomenthone formation.

Redox Transformations

As indicated in the preceding section, the presence of either limonene-3-hydroxylase or limonene-6-hydroxylase is the critical factor in determining whether a *Mentha* species produces a peppermint or spearmint type of essential oil, since the oxygenation pattern of a *p*-menthane monoterpene dictates its subsequent metabolism. For instance, in peppermint, the C3-oxygenated alcohol, (-)-*trans*-isopiperitenol, is oxidized to the corresponding ketone, (-)-isopiperitenone, which ultimately gives rise to the major products (-)-menthone and (-)-menthol by a series of reductive steps. By contrast, in spearmint, the C6-oxygenated alcohol, (-)-*trans*-carveol, is oxidized to (-)-carvone, the major monoterpene accumulated in this species, which then gives rise to small quantities of dihydrocarvones and dihydrocarveols.[23,26] These pathways appear to be quite different on a chemical level, yet the basic reaction types involved are virtually identical. In fact, as indicated previously, after the limonene hydroxylation step, the enzymes that catalyze the remaining reactions of monoterpene biosynthesis are very similar, regardless of whether peppermint or spearmint-type products are formed.[23] This metabolic similarity of the redox steps can be demonstrated by a simple experiment that bypasses the hydroxylation step. If isolated spearmint leaf disks are administered exogenous (-)-*trans*-isopiperitenol, they will produce (-)-menthol (Croteau and Wagschal, unpublished results). Conversely, when peppermint leaf disks are fed (-)-*trans*-carveol, they readily accumulate (-)-carvone. The enzymes responsible for these transformations are now described.

p-Menthadienol dehydrogenase (enzyme 4). After (-)-limonene is hydroxylated to either (-)-*trans*-isopiperitenol or (-)-*trans*-carveol, these allylic alcohols are enzymatically oxidized to the corresponding α,β−unsaturated ketones, (-)-isopiperitenone and (-)-carvone. The dehydrogenase responsible for these conversions has been found in the soluble supernatants of glandular trichome extracts[29,30] prepared from peppermint, native spearmint and Scotch spearmint.[23,26] The enzyme was demonstrated to have a molecular weight of about 66,000 by gel filtration, and a preference for NAD as oxidant.[61] (-)-*trans*-Isopiperitenol is a somewhat more efficient substrate than (-)-*trans*-carveol by a factor of three, and neither *cis*-isomer is oxidized at a measurable rate. An allylic double bond is required for catalysis, and substrates with endocyclic double bonds are preferred to those with exocyclic double bonds.[61]

p-Menthadienol dehydrogenase activity is always present at a level greatly exceeding that of either limonene-3-hydroxylase or limonene-6-hydroxylase, the immediately preceding enzymes.[23,26] For example, in cell-free extracts of native spearmint leaves, (-)-*trans*-carveol dehydrogenase is more than 50 times as active as limonene-6-hydroxylase,[26] while in peppermint glandular trichome extracts, (-)-*trans*-isopiperitenol dehydrogenase activity is over 1000 times as high as that of limonene-3-hydroxylase (Gershenzon and Croteau, unpublished results). These *in vitro* differences probably account for the failure of either (-)-*trans*-carveol or (-)-trans-isopiperitenol to accumulate in the essential oils of *Mentha* species. To date, no genes associated with *p*-menthadienol dehydrogenase activity have been identified.

p-Menthadienone-$\Delta^{1,2}$-reductase (enzyme 5). In peppermint, the product of dehydrogenase action, (-)-isopiperitenone, is converted to (+)-*cis*-isopulegone by a position-specific and stereospecific double bond reductase. This enzyme is extremely active in peppermint glandular trichome extracts, and has a molecular weight of roughly 60,000 and a strong preference for NADPH as a reductant.[62] $\Delta^{1,2}$-Reductase activity is also present in glandular trichome extracts of native and Scotch spearmint, but at lower levels than those found in peppermint.[23] The enzymes of all three species exhibit a very high degree of specificity for (-)-isopiperitenone, with neither piperitenone nor piperitone being effective substrates.[23,62] Curiously, (-)-carvone is not detectably reduced either, although it seems likely that this $\Delta^{1,2}$-reductase is responsible for producing the low levels of dihydrocarvone and dihydrocarveols observed in the essential oils of both native and Scotch spearmint.[23]

A dominant allele A that appears to encode p-menthadienone-$\Delta^{1,2}$-reductase was first described in 1960 by Murray.[18] This allele, which is equivalent to the more recently proposed X locus of Hendriks,[60] was once believed to cause the reduction of piperitenone to (+)-pulegone.[2,17,60,63] However, from a modern vantage point, the A locus is seen to be responsible for the conversion of (-)-isopiperitenone to (+)-cis-isopulegone,[62] an intermediate which is subsequently isomerized to (+)-pulegone. In the absence of $\Delta^{1,2}$-reductase activity, (-)-isopiperitenone can be diverted to piperitenone, piperitone or the corresponding epoxides (Fig. 1). In fact, Murray[18] demonstrated that these 1,2-unsaturated or 1,2-epoxy-metabolites accumulate in individuals of $M.$ $longifolia$, $M.$ $rotundifolia$, $M.$ $crispa$ and $M.$ $spicata$ with the recessive genotype aa. The A locus (or its equivalent) has also been implicated in the $\Delta^{1,2}$-reduction of the C6-oxygenated products, (-)-carvone and (-)-$trans$-carveol, to epimers of dihydrocarvone and dihydrocarveol, respectively.[2,60]

p-Menthenone isomerase (enzyme 6). The allylic isomerization of (+)-cis-isopulegone to (+)-pulegone and of (-)-isopiperitenone to piperitenone is catalyzed by an activity designated as p-menthenone isomerase. This enzyme has been previously detected in soluble extracts of peppermint, native spearmint and Scotch spearmint, and has a molecular weight of about 55,000 and no cofactor requirement.[23,61,62] (+)-cis-Isopulegone is the preferred substrate at neutral pH, with the isomerization of (-)-isopiperitenone and the corresponding 1,2-oxide occurring at only 15% of this rate. (-)-Carvone is not a substrate at all, indicating a strict requirement for a C3-carbonyl function on the p-menthane ring. However, the C3-ketone group is not in itself sufficient to promote catalysis, since (+)-$trans$-isopulegone, epimeric to (+)-cis-isopulegone at C4, is not a substrate either.[23,62] The mechanism of the reaction has been shown to involve an intramolecular 1,3-hydrogen transfer, reminiscent of the allylic isomerizations mediated by the ketosteroid isomerases.[25,61] Genes affecting the action of p-menthenone isomerase have not yet been discovered. However, this enzyme appears to be present at high levels in all *Mentha* taxa of known monoterpene composition, since neither of its two substrates, (-)-isopiperitenone nor (+)-*cis*-isopulegone, has ever been observed in greater than trace amounts.

p-*Menthenone*-$\Delta^{4,8}$-*reductases* (enzymes 7a and 7b). The isomerase products, (+)-pulegone and piperitenone, can have various metabolic fates in *Mentha* species, including oxidation [(+)-pulegone to (+)-menthofuran, piperitenone to (+)-piperitenone oxide] and reduction of the double bond ($\Delta^{4,8}$) of the isopropylidene moiety [(+)-pulegone to (-)-menthone or (+)-isomenthone, piperitenone to (+)- and (-)-piperitone] (Fig. 1). The NADPH-dependent reduction of the $\Delta^{4,8}$-double bond is accomplished by two distinct enzyme species of opposite stereospecificity. One enzyme (7a) generates the (4S)-isopropyl function, while the other (7b) gives the antipodal (4R)-isopropyl function.[23,62] Enzyme 7a is responsible for the conversion of (+)-pulegone to (-)-menthone, and is apparently also capable of reducing, at significantly slower rates, piperitenone to (+)-(4S)-piperitone, and (+)-piperitenone oxide to (-)-*trans*-piperitone oxide.[23] Enzyme 7b converts (+)-pulegone to (+)-isomenthone, and is almost certainly the same catalyst that transforms piperitenone to (-)-piperitone, and (+)-piperitenone oxide to (-)-*cis*-piperitone oxide, again at substantially slower rates.[23] Neither enzyme is able to reduce the $\Delta^{8,9}$-double bond found in (-)-carvone, (-)-isopiperitenone, dihydrocarvone, carveol or (-)-isopiperitenol. Indeed, $\Delta^{8,9}$-reduction has never been previously observed in a *Mentha* species, even in crude extracts supplied with both NADH and NADPH as reductants. The relative proportions of reductases 7a and 7b vary among the commercial *Mentha* species, ranging from over 10:1 in peppermint to roughly 3:1 in Scotch spearmint.[23,64] These catalysts are very similar in most of their basic properties, including molecular weight, kinetic constants and chromatographic behavior.[23,64]

Based on a series of crossing experiments among different *Mentha* species, Murray and collaborators postulated that the reactions catalyzed by the p-menthenone-$\Delta^{4,8}$-reductases are under the control of a single genetic locus with multiple alleles.[17,63] In their model, the P^S allele causes the reduction of (+)-pulegone to (-)-menthone (mediated by enzyme 7a), while the P^r allele leads to the conversion of (+)-pulegone to (+)-isomenthone (via enzyme 7b). P^r is "mostly dominant" over P^S, while both P^r and P^S are each completely dominant over the recessive allele p, which blocks the expression of both $\Delta^{4,8}$ reductases. Different combinations of these alleles are thought to be responsible for the variety of (-)-menthone to (+)-isomenthone ratios observed in *Mentha* essential oils.[63] Given the incomplete dominance of P^r over P^S and the fact that in polyploid species, such as *M. arvensis* and *M. x piperita*, these alleles will be present on more than one chromosome pair, a wide range of (-)-menthone to (+)-isomenthone ratios could theoretically be obtained. Lawrence[2] hypothesized that the *P* locus also controls the reductions of piperitenone to the piperitones, and piperitenone oxide to the piperitone oxides, a proposition consistent with the available biochemical data.[23]

Oxidation of (+)-pulegone to (+)-menthofuran. One of the most unusual monoterpene structures found in *Mentha* is menthofuran, a compound of special concern in commercial peppermint-type species because of its undesirable influence on oil quality. Little modern research has been carried out on the biosynthetic origin of menthofuran. Pulegone is widely believed to be an immediate precursor based on chemical plausibility, co-occurrence data,[9] preliminary feeding studies,[65] and parallels with mammalian metabolism where liver microsomal preparations have been demonstrated to transform pulegone to menthofuran in a process requiring cytochrome P-450 oxygenase activity.[66,67] A hypothetical pathway from pulegone to menthofuran can be formulated consisting of several discrete steps: (1) hydroxylation of the *cis*-methyl group of the isopropylidene moiety, (2) intramolecular cyclization to form a hemiketal, and (3) dehydration to generate the furan ring. However, it is not clear how many enzymes might be involved in these conversions.

Regardless of the biochemical details, the oxidation of pulegone to menthofuran appears to be regulated by a pair of alleles designated *F* and *f* occupying a single locus. *F* is incompletely dominant to *f*, and the maximum accumulation of menthofuran occurs in the recessive genotype *ff*.[56] Plants with the *ff* genotype contain menthofuran at 60-80% of total monoterpenes, while the heterozygote *Ff* contains 0.4-25%, and the homozygous dominant *FF* less than 0.1%. Since the oxidation of (+)-pulegone to (+)-menthofuran appears to take precedence over the $\Delta^{4,8}$-reduction of (+)-pulegone to (-)-menthone and (+)-isomenthone, the allelic pair *F* and *f* may be considered epistatic to the *P* gene discussed above.[17,56] In *ff* individuals, only trace amounts of menthone, isomenthone and derivatives are found, regardless of whether they have *PP*, *Pp* or *pp* genotypes.

The variable occurrence of menthofuran in certain mint species provides an excellent illustration of how developmental and environmental factors can modify genetic control. For example, the mature leaves of Black Mitcham peppermint (*M. x piperita* Huds. f. *rubescens* Camus) typically contain menthofuran at a level of 1-5% of total monoterpenes,[2,12,68,69] indicating this cultivar to be heterozygous at the *F* locus (*Ff* = 0.4-25% menthofuran, see above).[12] However, inflorescences of Black Mitcham peppermint usually have much higher levels of menthofuran (20-60%) than the leaves,[2,68,70,71] and under short-day photoperiods leaf menthofuran content can reach 80%.[16,69,70,72]

p-Menthanone-3-keto-reductases (enzymes **8a** and **8b**). In peppermint, the products of $\Delta^{4,8}$-reductase activity, the ketones (-)-menthone and (+)-isomenthone, are converted to their corresponding alcohols by a pair of stereospecific, NADPH-dependent keto-reductases.[73] One of these enzymes (**8a**) generates only products with the (3R)-configuration, reducing, at comparable rates, (-)-menthone to (-)-menthol, and (+)-isomenthone to (+)-neoisomenthol. The other enzyme (**8b**) produces only alcohols with the (3S)-configuration, transforming (-)-menthone to (+)-neomenthol, and (+)-isomenthone to (+)-isomenthol. Both enzymes have many properties in common, including molecular weight (approximately 35,000), cofactor preference (NADPH > NADH), pH optimum (about 7.5), kinetic constants, lack of significant reversibility, and sensitivity to thiol-directed reagents. These characteristics are shared by many other cytoplasmic keto-reductases as well.[73] α,β-Unsaturated ketones, such as carvone and pulegone, are rather poor substrates for both reductase **8a** and **8b**. Nevertheless, the conversion of (-)-carvone to (-)-(6R)-*trans*-carveol and (-)-(6S)-*cis*-carveol is probably attributable to these enzymes, as is the formation of trace levels of various other monoterpene alcohols in *Mentha* species, such as dihydrocarveol and isopulegol.[73] The relative activities of **8a** and **8b** vary among the commercial mint species. For example, in peppermint the (3R)-alcohol forming enzyme (**8a**) predominates, whereas in Scotch spearmint the (3S)-alcohol forming catalyst (**8b**) is the major activity.[23]

The expression of enzymes **8a** and **8b** has been shown to be under simple genetic control. In 1960, Murray established that the reduction of (-)-menthone to (-)-menthol in *M. arvensis* (activity **8a**) was regulated by a single locus designated R having both a dominant (R) and a recessive (r) allele.[20] Thus, the genotypes RR and Rr convert (-)-menthone to (-)-menthol, while rr individuals accumulate only (-)-menthone. The R locus was later also implicated in the conversion of carvone to carveols, and of dihydrocarvone to dihydrocarveols, that occurs in *M. crispa*, *M. spicata* and *M. x gracilis*.[55] More recently, based on oil analyses of many *Mentha* species and segregation patterns in several artificial hybrids, Lawrence[2,9] inferred the existence of two R-type genes, one named R_r for the production of alcohols with a (3R)-configuration (activity **8a**) and the other named R_s for the production of (3S)-alcohols (activity **8b**). The nature of the actual proteins encoded by these loci is unclear, since control over 3-keto reduction appears to be "leaky" in some cases. For example, in *M. arvensis* var. *piperascens*, individuals with dominant R_rR_r or R_rr_r

genotypes accumulate (-)-menthol to 40-80% of total monoterpenes, but recessive $r_r r_r$ genotypes also contain significant amounts (3-15%) of (-)-menthol.[20] Thus, the R_r locus may encode a regulatory protein rather than the enzyme itself, or the recessive allele r_r may simply code for a catalytically less-efficient form of the enzyme.

The genetic control of 3-keto-reductase activity often exhibits an interesting developmental aspect. For instance, young peppermint leaves contain large quantities of (-)-menthone and very little (-)-menthol, even though this species has an $R_r r_r$ genotype.[12,59] (-)-Menthol only appears in substantial amounts in peppermint after leaves have reached maturity,[16,69,72] indicating that the 3-keto-reductase activity in this species (predominantly 8a) is only expressed at high levels late in leaf development (Fig. 3). Analogous developmental changes involving the reduction of carvone to carveols, and of dihydrocarvone to dihydrocarveols, seem to occur in *Mentha* species with spearmint-type oils (*M. crispa*, *M. spicata* and *M. x gracilis*).[74] Curiously, other genes of monoterpene metabolism also seem to be maximally expressed late in leaf development, including the *A* gene (*p*-menthadienone-$\Delta^{1,2}$-reductase) in species with spearmint-type oils[74] and the *E* gene (acetyltransferase) discussed below.

Conjugation Reactions

The *p*-menthane alcohols of *Mentha* species are not stable end-products of metabolism, since they can form conjugates with acetate or glucose moieties, especially during the late stages of plant development. The enzyme-catalyzed reactions involved have only been examined in peppermint.[75,76] However, the widespread occurrence of acetate and glucose conjugates in *Mentha*,[9,77-81] indicates that the requisite enzymes have an equally wide distribution.

p-Menthanol acetyltransferase. (-)-Menthyl acetate may constitute up to 15% of the total monoterpenes present in the mature leaves of flowering peppermint.[75] The acetyltransferase responsible for the formation of this metabolite from (-)-menthol is an operationally soluble enzyme with a molecular weight of 37,000 as determined by gel filtration. This transferase is highly specific for acetyl-CoA, with propionyl-CoA and butyryl-CoA being very inefficient as acyl donors. However, much less selectivity is exhibited with regard to the alcohol utilized. The peppermint enzyme is capable of acetylating a variety of monoterpenoid and non-monoterpenoid alcohols, with *n*-decanol being the most efficient substrate.[75] Thus, the nature of the acetates produced in the plant may depend upon the type of alcohols available for reaction, rather than the

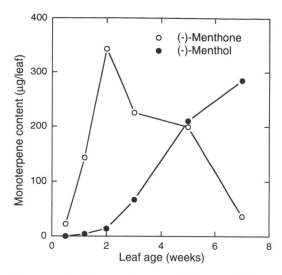

Fig. 3. Changes in (-)-menthone and (-)-menthol content in developing peppermint leaves. (-)-Menthone is reduced to (-)-menthol by p-menthanone-3-keto-reductase (enzyme **8**). Compared to other enzymes of monoterpene biosynthesis in peppermint, this activity is expressed much later in leaf development.

selectivity of the acetyltransferase itself. The availability of monoterpene alcohols may in turn be influenced by specific compartmentation phenomena. For example, mature peppermint leaves contain 40% (-)-menthol and 5% (+)-neomenthol, yet only (-)-menthyl acetate is found in this species.[82] No significant amount of (+)-neomenthyl acetate is present despite the fact that (-)-menthol and (+)-neomenthol are acetylated *in vitro* with nearly equal facility.[75] Hence, the enzyme must only have access to (-)-menthol *in vivo*.[82] Almost nothing is known concerning the genetic basis of monoterpene acetylation in *Mentha*. Rudimentary studies indicate that a single dominant allele, referred to as *E* for "esterase", appears to be responsible for the conversion of a variety of monoterpene alcohols into their corresponding acetates.[1,74]

   p-Menthanol glucosyltransferase. Monoterpene alcohols in *Mentha* can occur as conjugates with glucose as well as with acetate. For instance, glucosides of (-)-menthol, (+)-neomenthol, (-)-*trans*-carveol, (-)-*cis*-carveol, (-)-dihydrocarveol and (+)-neodihydrocarveol have been reported from various commercially important species of *Mentha*.[77-81] However, the enzymatic machinery for glucoside formation has only been investigated in peppermint.[76] A UDP-glucose:monoterpenol glucosyltransferase has been described from peppermint leaves that converts (+)-neomenthol to (+)-neomenthyl-β-D-glucoside. This enzyme is operationally soluble with a molecular weight of 46,000, and transferase activity can be stimulated by magnesium ions. Both (+)-neomenthol and (-)-menthol are glucosylated at equal rates *in vitro*, suggesting that the enzyme may have only a low specificity for particular monoterpene alcohols. However, as for acetylation, only one of the menthol isomers [(+)-neomenthol in this case] seems to be glucosylated *in vivo*, suggesting that strict compartmentation effects operate in the intact plant.[82] No information is currently available on the genetic control of monoterpene glucosylation, but the possible role of glycosides in monoterpene metabolism[4,77] should stimulate interest in the genetic factors underlying their formation.

GENETIC CONTROL IN OTHER MONOTERPENE-PRODUCING PLANTS

Many monoterpene-accumulating species of the Lamiaceae and Asteraceae show extensive infraspecific variation in monoterpene composition.[83,84] Numerous chemical races or chemotypes have been described based on differences in monoterpene profiles. Such differences appear to be under genetic rather than environmental control, since they are largely retained when plants are grown side by side in a common garden. In an effort to understand the factors that regulate the inheritance of monoterpene composition in these families, a large number of crosses have been made among individuals of different chemical races and the progeny analyzed. Although no genus has been examined in the same detail as *Mentha*, together these studies have revealed that several general features of the genetic control of monoterpene composition are common to many different groups of plants.

As we have already seen, the differences in monoterpene composition among *Mentha* taxa are usually attributable to the action of only a few genes. For example, the accumulation of linalool and linalyl acetate instead of cyclic monoterpenes in *M. citrata* is due to the presence of the dominant *I* allele rather than the recessive allele *i*.[9,45] In addition, the different compositional types in

Mentha usually show simple dominance relationships with each other, e.g. C6-oxygenation is dominant to C3-oxygenation.[1,18,24,55,56] Both of these characteristics are also observed in species of *Perilla*,[85-89] *Satureja*[57] and *Thymus*.[90,91] In *Thymus vulgaris* (garden thyme), for instance, six chemical races are native to the Mediterranean region of France, each of which contains a single major monoterpene constituent: geraniol, linalool, α-terpineol, thujanol, thymol or carvacrol.[90] These races form a strict dominance hierarchy: geraniol > α-terpineol > thujanol > linalool > carvacrol > thymol (meaning that the geraniol race is dominant to all the others, the α-terpineol race is dominant to all but the geraniol race, etc.), and are thought to differ from each other at just six loci.[91] In *Perilla frutescens*, an annual mint cultivated extensively in Japan, six chemical races have also been described: one that accumulates cyclohexanoid monoterpenes, three that produce different types of furanoid monoterpenes, one that contains only acyclic monoterpenes, and one that does not produce any monoterpenes in its essential oil but accumulates phenylpropanoids instead.[85,87] The genetic differences among these races are very small. For example, the production of cyclohexanoid monoterpenoids instead of any of the furanoid skeletal types is due to genetic differences at just two loci, designated *G* and *H*. The cyclohexanoid chemical race is dominant over all the others.

Strict dominance relationships among chemical races are not evident in all monoterpene-containing species. Crosses between plants of varying monoterpene composition sometimes give rise to progeny whose chemistry resembles that of both parents.[92,93] In *Hedeoma*, for instance, hybrids between plants possessing principally acyclic monoterpenes, such as geranial and neral, and plants possessing mainly monocyclic monoterpenes, such as pulegone and isomenthone, produce an "additive" chemistry comprising monoterpenes of both major types. This genus also differs from others in the Lamiaceae in that multiple loci seem to control the inheritance of monoterpene composition,[92] although interpretation of the crossing results is complicated by the lack of biosynthetic knowledge. Indeed, in all these investigations, the assignment of genes to specific chemical transformations must be regarded as extremely tentative because of uncertainties about the sequence of intermediates in the biosynthetic pathway. A number of the genes implicated in the control of monoterpene profiles may actually encode regulatory proteins rather than the biosynthetic enzymes themselves, since control is sometimes leaky[20,86,92] and only poorly correlated with changes in enzyme activity.[88]

Like herbaceous plants, many conifers also exhibit pronounced infraspecific variation in monoterpene composition, much of which has been demonstrated to be genetically controlled.[83,84,94] Chemical analyses of families obtained from controlled pollinations have suggested that the inheritance of individual monoterpenes is often under the control of a single gene having a dominant/recessive pair of alleles.[95-103] The allele specifying a high monoterpene content is usually observed to be dominant to the allele specifying a low monoterpene content.[95,96,98,100] For example, in *Pinus elliottii* (slash pine), variation in the amount of β-pinene in cortical oleoresin was reported to be under single gene control, with a high amount of this compound (21-74% of total resin monoterpenes) being dominant over a low amount (2-8% of resin monoterpenes).[100] In this study, the myrcene content of *P. elliottii* was also shown to be under monogenic control, and the segregation of the alleles regulating myrcene content was found to be independent of the alleles regulating β-pinene content. However, in other conifers the inheritance of different monoterpenes is sometimes linked, e.g., α-pinene, β-pinene and 3-carene in *Picea abies*.[96] Such linkage may be due to the formation of multiple products by a single biosynthetic enzyme.[50,51,104-107] In *Abies grandis*, for instance, a monoterpene cyclase induced by wounding produces both α–pinene (40%) and β-pinene (60%).[106]

Considerable criticism has recently been directed against investigations of the genetic control of monoterpene composition in conifers because of the use of inappropriate statistical methods and the fact that quantitation of specific compounds is nearly always expressed as a percent of total resin rather than as an absolute amount.[108] Thus, the evidence cited in support of monogenic control may actually be quite weak. In fact, Birks and Kanowski[108] assert that nearly all published models of gene control mechanisms for regulating monoterpene composition in conifers are fundamentally flawed. Improved knowledge of the pathways of monoterpene biosynthesis in this plant group should greatly assist in understanding the genetic control of monoterpene formation.

CONCLUSION

This chapter has correlated the results of the classical crossing experiments of Murray with more recent biosynthetic findings to assemble an up-to-date picture of the genetic control of monoterpene formation in *Mentha* species. For the first time, classical genetic and biochemical approaches to the study of monoterpene biosynthesis have been formally integrated. Nearly all the genes described by Murray and later workers have been assigned to specific, well-characterized transformations in the biosynthesis of *Mentha* monoterpenes, although not every step of monoterpene metabolism in this genus can yet be linked to a particular gene.

The first committed reaction of monoterpene biosynthesis in the commercial mint species is the cyclization of the ubiquitous intermediate, geranyl pyrophosphate, to (-)-limonene, catalyzed by geranyl pyrophosphate:(-)-limonene cyclase. No gene associated with the expression of this extensively-investigated enzyme was detected in Murray's work, but modern molecular methods have led to the isolation and sequencing of a cDNA clone that encodes geranyl pyrophosphate:(-)-limonene cyclase. The loci I and I_S appear to control alternate reactions of geranyl pyrophosphate, although no biochemical research has been carried out on these transformations so far. The (-)-limonene produced in *Mentha* is a substrate for two cytochrome P450-dependent hydroxylases that determine the basic types of monoterpenes formed in most species of the genus by regulating the oxygenation pattern (C3 versus C6). The action of these well-studied enzymes is controlled by two closely-linked diallelic loci designated C and Lm. Subsequent redox transformations may also have a significant influence on monoterpene composition. These reactions have been well-characterized biochemically, and specific loci (A, F, P and R) have been assigned to nearly all of them.

Most of the genes implicated in the control of monoterpene biosynthesis in mints are diallelic with simple dominant and recessive alleles that segregate independently in standard Mendelian fashion. However, a few loci are multiallelic (e.g., P), and others possess alleles that are incompletely dominant to each other (e.g., F and f). The nature of the actual products of these genes is still uncertain, though some may code for regulatory proteins rather than the biosynthetic enzymes themselves. Clearly, further molecular genetic investigations are needed to decipher the various levels of genetic interaction that determine the pattern of inheritance of the major monoterpenes in *Mentha*. Additional research on both the genes and the corresponding enzymes should permit the modification of monoterpene yield and composition in commercial mint species, and facilitate the development of entirely new varieties that produce enhanced levels of monoterpenes of particular value in flavor, fragrance and pharmaceutical applications.

ACKNOWLEDGEMENTS

The work by the authors described herein was supported in part by grants from the U.S. National Institutes of Health, National Science Foundation, Department of Energy and Washington Mint Commission/Mint Industry Research Council, and by Project 0268 from the Agricultural Research Center,

Washington State University. We thank Joyce Tamura-Brown for typing the manuscript, Roy LaFever for preparation of the artwork, and Eran Pichersky for critical discussions.

REFERENCES

1.   HEFENDEHL, F.W., MURRAY, M.J. 1976. Genetic aspects of the biosynthesis of natural odors. Lloydia 39:39-52.

2.   LAWRENCE, B.M. 1981. Monoterpene interrelationships in the *Mentha* genus: A biosynthetic discussion. In: Essential Oils. (B.D. Mookherjee, C.J. Mussinan, eds.) Allured Publ. Co., Wheaton, IL, pp. 1-81.

3.   CROTEAU, R. 1981. Biosynthesis of monoterpenes. In: Biosynthesis of Isoprenoid Compounds. Vol. 1. (J.W. Porter, S.L. Spurgeon, eds.) John Wiley and Sons, New York, pp. 225-282.

4.   CROTEAU, R. 1987. Biosynthesis and catabolism of monoterpenoids. Chem. Rev. 87:929-954.

5.   CROTEAU, R. 1991. Metabolism of monoterpenes in mint (*Mentha*) species. Planta Med. 57(Suppl.):S10-S14.

6.   COLBY, S.M., ALONSO, W.R., KATAHIRA, E.J., MCGARVEY, D.J., CROTEAU, R. 1993. 4S-Limonene synthase from the oil glands of spearmint (*Mentha spicata*): cDNA isolation, characterization and bacterial expression of the catalytically active monoterpene cyclase. J. Biol. Chem., 268:23016-23024.

7.   TUCKER, O.A., HARLEY, R.M., FAIRBROTHERS, D.E. 1980. The Linnaean types of *Mentha* (Lamiaceae). Taxon 29:233-255.

8.   KOKKINI, S. 1991. Chemical races within the genus *Mentha* L. In: Essential Oils and Waxes. (H.F. Linskens, J.F. Jackson, eds.) Springer-Verlag, Berlin, pp. 63-78.

9.   LAWRENCE, B.M. 1978. A study of the monoterpene interrelationships in the genus *Mentha* with special reference to the origin of pulegone and menthofuran, Ph.D. Thesis, Groningen State University, Groningen.

10. HARLEY, R.M., BRIGHTON, C.A. 1977. Chromosome numbers in the genus *Mentha*. Bot. J. Linn. Soc., 74:71-96.

11. LAWRENCE, B.M. 1985. A review of the world production of essential oils. Perfum. Flav. 13:2-16.

12. MURRAY, M.J., LINCOLN, D.E., MARBLE, P.M. 1972. Oil composition of *Mentha aquatica* x *M. spicata* F_1 hybrids in relation to the origin of *M.* x *piperita*. Can. J. Genet. Cytol. 14:13-29.

13. TUCKER, A.O., FAIRBROTHERS, D.E. 1990. The origin of *Mentha* x *gracilis* (Lamiaceae). I. Chromosome numbers, fertility, and three morphological characters. Econ. Bot. 44:183-213.

14. TUCKER, A.O., HENDRIKS, H., BOS, R., FAIRBROTHERS, D.E. 1991. The origin of *Mentha* x *gracilis* (Lamiaceae). II. Essential oils. Econ. Bot. 45:200-215.

15. KJONAAS, R., CROTEAU, R. 1983. Demonstration that limonene is the first cyclic intermediate in the biosynthesis of oxygenated *p*-menthane monoterpenes in *Mentha piperita* and other *Mentha* species. Arch. Biochem. Biophys. 220:79-89.

16. BURBOTT, A.J., LOOMIS, W.D. 1967. Effects of light and temperature on the monoterpenes of peppermint. Plant Physiol. 42:20-28.

17. LINCOLN, D.E., MURRAY, M.J. 1978. Monogenic basis for reduction of (+)-pulegone to (-)-menthone in *Mentha* oil biogenesis. Phytochemistry 17:744-753.

18. MURRAY, M.J. 1960. The genetic basis for a third ketone group in *Mentha spicata* L. Genetics 45:931-937.

19. MURRAY, M.J., LINCOLN, D.E. 1970. The genetic basis of acyclic constituents in *Mentha citrata* Ehrh. Genetics 65:457-471.

20. MURRAY, M.J. 1960. The genetic basis for the conversion of menthone to menthol in Japanese mint. Genetics 45:925-929.

21. MURRAY, M.J., REITSEMA, R.H. 1954. The genetic basis of the ketones, carvone and menthone, in *Mentha crispa* L. J. Am. Pharm. Assoc. Sci. Ed. 43:612-613.

22. KARP, F., MIHALIAK, C.A., HARRIS, J.L., CROTEAU, R. 1990. Monoterpene biosynthesis: Specificity of the hydroxylations of (-)-limonene by enzyme preparations from peppermint (*Mentha piperita*), spearmint (*Mentha spicata*), and perilla (*Perilla frutescens*) leaves. Arch. Biochem. Biophys. 276:219-226.

23. CROTEAU, R., KARP, F., WAGSCHAL, K.C., SATTERWHITE, D.M., HYATT, D.C., SKOTLAND, C.B. 1991. Biochemical characterization of a spearmint mutant that resembles peppermint in monoterpene content. Plant Physiol. 96:744-753.

24. LINCOLN, D.E., MARBLE, P.M., CRAMER, F.J., MURRAY, M.J. 1971. Genetic basis for high limonene-cineole content of exceptional *Mentha citrata* hybrids. Theor. Appl. Genet. 41:365-370.

25. TALALAY, P., BENSON, A.M. 1972. Ketosteroid isomerases. In: The Enzymes. 3rd Ed., Vol. 6. (P.D. Boyer, ed.) Academic Press, New York, pp. 591-618.

26. GERSHENZON, J., MAFFEI, M., CROTEAU, R. 1989. Biochemical and histochemical localization of monoterpene biosynthesis in the glandular trichomes of spearmint (*Mentha spicata*). Plant Physiol. 89:1351-1357.

27. GERSHENZON, J., MCCASKILL, D.G., RAJAONARIVONY, J., MIHALIAK, C., KARP, F., CROTEAU, R. 1991. Biosynthetic methods for plant natural products: New procedures for the study of glandular trichome constituents. Rec. Adv. Phytochem. 25:347-370.

28. MCCASKILL, D., GERSHENZON, J., CROTEAU, R. 1992. Morphology and monoterpene biosynthetic capabilities of secretory cell clusters isolated from glandular trichomes of peppermint (*Mentha piperita* L.). Planta 187:445-454.

29. GERSHENZON, J., DUFFY, M.A., KARP, F., CROTEAU, R. 1987. Mechanized techniques for the selective extraction of enzymes from plant epidermal glands. Anal. Biochem. 163:159-164.

30. GERSHENZON, J., MCCASKILL, D., RAJAONARIVONY, J.I.M., MIHALIAK, C., KARP, F., CROTEAU, R. 1992. Isolation of secretory cells from plant glandular trichomes and their use in biosynthetic studies of monoterpenes and other gland products. Anal. Biochem. 200:130-138.

31. ALONSO, W.R., RAJAONARIVONY, J.I.M., GERSHENZON, J., CROTEAU, R. 1992. Purification of 4S-limonene synthase, a monoterpene cyclase from the glandular trichomes of peppermint (*Mentha* x *piperita*) and spearmint (*M. spicata*). J. Biol. Chem. 267:7582-7587.

32. RAJAONARIVONY, J.I.M., GERSHENZON, J., CROTEAU, R. 1992. Characterization and mechanism of 4S-limonene synthase, a monoterpene cyclase from the glandular trichomes of peppermint (*Mentha* x *piperita*) Arch. Biochem. Biophys. 296:49-57.

33. RAJAONARIVONY, J.I.M., GERSHENZON, J., MIYAZAKI, J., CROTEAU, R. 1992. Evidence for an essential histidine residue in 4S-limonene synthase and other terpene cyclases. Arch. Biochem. Biophys. 299:77-82.

34. CROTEAU, R., ALONSO, W.R., KOEPP, A.E., SHIM, J.-H., CANE, D.E. 1993. Irreversible inactivation of monoterpene cyclases by a mechanism-based inhibitor. Arch. Biochem. Biophys., in press.

35. ALONSO, W.R., CROCK, J.E., CROTEAU, R. 1993. Production and characterization of polyclonal antibodies in rabbits to 4S-limonene synthase from spearmint (*Mentha spicata*). Arch. Biochem. Biophys. 301:58-63.

36. GERSHENZON, J., ALONSO, W.R., CROTEAU, R. 1993. Subcellular localization of (4S)-limonene synthase in glandular trichome secretory cells of peppermint (*Mentha* x *piperita*). submitted.

37. FACCHINI, P.J., CHAPPELL, J. 1992. Gene family for an elicitor-induced sesquiterpene cyclase in tobacco. Proc. Natl. Acad. Sci. USA 89:11088-11092.

38. CANE, D.E. 1992. Terpenoid cyclases. Design and function of electrophilic catalysts. In: Secondary Metabolites: Their Function and Evolution. CIBA Foundation Symposium 171. (D.J. Chadwick, J. Whelan, eds.) John Wiley and Sons, West Sussex, UK, pp. 163-183.

39. POULTER, C.D., RILLING, H.C. 1981. Prenyl transferases and isomerases. In: Biosynthesis of Isoprenoid Compounds. Vol. 1. (J.W. Porter, S.L. Spurgeon, eds.) John Wiley and Sons, New York, pp. 161-224.

40. ASHBY, M.N., EDWARDS, P.A. 1990. Elucidation of the deficiency in two yeast coenzyme Q mutants. Characterization of the structural gene encoding hexaprenyl pyrophosphate synthase. J. Biol. Chem. 265:13157-13164.

41. GERSHENZON, J., CROTEAU, R. 1990. Regulation of monoterpene biosynthesis in higher plants. Rec. Adv. Phytochem. 24:99-160.

42. CLASTRE, M., BANTIGNIES, B., FERON, G., SOLER, E., AMBID, C. 1993. Purification and characterization of geranyl diphosphate synthase from *Vitis vinifera* L. cv Muscat de Frontignan cell cultures. Plant Physiol. 102:205-211.

43. CROTEAU, R., PURKETT, P.T. 1989. Geranyl pyrophosphate synthase: Characterization of the enzyme and evidence that this chain-length specific prenyltransferase is associated with monoterpene biosynthesis in sage (*Salvia officinalis*). Arch. Biochem. Biophys. 271:524-535.

44. MALINGRE, T.M. 1971. Chemotaxonomisch onderzock von *Mentha arvensis* L. Pharm. Weekblad 106:165-171.

45. TODD, W.A., MURRAY, M.J. 1968. New essential oils from hybridization of *Mentha citrata* Ehrh. Perfum. Essent. Oil Rec. 59:97-102.

46. HEFENDEHL, F.W., MURRAY, M.J. 1973. Monoterpene composition

of a chemotype of *Mentha piperita* having high limonene. Planta Med. 23:101-109.

47.   MURRAY, M.J., HEFENDEHL, F.W. 1973. Changes in monoterpene composition of *Mentha aquatica* produced by gene substitution from a high limonene strain of *M. citrata*. Phytochemistry 12:1875-1880.

48.   LINCOLN, D.E., MURRAY, M.J., LAWRENCE, B.M. 1986. Chemical composition and genetic basis for the isopinocamphone chemotype of *Mentha citrata* hybrids. Phytochemistry 25:1857-1863.

49.   CROTEAU, R., ALONSO, W.R., KOEPP, A.E., JOHNSON, M.A. 1993. Biosynthesis of monoterpenes: Characterization and mechanism of action of 1,8-cineole synthase. submitted.

50.   GAMBLIEL, H., CROTEAU, R. 1984. Pinene cyclases I and II: Two enzymes from sage (*Salvia officinalis*) which catalyze stereospecific cyclizations of geranyl pyrophosphate to monoterpene olefins of opposite configuration. J. Biol. Chem. 259:740-748.

51.   HALLAHAN, T.W., CROTEAU, R. 1988. Monoterpene biosynthesis: Demonstration of a geranyl pyrophosphate:sabinene hydrate cyclase in soluble enzyme preparations from sweet marjoram (*Majorana hortensis*). Arch. Biochem. Biophys. 264:618-631.

52.   HALLAHAN, T.W., CROTEAU, R. 1989. Monoterpene biosynthesis: Mechanism and stereochemistry of the enzymatic cyclization of geranyl pyrophosphate to (+)-*cis*- and (+)-*trans*-sabinene hydrate. Arch. Biochem. Biophys. 269:313-326.

53.   KARP, F., CROTEAU, R. 1992. Hydroxylation of (-)-β-pinene and (-)-α-pinene by a cytochrome P-450 system from hyssop (*Hyssopus officinalis*). In: Secondary Metabolite Biosynthesis and Metabolism. (R.J. Petroski, S.P. McCormick, eds.) Plenum Press, New York, pp. 253-260.

54.   FUNK, C., CROTEAU, R. 1993. Induction and characterization of a cytochrome P450-dependent camphor hydroxylase in tissue cultures of common sage (*Salvia officinalis*). Plant Physiol. 101:1231-1237.

55.   HEFENDEHL, F.W., MURRAY, M.J. 1972. Changes in monoterpene composition in *Mentha aquatica* produced by gene substitution. Phytochemistry 11:189-195.

56.   MURRAY, M.J., HEFENDEHL, F.W. 1972. Changes in monoterpene composition of *Mentha aquatica* produced by gene substitution from *Mentha arvensis*. Phytochemistry 11:2469-2474.

57.   LINCOLN, D.E., LANGENHEIM, J.H. 1981. A genetic approach to monoterpenoid compositional variation in *Satureja douglasii*. Biochem. Syst. Ecol. 9: 153-160.

58. CAPDEVILA, J., SAEKI, Y., FALCK, J.R. 1984. The mechanistic plurality of cytochrome P-450 and its biological ramifications. Xenobiotica 14:105-118.

59. MURRAY, M.J., LINCOLN, D.E. 1972. Oil composition of *Mentha aquatica* - *M. longifolia* F_1 hybrids and *M. dumetorum*. Euphytica 21: 337-343.

60. HENDRIKS, H., VAN OS, F.H.L., FEENSTRA, W. J. 1976. Crossing experiments between some chemotypes of *Mentha longifolia* and *Mentha suaveolens*. Planta Med. 30:154-162.

61. KJONAAS, R.B., VENKATACHALAM, K.V., CROTEAU, R. 1985. Metabolism of monoterpenes: Oxidation of isopiperitenol to isopiperitenone, and subsequent isomerization to piperitenone, by soluble enzyme preparations from peppermint (*Mentha piperita*) leaves. Arch. Biochem. Biophys. 238:49-60.

62. CROTEAU, R., VENKATACHALAM, K.V. 1986. Metabolism of monoterpenes: Demonstration that (+)-*cis*-isopulegone, not piperitenone, is the key intermediate in the conversion of (-)-isopiperitenone to (+)-pulegone in peppermint (*Mentha piperita*). Arch. Biochem. Biophys. 249:306-315.

63. MURRAY, M.J., LINCOLN, D.E., HEFENDEHL, F.W. 1980. Chemogenetic evidence supporting multiple allele control of the biosynthesis of (-)-menthone and (+)-isomenthone stereoisomers in *Mentha* species. Phytochemistry 19:2103-2110.

64. BATTAILE, J., BURBOTT, A.J., LOOMIS, W.D. 1968. Monoterpene interconversions: Metabolism of pulegone by a cell-free system from *Mentha piperita*. Phytochemistry 7:1159-1163.

65. BATTAILE, J., LOOMIS, W.D. 1961. Biosynthesis of terpenes. II. The site and sequence of terpene formation in peppermint. Biochim. Biophys. Acta 51:545-552.

66. GORDON, W.P., HUITRIC, A.C., SETH, C.L., MCCLANAHAN, R.H., NELSON, S.D. 1987. The metabolism of the abortifacient terpene, (*R*)-(+)-pulegone, to a proximate toxin, menthofuran. Drug Metab. Dispos. 15:589-595.

67. NELSON, S.D., MCCLANAHAN, K., KNEBEL, N., THOMASSEN, D., GORDON, W.P., OISHI, S. 1992. The metabolism of (*R*)-(+)-pulegone, a toxic monoterpene. In: Secondary Metabolite Biosynthesis and Metabolism. (R.J. Petroski, S.P. McCormick, eds.) Plenum Press, New York, pp. 287-296.

68. MAFFEI, M., SACCO, T. 1987. Chemical and morphometrical

comparison between two peppermint notomorphs. Planta Med. 53:214-216.

69. VOIRIN, B., BRUN, N., BAYET, C. 1990. Effects of daylength on the monoterpene composition of leaves of *Mentha* x *piperita*. Phytochemistry 29:749-755.

70. GRAHLE, A., HOLTZEL, C. 1963. Photoperiodische Abhangigkeit der Bildung des ätherischen Öls bei *Mentha piperita* L. Naturwiss. 50:552-555.

71. LAWRENCE, B.M., SHU, C.-K., HARRIS, W.R. 1989. Peppermint oil differentiation. Perfum. Flav. 14: 21-30.

72. CLARK, R.J., MENARY, R.C. 1980. Environmental effects on peppermint (*Mentha piperita* L.). I. Effect of daylength, photon flux density, night temperature and day temperature on the yield and composition of peppermint oil. Aust. J. Plant Physiol. 7: 685-692.

73. KJONAAS, R., MARTINKUS-TAYLOR, C., CROTEAU, R. 1982. Metabolism of monoterpenes: Conversion of *l*-menthone to *l*-menthol and *d*-neomenthol by stereospecific dehydrogenases from peppermint (*Mentha piperita*) leaves. Plant Physiol. 69:1013-1017.

74. MURRAY, M.J., FAAS, W., MARBLE, P. 1972. Effects of plant maturity on oil composition of several spearmint species grown in Indiana and Michigan. Crop Sci. 12:723-728.

75. CROTEAU, R., HOOPER, C.L. 1978. Metabolism of monoterpenes. Acetylation of (-)-menthol by a soluble enzyme preparation from peppermint (*Mentha piperita*) leaves. Plant Physiol. 61:737-742.

76. MARTINKUS, C., CROTEAU, R. 1981. Metabolism of monoterpenes. Evidence for compartmentation of *l*-menthone metabolism in peppermint (*Mentha piperita*) leaves. Plant Physiol. 68:99-106.

77. CROTEAU, R., MARTINKUS, C. 1979. Metabolism of monoterpenes. Demonstration of (+)-neomenthyl–β-D-glucoside as a major metabolite of (-)-menthone in peppermint (*Mentha piperita*). Plant Physiol. 64:169-175.

78. SAKATA, I., MITSUI, T. 1975. Isolation and identification of *l*-menthyl-β-D-glucoside from shubi. Agric. Biol. Chem. 39:1329-1330.

79. SAKATA, I., KOSHIMIZU, K. 1978. Occurrence of *l*-menthyl-β-D-glucoside and methyl palmitate in rhizoma of Japanese peppermint. Agric. Biol. Chem. 42:1959-1960.

80. SHIMIZU, S., SHIBATA, H., MAEJIMA, S. 1990. A new monoterpene glucoside, *l*-menthyl-6'-*O*-acetyl glucoside. J. Ess. Oil Res. 2:21-24.

81. SHIMIZU, S., SHIBATA, H., KARASAWA, D., KOZAKI, T. 1990. Carvyl- and dihydrocarvyl-β-D-glucosides in spearmint (Studies on terpene glycosides in *Mentha* plants, Part II). J. Ess. Oil Res. 2:81-86.

82. CROTEAU, R., WINTERS, J.N. 1982. Demonstration of the intercellular compartmentation of *l*-menthone metabolism in peppermint (*Mentha piperita*) leaves. Plant Physiol. 69:975-977.

83. GIANNASI, D.E., CRAWFORD, D.J. 1986. Biochemical systematics II. A reprise. In: Evolutionary Biology, Vol. 20. (M.K. Hecht, B. Wallace, G.T. Prance, eds.) Plenum Press, New York, pp. 25-248.

84. HARBORNE, J.B., TURNER, B.L. 1984. Plant Chemosystematics. Academic Press, London, 562 pp.

85. KOEZUKA, Y., HONDA, G., TABATA, M. 1986. Genetic control of the chemical composition of volatile oils in *Perilla frutescens*. Phytochemistry 25:859-863.

86. KOEZUKA, Y., HONDA, G., TABATA, M. 1986. Genetic control of isoegomaketone formation in *Perilla frutescens*. Phytochemistry 25:2656-2657.

87. NISHIZAWA, A., HONDA, G., TABATA, M. 1990. Genetic control of perillene accumulation in *Perilla frutescens*. Phytochemistry 29:2873-2875.

88. NISHIZAWA, A., HONDA, G., TABATA, M. 1992. Genetic control of the enzymatic formation of cyclic monoterpenoids in *Perilla frutescens*. Phytochemistry 31:139-142.

89. NISHIZAWA, A., HONDA, G., TABATA, M. 1989. Determination of final steps in biosyntheses of essential oil components in *Perilla frutescens*. Planta Med. 55:251-253.

90. GRANGER, R., PASSET, J. 1973. *Thymus vulgaris* spontane de France: races chimiques et chemotaxonomie. Phytochemistry 12:1683-1691.

91. VERNET, P., GOUYON, P.H., VALDEYRON, G. 1986. Genetic control of the oil content in *Thymus vulgaris* L.: A case of polymorphism in a biosynthetic chain. Genetica 69:227-231.

92. IRVING, R.S., ADAMS, R.P. 1973. Genetic and biosynthetic relationships of monoterpenes. Rec. Adv. Phytochem. 6:187-214.

93. LOKKI, J., SORSA, M., FORSEN, K., SCHANTZ, M.V. 1973. Genetics of monoterpenes in *Chrysanthemum vulgare*. I. Genetic control and inheritance of some of the most common chemotypes. Hereditas 74:225-232.

94.   VON RUDLOFF, E. 1975. Volatile leaf oil analysis in chemosystematic studies of North American conifers. Biochem. Syst. Ecol. 2:131-167.

95.   BERNARD-DAGAN, C., PAULY, G., MARPEAU, A., GLEIZES, M., CARDE, J.-P., BARADAT, P. 1982. Control and compartmentation of terpene biosynthesis in leaves of *Pinus pinaster*. Physiol. Veg. 20:775-795.

96.   ESTEBAN, I., BERGMANN, F., GREGORIUS, H.-R., HUHTINEN, O. 1976. Composition and genetics of monoterpenes from cortical oleoresin of Norway spruce and their significance for clone identification. Silv. Genet. 25:59-66.

97.   GANSEL, C.R., SQUILLACE, A.E. 1976. Geographic variation of monoterpenes in cortical oleoresin of slash pine. Silv. Genet. 25:150-154.

98.   HANOVER, J.W. 1966. Inheritance of 3-carene concentration in *Pinus monticola*. For. Sci. 12:447-450.

99.   MEIER, R.J., GOGGANS, J.F. 1978. Heritabilities and correlations of the cortical monoterpenes of Virginia pine (*Pinus virginiana* Mill.). Silv. Genet. 27:79-84.

100. SQUILLACE, A.E. 1971. Inheritance of monoterpene composition in cortical oleoresin of slash pine. For. Sci. 17:381-387.

101. SQUILLACE, A.E., SWINDEL, B.F. 1986. Linkage among genes controlling monoterpene constituent levels in loblolly pine. For. Sci. 32:97-112.

102. WHITE, E.E. 1984. Mode of genetic control of monoterpenes in foliage of controlled crosses of *Pinus contorta*. Silv. Genet. 33:115-119.

103. YAZDANI, R., RUDIN, D., ALDEN, T., LINDGREN, D., HARBOM, B., LJUNG, K. 1982. Inheritance pattern of five monoterpenes in Scots pine (*Pinus sylvestris* L.). Hereditas 97:261-272.

104. ALONSO, W.R., CROTEAU, R. 1991. Purification and characterization of the monoterpene cyclase γ-terpinene synthase from *Thymus vulgaris*. Arch. Biochem. Biophys. 286:511-517.

105. CROTEAU, R.B., WHEELER, C.J., CANE, D.E., EBERT, R., HA, H.-J. 1987. Isotopically sensitive branching in the formation of cyclic monoterpenes: Proof that (-)-α-pinene and (-)-β-pinene are synthesized by the same monoterpene cyclase via deprotonation of a common intermediate. Biochemistry 26:5383-5389.

106. LEWINSOHN, E., GIJZEN, M., CROTEAU, R. 1992. Wound-inducible pinene cyclase from grand fir: Purification, characterization and renaturation after SDS-PAGE. Arch. Biochem. Biophys. 293:167-173.

107. WAGSCHAL, K., SAVAGE, T.J., CROTEAU, R. 1991. Isotopically sensitive branching as a tool for evaluating multiple product formation by monoterpene cyclases. Tetrahedron 47:5933-5944.
108. BIRKS, J.S., KANOWSKI, P.J. 1988. Interpretation of the composition of coniferous resin. Silv. Genet. 37:29-39.

Chapter Nine

GENETIC MANIPULATION OF TERPENOID PHYTOALEXINS IN GOSSYPIUM: EFFECTS ON DISEASE RESISTANCE

Alois A. Bell, Robert D. Stipanovic, Marshall E. Mace, and Russell J. Kohel

USDA, ARS, Southern Crops Research Laboratory
College Station, Texas

INTRODUCTION

There is considerable evidence that the rates of phytoalexin synthesis relative to those of secondary colonization by pathogens is a critical determinant of disease resistance in most plants.[1] For example, resistance of cotton cultivars to fungal wilt pathogens, the bacterial blight pathogen, and nematodes is directly

Genetic Engineering of Plant Secondary Metabolism,
Edited by B.E. Ellis *et al.*, Plenum Press, New York, 1994

related to how rapidly terpenoid phytoalexins accumulate in infected tissues.[2] HMGR (3-hydroxy-3-methylglutaryl-CoA reductase) is one of the first enzymes in the terpenoid pathway and often plays a regulatory role for biosynthesis of steroids and terpenoids in plants and animals. Increases in activity of HMGR and accumulation of mRNA for HMGR slightly precede and occur parallel to the accumulation of terpenoid phytoalexins in cotton cultivars infected with *Verticillium dahliae*.[3,4] Genetic manipulation of the HMGR genes or regulatory genes affecting them to give more rapid synthesis of phytoalexins is one approach for improving disease resistance in cotton and other plant species that have terpenoid phytoalexins.

Another approach for improving disease resistance might be to alter biosynthetic pathways through the introduction of genes from foreign species to cause synthesis of more toxic phytoalexins. The feasibility of this approach is supported by the fact that even optical isomers of a phytoalexin can have significant differences in toxicity.[5,6] However, data to show that a change in toxicity has a significant effect on disease resistance is lacking. In this chapter we report studies that were designed to determine whether the specific phytoalexin composition in Upland cotton (*Gossypium hirsutum* L.) can be genetically altered to change resistance to Verticillium wilt. The approach was to identify structural variations in phytoalexins among *Gossypium* species and

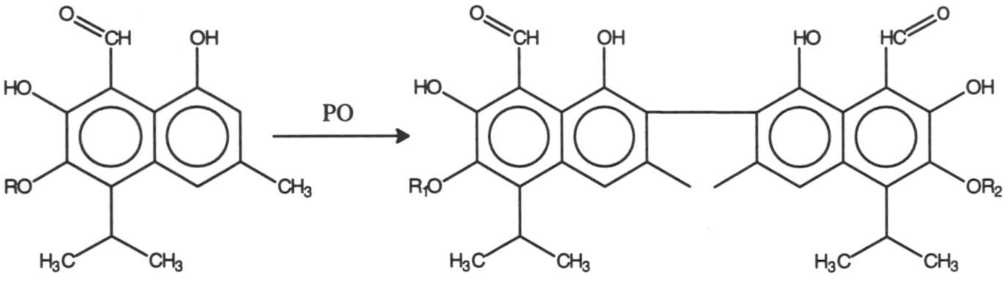

HG (R = H)
MHG (R = CH₃)

G (R₁, R₂ = H)
MG (R₁ = H; R₂ = CH₃)
DMG (R₁, R₂ = CH₃)

Fig.1. Structures and biosynthetic relationships of hemigossypol (HG), gossypol (G) and their methyl ethers (MHG, MG, DMG). Dimerization of HG to G is catalyzed by a specialized peroxidase (PO). Methylation occurs prior to synthesis of the terpenoid aldehydes as shown in Fig. 3.

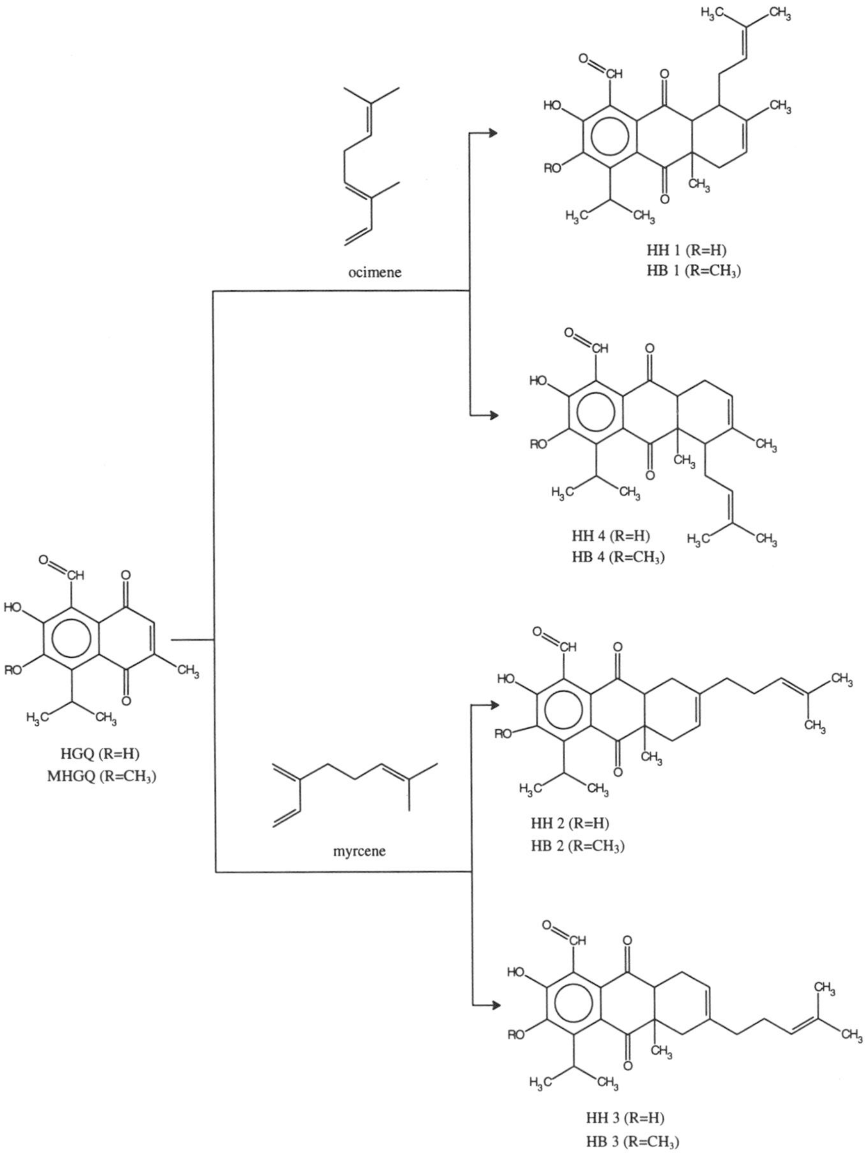

Fig. 2. Structures and biosynthetic relationships among hemigossypolone (HGQ), heliocides H1, H2, H3, and H4 (HH1, HH2, HH3, and HH4), and their methyl ethers (MHGQ, HB1, HB2, HB3, and HB4). The reactions between the quinones and ocimene or myrcene involves a Diels-Alder reaction and does not require an enzyme catalyst. Methylation occurs prior to quinone synthesis as shown in Fig. 3.

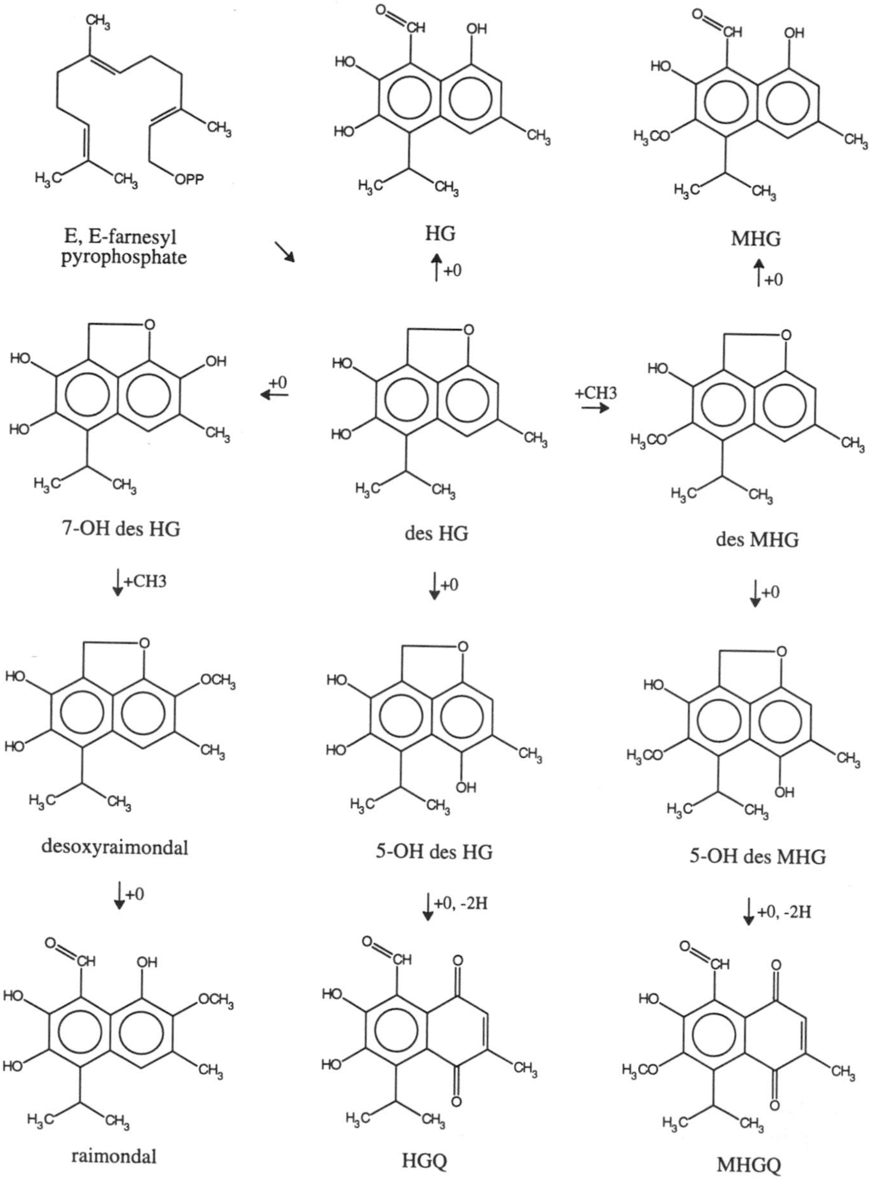

E, E-farnesyl
pyrophosphate

HG
↑+0

MHG
↑+0

7-OH des HG

←+0

des HG

+CH3→

des MHG

↓+CH3

↓+0

↓+0

desoxyraimondal

5-OH des HG

5-OH des MHG

↓+0

↓+0, -2H

↓+0, -2H

raimondal

HGQ

MHGQ

determine the effects of these variations on toxicity to several fungal pathogens *in vitro*. Genetic hybridization schemes then were developed, using selected variations as genetic characters, to transfer variations from wild *Gossypium* species into the Upland cotton cultivar `Tamcot CAMD-E', and the genetic control of the variations in this background was determined. Finally the effects of the foreign genes on terpenoid composition and on resistance to Verticillium wilt were determined.

STRUCTURAL VARIATIONS OF TERPENOID PHYTOALEXINS

Variations Among Tissues in Upland Cotton

The plant tribe Gossypieae in the family Malvaceae, which includes *Gossypium*, is characterized by the production of lysigenous pigment glands that contain gossypol (G; Fig. 1) in cotyledons of the seed. The glands are located below the epidermis among the mesophyll cells in cotyledons, leaves, and bracts and are scattered throughout the cortex of the stem and upper root bark. Glands are absent from stele tissue and young roots. Although glands in all tissues contain some gossypol, those in young leaves contain predominantly hemigossypolone (HGQ) and those in old leaves contain mostly heliocides H1, H2, H3, and H4 (HH1, HH2, HH3 and HH4, respectively) derived from hemigossypolone (Fig. 2). These same terpenoids predominate in pigment glands located in other organs and tissues, such as the bract, capsule, and stem surface, where the surrounding cells contain differentiated plastids. The heliocides

←————————————————————

Fig. 3. Structures and probable biosynthetic relationships among the major terpenoid phytoalexins formed by *Gossypium* species. Key to compounds: HG = hemigossypol, MHG = hemigossypol 3-methyl ether, 7-OH desHG = 7-hydroxydesoxyhemigossypol, desHG = desoxyhemigossypol, desMHG = desoxyhemigossypol 3-methyl ether, 5-OH desHG = 5-hydroxydesoxyhemigossypol, 5-OH desMHG = 5-hydroxydesoxyhemigosspol 3-methyl ether, HGQ = hemigossypolone, and MHGQ = hemigossypolone 3-methyl ether. See Figures 1 & 2 for further reactions forming gossypol and heliocides from HG, MHG, HGQ, and MHGQ

are formed from a Diels-Alder-type cyclization of hemigossypolone with the volatile monoterpenes ocimene and myrcene, which along with various other volatile terpenes also are stored in the glands (Fig. 2).[7] The only other cells (besides glands) that contain gossypol in healthy plants are the epidermal cells of roots, the outer dead cells of root bark, and occasionally xylem ray cells in stems of mature plants. Gossypol and the heliocides have been studied extensively because of their toxicity to various herbivores and their probable role in pest resistance.[8]

Intermediates in the biosynthetic pathways of gossypol and the heliocides in Upland cotton are also phytoalexins. Cells without differentiated chloroplasts synthesize primarily hemigossypol (HG; Figs. 1, 3) and its precursor desoxyhemigossypol (desHG; Fig 3) along with various amounts of gossypol in response to fungal infections. These compounds are accompanied by relatively small concentrations of their methyl ethers (desMHG, MHG, MG, and DMG; Figs. 1, 3). Cells with differentiated chloroplasts synthesize primarily hemigossypolone and its methyl ether (MHGQ; Fig. 2) as phytoalexins.[9] In addition, green cells synthesize 2,7-dihydroxycadalene and various derivatives, which have bactericidal activity but only weak antifungal activity.[5,9] The probable importance of all of these phytoalexins in the resistance of cotton to various diseases has been reviewed.[2,9,10,11]

Variations Among *Gossypium* Species

In surveys of 24 *Gossypium* species,[12,13] four qualitative variations were found in the biosynthetic pathways of terpenoid phytoalexins and of derived biocides accumulated in glands (Fig. 3). First, the percentage of the terpenoids with 3-methyl ethers varied from undetectable levels to more than 70% of the total terpenoids. Differences in methylation occurred in all parts of the plant but were most pronounced in leaves. The 3-methyl group apparently is introduced into the pathway only at the point where desHG is methylated to form desMHG (Fig. 3), because all comparable compounds (HG and MHG; HGQ and MHGQ; etc.) beyond this point show the same ratio of methylation as desHG and desMHG. Second, all American wild diploid cottons (D genome), except *Gossypium gossypioides*, do not accumulate hemigossypolone or heliocides in either diseased tissues or normal glands. This apparently is due to the absence of an enzyme required to oxidize the C-5 of desHG to form 5-hydroxy desHG (Fig.

3). Third, the American wild cotton *Gossypium raimondii* accumulates mostly raimondal (Fig. 3) in leaves but only traces of raimondal in infected steles. Thus, green cells of this species apparently contain an enzyme that oxidizes C-7 of desHG to form 7-hydroxy desHG, which is methylated and oxidized to form raimondal as shown in Figure 3. Finally, certain species do not accumulate heliocides H2, H3, B2, or B3 in glands because they lack the ability to synthesize myrcene (Fig. 2). However, this variation does not affect the quality of phytoalexins accumulated in diseased tissues. We were able to devise interspecific genetic crossing schemes to transfer the 3-methylation and the raimondal variations into a common Upland cultivar background.

TOXICITY OF TERPENOID PHYTOALEXINS TO FUNGI

Methylation of the 3-hydroxyl group of phytoalexins from cotton stem stele substantially reduces toxicity (Table 1). Addition of this methyl group approximately doubles the amount of phytoalexin required for the same inhibitory effect on all fungi tested except *Candida albicans*. This effect of 3-methylation was true for both the naphthofuran desHG and the terpenoid aldehyde HG. Addition of the 3-methyl group also reduces the water solubility of the phytoalexins. The solubility of desHG, desMHG, HG and MHG in 0.15M potassium phosphate buffer at pH 6.3 and 24° C is 50.2, 4.3, 2.9 and 2.0 μg/ml, respectively.[11] Thus, desHG but not desMHG is sufficiently toxic and soluble to completely inhibit most fungi without the aid of a surfactant. Even with 2% DMSO it is sometimes impossible to get effective concentrations of MHG into solution.[14]

The effects of the presence of a methoxyl group at C-7 on toxicity were determined by comparing the toxicity of raimondal with that of HG against two defoliating and two nondefoliating strains of *V. dahliae* using the bioassay of Mace et al.[17] The toxicity of both compounds to the defoliating strain V44 was dose dependent, and about twice as much raimondal was always required to cause the same amount of inhibition as HG (Fig.4). At concentrations of 10 and 20 μg/ml, HG was significantly more toxic than raimondal against three of the four *V. dahliae* strains (Table 2). Thus, the presence of the 7-methoxyl group in raimondal reduces the toxicity to *V. dahliae* compared to that of HG to about the .same extent as does presence of the 3-methoxyl group in MHG (Table 1)

Table 1. Effects of 3-methyl ether formation on toxicity of cotton phytoalexins.[a]

Fungal Species[b]	Toxic Measure[c]	Phytoalexin Concentration (μg/ml)			
		desHG	desMHG	HG	MHG
V.d.	EC_{50}	5-15[d]	10-20	5-25	20-80
	EC_{100}	10-20	20-50	40	90
F.o.v.	EC_{50}	9	13	29	>35
C.a.	MIC	32-64	32-50	-	-
C.n.	MIC	8	16	-	-

[a]Adapted from references 14, 15, and 16.
[b]V.d. = *Verticillium dahliae*; F.o.v. = *Fusarium oxysporum* f.sp. *vasinfectum*; C.a. = *Candida albicans*; C.n. = *Cryptococcus neoformans*.
[c]EC_{50} = Minimum concentration that reduces growth to one-half; EC_{100} = minimum concentration that completely inhibits growth; MIC = minimum concentration that significantly inhibits growth.
[d]Where a range is given, effective concentrations varied depending on the type of microbial cell (conidia vs. mycelia) and the strain of the fungus.

Table 2. Percent inhibition of *Verticillium dahliae* growth by hemigossypol (HG) and raimondal (RA) assayed by the procedure of Mace et al.[17]

Verticillium dahliae Strain	Concentration			
	10 ppm HG	10 ppm RA	20 ppm HG	20ppm RA
	(% Inhibition of Growth)			
V-C 1 (Defoliating Strains):				
V44	37	10*	78	34*
V76	32	25	53	38*
V-C 2 (Nondefoliating Strains):				
TS2	36	2*	54	18*
PH	34	42	58	53

*Significantly less toxic than HG for the same strain and concentration. (LSD; $P = 0.05$)

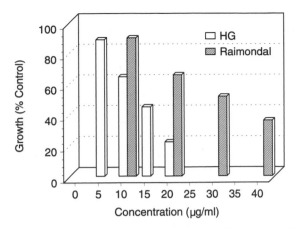

Fig. 4. Growth inhibition of *Verticillium dahliae* strain V44 by hemigossypol and raimondal. Toxicity was measured using the bioassay of Mace *et al.*[17]

DEVELOPMENT OF LINES WITH STRUCTURAL VARIATIONS

Transfer from Foreign Species of 3-Methylation and Raimondal (7- Hydroxylation and 7-Methylation) Traits

The transfer of the structural variation traits into the Upland cotton (*G. hirsutum*) cultivar 'Tamcot CAMD-E' was carried out using the crossing schemes described by Bell.[18] The approximate ratio of hemigossypolone methyl ether (MHGQ, Fig. 3) to hemigossypolone (HGQ, Fig. 3) as a measure of 3-methylation, and the presence or absence of raimondal as a measure of 7-hydroxylation followed by 7-methylation (Fig. 3), were monitored with the TLC fingerprint method of Bell and Stipanovic.[19] When Upland cotton was crossed with various *Gossypium* species containing high concentrations of methylated terpenoids in pigment glands, methylation in the F_1 generation was inherited as a recessive trait from all species, except *G. sturtianum,* which expresses it as a completely dominant trait.[20] Similarly, crosses between *G. raimondii* and various other species showed that accumulation of raimondal was a dominant trait.[20] Enhanced methylation was transferred from *G. barbadense* as a recessive trait and from a *G. hirsutum* X *G. sturtianum* hexaploid as a dominant trait. The raimondal trait was transferred from a *G. arboreum* X *G. raimondii* synthetic amphidiploid. In each case, the desired trait was transferred by successive backcrosses to 'Tamcot CAMD-E'. Test crosses to *G. barbadense* were used to detect genes for the recessive methylation trait in backcross progeny.

Progeny from the fourth backcross were self-pollinated and seeds were grown to give segregating populations. Test crosses then were used to identify and select three pairs of phenotypically similar sister plants, of which one was homozygous for the presence, and the other homozygous for the absence, of the desired trait. Backcrosses for each trait were carried out simultaneously in four separate family lines to minimize chance correlations between changes in phytoalexin composition and changes in disease resistance. For each trait, comparisons of terpenoid composition and disease severity were based on six progeny from each sister plant of three pairs from each of the four family lines in each experiment. Experiments were repeated three times; data as means of all observations are shown in Figures 5-8.

Genetics of Variations

A single line with the raimondal trait (7-hydroxylation folowed by 7-methylation) and a single line with the 3-methylation trait from *G. barbadense* were crossed with the *G. hirsutum* Texas marker stock 'TM-1', and F_1, F_2, and backcross progeny were analyzed for the segregation of these characters. Chi-square analyses of the segregation of the raimondal trait (Table 3) indicate that it

Table 3.Chi-square analyses of genetic control of raimondal.

Theory:	Two Dominant Genes with epistasis (Ra_1Ra_2)
F_2:	Theoretical - 9 raimondal: 7 normal
	Observed - 158 raimondal: 130 normal
	Chi-square 0.22 (P = 0.7 - 0.5)
Backcross:	Theoretical - 1 raimondal: 3 normal
	Observed -51 raimondal: 137 normal
	Chi-square 0.45 (P = 0.7 - 0.5)

is controlled by two dominant genes with epistasis. This result is consistent with our proposed biosynthetic scheme (Fig. 3), which requires separate enzymes for the oxidation and methylation at C-7 during raimondal synthesis. Although our genetic studies of methylation are not complete, they appear to support the preliminary conclusion that the high levels of methylation in *G. barbadense* are due to a single homozygous recessive gene.[20] The segregation of the methylation trait from *G. sturtianum* has not been consistent among the four family lines and deviates significantly from either a one or two dominant gene model, although it behaves most like a single dominant gene.

EFFECTS OF FOREIGN GENES ON TERPENOID PHYTOALEXINS

Terpenoid Composition in Leaves

The effects of the foreign genes on the composition of terpenoids in glands of leaves of healthy plants is shown in Table 4. Healthy, rather than diseased leaves, were used because the terpenoids can be extracted from the pigment glands with much greater efficiency than from diseased tissues. Hemigossypolone (HGQ) and its 3-methyl ether (MHGQ), the predominant compounds formed as phytoalexins in leaves[22], are extremely reactive with proteins and are often lost during extraction, especially if cells are disrupted to release proteins. Neither methylated terpenoids (MHGQ and HB1 to HB4) nor raimondal were found in leaves of the 'Tamcot CAMD-E', cultivar of *G. hirsutum* which is typical of commercially grown Upland cotton. In family lines with genes from *G. barbadense* and *G. sturtianum*, 40 and 59%, respectively, of the terpenoids in leaves were methylated; the difference in the percentages of methylation was highly significant (LSD, P = 0.01). Thus, the gene(s) from *G. sturtianum* more effectively increase methylation. In lines with genes from *G. raimondii* 42% of the total terpenoid content in leaves was raimondal; there also was a significant (LSD, P = 0.05) increase in gossypol concentration and decrease in total terpenoids. Similar observations were made by Altman et al.[23] in their more detailed observations of raimondal inheritance in F_1 hybrids between species. It is possible that some intermediate, such as 7-hydroxyhemigossypol or its ortho-quinone derivative, may cause feedback regulation of the pathway.

Table 4. Terpenoid concentrations in leaves of cotton breeding lines containing genes for increased 3-methyl ether (MHGQ) and raimondal (7-methyl ether) synthesis.[a]

| Terpenoid[b] | Breeding Line or Cultivar[c] | | | |
	CE	CE-MEB	CE-MES	CE-RA
	(μg/g Fresh Weight)			
Gossypol	22	18	19	50
HGQ	16	30	37	7
HH_1 - HH_4	983	181	149	88
MHGQ	0	37	112	0
HB_1 - HB_4	0	103	156	0
Raimondal	0	0	0	106
% 3 Methyl-Ethers	0	40	59	0
% Raimondal	0	0	0	42

[a]Leaves from plants with 8 to 10 true leaves were freeze-dried and analyzed for terpenoids using the methods of Stipanovic et al.[21]
[b]HGQ = hemigossypolone; HH_1-HH_4 = heliocides H_1, H_2, H_3, H_4; MHGQ = hemigossypolone 3-methyl ether; HB_1-HB_4 = heliocides B_1, B_2, B_3, B_4.
[c]CE = 'Tamcot CAMD-E' cultivar of *G hirsutum*; CE-MEB = CE breeding line with recessive gene for enhanced methylation from *G. barbadense*; CE-MES = CE breeding line with dominant gene(s) for enhanced methylation from *G. sturtianum*; CE-RA = CE breeding line with genes for raimondal synthesis from *G. raimondii*.

Terpenoid Composition in Stem Stele

The patterns of terpenoid composition in diseased stelar tissue 2 weeks after inoculation with *V. dahliae* are shown in Figures 5 and 6. Lines with the 3-methylation gene(s) from *G. sturtianum* had significantly (LSD, P=O.O5) higher concentrations of 3-methylated terpenoid phytoalexin than sister lines without the genes. Less than 8% of the total terpenoids, however, were 3-methylated in any line. There also was a significant decrease in desHG and HG and a corresponding increase in gossypol associated with the *G. sturtianum* gene(s). These changes, however, may reflect a greater loss of redox potential or increased peroxidase activity in plants with the methylation trait, which were more severely diseased than their corresponding sister plants that lacked methylation (Fig. 7). Lines with the 3-methylation gene from *G. barbadense* had similar or even less pronounced changes in terpenoid composition (not shown) than those

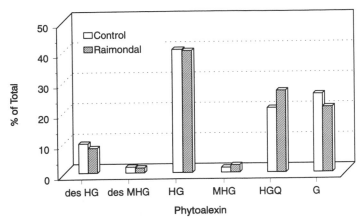

Fig. 5 Percent distribution of terpenoid phytoalexins in stem stele of sister plants homozygous for 3-methylation genes from *Gossypium sturtianum* (methylation) and lacking 3-methylation genes (control) at 2 weeks after inoculation with *Verticillium dahliae*.

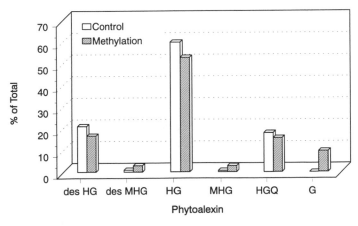

Fig. 6. Percent distribution of terpenoid phytoalexins in stem stele of sister plants homozygous for raimondal genes from *Gossypium raimondii* (raimondal) and lacking raimondal genes (control) at 2 weeks after inoculation with *Verticillium dahliae*.

from *G. sturtianum*. Lines with genes for raimondal synthesis from *G. raimondii* also had only small changes in the terpenoid composition of the stele. Raimondal was not detected by the HPLC analytical procedure, and only traces were found with TLC. There was a small but significant increase of hemigossypolone and a decrease of gossypol in stele of lines that produced raimondal in leaves However, concentrations of the most toxic phytoalexins (desHG, desMHG, and HG) were not significantly affected by the raimondal genes.

EFFECTS OF FOREIGN GENES ON DISEASE RESISTANCE

Significant increases in disease severity were associated with all of the foreign genes, when plants were inoculated separately with two isolates of defoliating and nondefoliating pathotypes. Plants possessing the *G. sturtianum* and *G. raimondii* genes for synthesis of 3-methylated terpenes and raimondal (7-hydroxylated and 7-methylated), respectively, had a greater defoliation index (Figures 7 and 8) and a lower shoot weight than corresponding sister plants without these traits. These differences were consistent within each of the four family lines that were developed for each of the different foreign genes. Plants with the 3-methylation trait from *G. barbadense* also were more severely diseased than their sister plants (not shown). Thus, the differences in disease severity probably are due to the genes affecting the quality of phytoalexins, although closely linked genes might also be responsible. The latter seems unlikely, especially for 3-methylation, because this trait introduced with recessive and dominant genes from two different species gave consistent changes.

CONCLUSION

The toxicity of phytoalexins in cotton to *V. dahliae* apparently is a critical determinant of the plant's resistance to Verticillium wilt. Introducing genes that cause 40-60% of the terpenoid synthesis in leaves to be diverted to the synthesis of compounds less toxic than those normally formed resulted in significant increases in susceptibility to Verticillium wilt, but these genes did not confer appreciable changes in terpenoid composition in diseased stems. However, various studies have shown that extensive colonization of the leaf is essential for symptom production, and that damage to the leaves is the major cause of plant yield losses.[24,25] Thus, significant changes in resistance would be expected from altering terpenoid composition only in leaves.

The toxicity of the terpenoid phytoalexin desHG to fungi is due to the free radicals and active oxygen species, especially hydrogen peroxide, formed

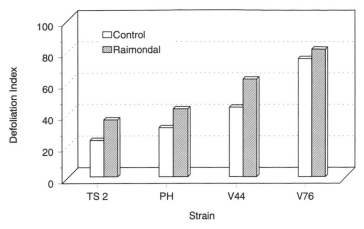

Fig. 7. Defoliation index of sister plants homozygous for 3-methylation genes from *Gossypium sturtianum* (methylation) and lacking 3-methylation genes (control) at 2 weeks after inoculation with various strains of *Verticillium dahliae*. The defoliation index is the percentage of reduction of the leaf weight to stem weight ratio; thus, a defoliation index of 100 = complete defoliation.

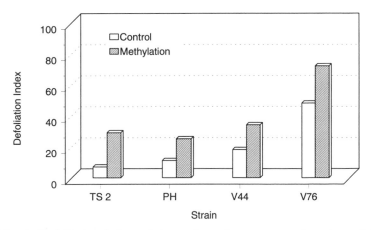

Fig. 8. Defoliation index of sister plants homozygous for raimondal genes from *Gossypium raimondii* (raimondal) and lacking raimondal genes (control) at 2 weeks after inoculation with various strains of *Verticillium dahliae*. The defoliation index is the percentage of reduction of the leaf weight to stem weight ratio; thus, a defoliation index of 100 = complete defoliation.

during oxidation.[26,27] If other cotton phytoalexins also depend on free radicals for toxic activity, it is not surprising that 3- or 7-methoxyl groups diminish activity. At room temperature in pH 6.3 phosphate buffer and air, 58 and 87% of desHG, compared to only 6 and 8% of desMHG (unpublished), are destroyed autoxidatively after 24 and 48 hours, respectively. Thus, 3-methylation stabilizes the terpenoids against oxidation that gives rise to free radicals. The major free radical formed from HG apparently involves removal of the proton at C-7, since gossypol is the major product of oxidation of HG by peroxidase.[28] Thus, addition of the 7-methoxyl group to HG to form raimondal also would stabilize the molecule against oxidation and free radical formation.

Methylation of hydroxyl groups also affects the toxic activity of leguminous phytoalexins. Demethylation of pisatin or sativan causes a decrease in toxicity to *Aphanomyces euteiches* and *Fusarium solani* f. sp. *cucurbitate*, whereas it increases the toxicity of dehydropisatin.[29] The mechanism of toxicity of these compounds has not been determined, but free radicals probably are involved.[26]

Genetically diverting phytoalexin synthesis toward production of phytoalexins with greater toxicity to *V. dahliae* should increase resistance of cotton to Verticillium wilt. However, enzymes required for synthesis of terpenoids with greater toxicity than those in commercial cotton cultivars do not appear to be present within *Gossypium* species. Therefore, we have begun to search for potent phytoalexins in other closely related genera in the family Malvaceae. Kenaf (*Hibiscus cannabinus* L.) recently was found to be resistant to Verticillium wilt and to produce the phytoalexin hibiscanone (3,8-dimethyl-1,2-naphthoquinone) which is about four times more toxic than desHG, the most toxic phytoalexin in cotton.[26] Because hibiscanone, like the terpenoids in cotton, apparently is derived from farnesyl pyrophosphate, it may be possible to introduce genes from kenaf to cotton to divert terpenoid synthesis to the production of more toxic phytoalexins.

REFERENCES

1.   BELL, A.A. 1980. The time sequence of defense. In: Plant Diseases: An Advanced Treatise, vol. 5, How Plants Defend Themselves, (J.G. Horsfall and E.B. Cowling, eds.), Academic Press, New York, pp. 53-73.

2.   BELL, A.A., MACE, M.E., STIPANOVIC, R.D. 1986. Biochemistry of cotton (*Gossypium*) resistance to pathogens. In: Natural Resistance

of Plants to Pests, ACS Symposium Series, No. 296, (M.B. Green and P.A. Hedin, eds.), American Chemical Society, Washington, D. C.

3.   BENEDICT, C.R., JI, W., BELL, A.A., STIPANOVIC, R.D. 1992. The regulation of the pathway for biosynthesis of sesquiterpenoid phytoalexins by 3-hydroxy-3-methylglutaryl-CoA reductase (HMGR). Plant Physiol. (Suppl) 99: 36.

4.   JOOST, O. 1993. Early genetic events in the interaction of *Verticillium dahliae* and *Gossypium* species. Ph.D. Thesis, Texas A&M University, College Station, 78 pp.

5.   ESSENBERG, M., GROVER, P.B., JR., COVER, E.C. 1990. Accumulation of antibacterial sesquiterpenoids in bacterially inoculated *Gossypium* leaves and cotyledons. Phytochemistry 29: 3107-3113.

6.   DELSERONE, L.M., MATTHEWS, D.E., VANETTEN, H.D. 1992. Differential toxicity of enantiomers of maackiain and pisatin to phytopathogenic fungi. Phytochemistry 31: 3813-3819.

7.   ELZEN, G.W., WILLIAMS, H.J., BELL, A.A., STIPANOVIC, R.D., VINSON, S.B. 1985. Quantification of volatile terpenes of glanded and glandless *Gossypium hirsutum* L. cultivars and lines by gas chromatography. J. Agric. Food Chem. 33: 1079-1082.

8.   STIPANOVIC R.D., BELL, A.A., LUKEFAHR, M.J. 1977. Natural insecticides from cotton (*Gossypium*). In: Host Plant Resistance to Pests, ACS Symposium Series, No. 62, (P.A. Hedin, ed.), American Chemical Society, Washington, D.C., pp. 197-214.

9.   BELL, A.A., STIPANOVIC, R.D., AND MACE, M.E. 1993. Cotton phytoalexins: A review. In: Proc. Beltwide Cotton Prod. Res. Conf., (D.J. Herber and D.A. Richter, eds.), National Cotton Council of America, Memphis, TN, pp. 197-201.

10. BELL, A.A., STIPANOVIC, R.D. 1978. Biochemistry of disease and pest resistance in cotton. Mycopathologia 65: 91-106.

11. STIPANOVIC, R.D., MACE, M.E., ALTMAN, D.W., BELL, A.A. 1988. Chemical and anatomical response in *Gossypium* spp. challenged by *Verticillium dahliae*. In: Biologically Active Natural Products: Potential Use in Agriculture, ACS Symposium Series, No. 380, (H.G. Cutler, ed.), American Chemical Society, Washington, D.C., pp. 262-272.

12. BELL, A.A., STIPANOVIC, R.D., HOWELL, C.R., FRYXELL, P.A. 1975. Antimicrobial terpenoids of *Gossypium*: Hemigossypol, 6-

methoxyhemigossypol and 6-desoxyhemigossypol. Phytochemistry 14: 225-231.

13. BELL, A.A., STIPANOVIC, R.D., O'BRIEN, D.H., FRYXELL, P.A. 1978. Sesquiterpenoid aldehyde quinones and derivatives in pigment glands of *Gossypium*. Phytochemistry 17: 1297-1305.

14. ZHANG, J., MACE, M.E., STIPANOVIC, R.D., BELL, A.A. 1993. Production and fungitoxicity of the terpenoid phytoalexins in cotton inoculated with *Fusarium oxysporum* f. sp. *vasinfectum*. J. Phytopathology. (in press)

15. MACE, M.E., STIPANOVIC, R. D., BELL, A. A. 1990. Relation between sensitivity to terpenoid phytoalexins and virulence to cotton of *Verticillium dahliae* strains. Pest. Biochem. Physiol. 36: 79-82.

16. MACE, M.E., STIPANOVIC, R.D., BELL, A.A. 1993. Toxicity of cotton phytoalexins to zoopathogenic fungi. Natural Toxins 1: 294-295.

17. MACE, M.E., STIPANOVIC, R.D., ZHANG, J., BELL, A.A. 1991. A quantitative assay of the terpenoid phytoalexin desoxyhemigossypol to *Verticillium dahliae* strains. In: Proc. Beltwide Cotton Prod. Res. Conf., (D.J. Herber and D.A. Richter, eds.), National Cotton Council of America, Memphis, TN, p.186.

18. BELL, A.A. 1984. Morphology, chemistry, and genetics of *Gossypium* adaptations to pests. In: Phytochemical Adaptations to Stress, Recent Advance in Phytochemistry, vol. 18, (B.N. Timmermann, C. Steelink, and F.A. Loewus, eds.), Plenum Press, New York, pp. 197-229.

19. BELL, A.A., STIPANOVIC, R.D. 1977. The chemical composition, biological activity, and genetics of pigment glands in cotton. In: Proc. Beltwide Cotton Prod. Res. Conf., (J.M. Brown, ed.), National Cotton Council of America, Memphis, TN, pp. 244-258.

20. BELL, A.A., STIPANOVIC, R.D., ELZEN, G.W., WILLIAMS, H.J. 1986. Structural and genetic variation of natural pesticides in pigment glands of cotton (*Gossypium*). In: Allelochemicals: Role in Agriculture, Forestry, and Ecology, ACS Symposium Series, No. 330, (G.R. Waller, ed.), American Chemical Society, Washington, D.C., pp. 477-490.

21. STIPANOVIC, R.D., ALTMAN, D.W., BEGIN, D.L., GREENBLATT, G.A., BENEDICT, J.H. 1988. Terpenoid aldehydes in Upland cottons: Analysis by aniline and HPLC methods. J. Agric. Food Chem. 36: 509-515.

22. BELL, A.A., STIPANOVIC, R.D. 1987. Association of spontaneous phytoalexin synthesis with lethal reactions in interspecific hybrids of *Gossypium* spp. In: Proc. Beltwide Cotton Prod. Res. Conf., (T.C. Nelson, ed.), National Cotton Council of America, Memphis, TN, pp. 555-556.

23. ALTMAN, D W., STIPANOVIC, R.D., BELL, A.A. 1990. Terpenoids in foliar pigment glands of A, D, and AD genome cottons: Introgression potential for pest resistance. J. Heredity 81: 447-454.

24. BELL, A.A. 1992. Verticillium wilt. In: Cotton Diseases, (R.J. Hillocks, ed.), CAB Int., London, England, pp. 87-126.

25. BELL, A.A. 1993. Biology and ecology of *Verticillium dahliae*. In: Biology of Sclerotial-Forming Fungi, (S.D. Lyda and C.M. Kenerley, eds.), Texas A&M University Press, College Station, TX, pp. 147-210.

26. STIPANOVIC, R.D. MACE, M.E, ELISSADE, M.H., BELL, A.A. 1991 Desoxyhemigossypol, a cotton phytoalexin: Structure-activity relationship. In: Naturally Occurring Pest Bioregulators, ACS Symposium Series, No. 449, (P.A.Hedin, ed.) American Chemical Society, Washington, D.C., pp. 336-351.

27. STIPANOVIC, R.D. MACE, M.H., BELL, A.A., BEIRI, R.C. The role of free radicals in the decomposition of the phytoalexin desoxyhemigossypol. J. Chem. Soc. Perkin Trans 1, pp. 3189-3192.

28. VEECH, J..A., STIPANOVIC, R.D., BELL, A.A. 1976 Peroxidative conversion of hemigossypol. A revised structure for isohemigossypol. J. Chem. Soc. Chem. Commun., pp. 144-145.

29. VANETTEN, H.D. 1976 Antifungal activity of pterocarpans and other selected isoflavonoids. Phytochemistry 15:655-669.

30. BELL, A.A., C., ZHANG, J., REIBENSPIES, J., MACE, M.E. 1993. Identification and toxicity of phytoalexins from kenaf (*Hibiscus cannabinus* L.). Phytopathology 83: (in press).

Chapter Ten

GENETIC CHARACTERIZATION OF SECONDARY METABOLISM IN *ARABIDOPSIS*

Clint Chapple

Department of Biochemistry, Purdue University, West Lafayette, IN 47907-1153

INTRODUCTION

Over the last century, many thousands of plant species have been examined for their content of secondary metabolites, and tens of thousands of secondary metabolites have thus been identified. Unfortunately, many of the species from which these compounds have been isolated are intractable to genetic approaches, and the cloning of genes involved in their secondary metabolism will depend upon classical biochemical approaches. In contrast, a variety of classical- and molecular-genetic approaches are available for the study of biochemical, physiological, and developmental processes in *Arabidopsis*. The desirable attributes of this species have been described in numerous review

Genetic Engineering of Plant Secondary Metabolism,
Edited by B.E. Ellis *et al.*, Plenum Press, New York, 1994

articles,[1,2,3] and include its short life cycle and small genome,[4] the ease with which mutants can be isolated,[5] availability of *Agrobacterium*-mediated transformation,[6,7] and novel gene cloning strategies such as T-DNA tagging,[8] map-based cloning[9,10] and genomic subtraction.[11] The combined use of all of these techniques has produced rapid advances in the understanding of the developmental biology, physiology and biochemistry of *Arabidopsis*.

A systematic effort to dissect the secondary metabolic pathways of *Arabidopsis* will be a very productive way to increase our understanding of plant secondary metabolism. For example, by concentrating on a single organism, it will be possible to identify all of the major, and many of the minor, secondary metabolites that are accumulated by Arabidopsis. By identifying mutants that fail to accumulate these metabolites, it will be possible to assess the role(s) of each of these compounds in the life cycle of *Arabidopsis*. It may also be possible to clone all of the genes involved in secondary metabolism in Arabidopsis and to evaluate the genetic complexity of these pathways. How many structural genes are involved in the biosynthesis of secondary metabolites in *Arabidopsis*? To what extent are these genes related or duplicated, and how many regulatory loci impinge upon their expression? The identification of these genes and regulatory circuits in this model specieswill make future efforts toward the characterization of secondary metabolism in other plant species significantly easier.

GENE ISOLATION METHODS IN *ARABIDOPSIS*

Classic techniques of gene isolation depend upon the isolation of the gene's product, and the use of the isolated protein for the generation of oligonucleotide or antibody probes. Clearly, this approach has been successful in many cases, but it has its limitations. Cloning a gene in this way depends upon the abundance and stability of the encoded protein, and a suitable assay for the gene product. In a number of cases, such as the isolation of fatty acid desaturases and cytochrome P-450 dependent monooxygenases from plants, these issues have been significant obstacles to the isolation of their corresponding genes. While *Arabidopsis* does not offer any direct solutions to problems of enzyme stability or abundance, other cloning techniques make it an ideal source of genes of secondary metabolism.

Many of the gene isolation techniques used with *Arabidopsis* capitalize upon the fact that mutants can be readily obtained. Because *Arabidopsis* is self-fertile, mutations induced in alleles of single-copy dispensable genes in M1 seeds will become homozygous in a proportion of the M2 generation.[12] Given a suitable screening procedure, these mutants can be identified and backcrossed to wild type to eliminate other background mutations. Once identified, the chromosomal position of the locus affected in the mutant can be mapped using classical genetic markers[13] or molecular techniques such as RFLP mapping[14,15] or RAPD mapping combined with the use of recombinant inbred lines.[16] One application of this approach is to match the map position of a previously identified mutation with the map position of cDNAs that have been identified through standard biochemical techniques, heterologous probing, or random sequencing. For example, after low-stringency probing to identify a family of MADS-box-containing genes, the *AGL7* (*AGAMOUS*-like) gene was identified, and by RFLP mapping, was found to reside on chromosome 1, very close to the *APETALA1* (*AP1*) locus. The identity of the *AGL7* clone was subsequently confirmed by *Agrobacterium*-mediated complementation of an *ap1* mutant.[17]

More recently, it has become possible to conduct map-based cloning that permits the isolation of a gene based upon a mutant phenotype alone. In this procedure, the mutant gene is positioned on the *Arabidopsis* genome map by RFLP or RAPD mapping, and these same molecular markers are used to identify yeast artificial chromosome (YAC) or cosmid clones that map near the gene of interest. The DNA from these clones can then be used to generate new probes in order to isolate overlapping DNA from new YACs or cosmids. In this way, a contiguous series of clones is constructed, covering the map position of the gene that is to be cloned. Clones representing cDNAs endoded in this region can then be identified, sequenced, and used to complement the defect in the original mutant. Two examples of this techniquue are the cloning of the *ABI3* gene,[10] and the work of Arondel et al.[9] who cloned the omega-3 fatty acid desaturase gene from a cDNA library of *Brassica napus* using an *Arabidopsis* YAC that corresponded to the locus affected in the *fad3* mutant.

Genomic subtraction has not yet been applied to genes of secondary metabolism, but appears to be a powerful technique for the isolation of genes where deletion mutations can be identified. This technique has been applied to the isolation of the *GA1* locus of *Arabidopsis*.[11] In this procedure, biotinylated mutant DNA is repeatedly hybridized to wild type DNA, and avidin-coated beads are used to remove common sequences. After 4 to 5 rounds of this subtraction, the remaining wild type DNA, consisting of sequences missing from the deleted mutant DNA, is ligated to primers, amplified by PCR, and cloned into a suitable vector.[18] Currently this approach is limited by the availability of suitable deletion-mutant strains as starting material.

One of the most powerful techniques currently available for cloning

Arabidopsis genes is T-DNA tagging.[8] This technique uses the T-DNA of *Agrobacterium tumefaciens* to generate mutations by insertional gene disruption. Once identified, a T-DNA tagged mutant allele can be retrieved by several methods. The *GLABROUS 1* gene was cloned from a T-DNA tagged mutant by the construction of a mutant genomic library which was then screened with probes derived from the T-DNA.[19,20] An alternate approach is plasmid rescue in which DNA from the tagged line is digested, self-ligated and electroporated into competent *E. coli.*[21] A replication origin and β-lactamase gene in the T-DNA vector permit the recovery of plasmids that contain T-DNA and *Arabidopsis* genomic sequences. *A. thaliana* genes that have been recovered by plasmid rescue include those coding for chloroplast proteins,[22] transcription factors involved in floral development,[23] and photomorphogenesis regulatory proteins.[24]

Until recently, the ability to clone genes from *Arabidopsis* was limited by the spectrum of T-DNA tagged lines that were available. It had previously been estimated that 50,000 to 100,000 tagged lines would be required for saturation mutagenesis of the *Arabidopsis* genome with T-DNA inserts,[25,8] but only a few thousand lines were available. Because the seed transformation technique[25] is inefficient and difficult for other laboratories to reproduce, a new method has recently been developed for the transformation of adult flowering *Arabidopsis* plants by vacuum infiltration in *Agrobacterium* suspensions.[7] Combined with the use of a new vector encoding Basta (phosophinothricin) resistance, which permits selection of transformed plants under low-cost greenhouse conditions rather than under sterile kanamycin selection conditions,[26] this technique promises to provide a full complement of T-DNA insertional mutant lines. In addition to its contribution to the cloning of new genes by T-DNA tagging, this new transformation method will also aid in the identification of cloned genes by facilitating the complementation of EMS-induced mutants.

A number of *Arabidopsis* genes have been cloned by complementation of mutants in yeast. A gene encoding a potassium transport channel protein was recently cloned by transformation of an *Arabidopsis* cDNA library into a K^+-transport defective mutant of *Saccharomyces cerevisiae,* followed by selection of transformants that had recovered the ability to grow on media containing low concentrations of potassium.[27,28] Although this method can be effective for cloning genes that are common between *Arabidopsis* and yeast, it is not clear that it will be useful in cloning genes of secondary metabolism, although it could be used to clone primary metabolic genes that have homologues in secondary metabolism. For example, this technique could be used to clone the genes of leucine metabolism from *Arabidopsis*. These clones could then be used as heterologous probes for the genes of the chain extension pathway involved in glucosinolate biosynthesis, which appears to be catalyzed by a series of enzymes homologous to those involved in branched-chain amino acid biosynthesis.[29]

New methods for efficient large scale sequencing have led to efforts to sequence cDNA libraries representing the entire transcribed genomes of humans[30] and plants.[31,32] The *Arabidopsis* genome project has identified a number of Arabidopsis clones encoding genes of secondary metabolism, including chalcone synthase, myrosinase, several flavonoid-*O*-glucosyl-transferases, cinnamate-4-hydroxylase, and numerous other cytochrome P450-dependent monooxygenases. These projects have also identified genes that show high homology to previously sequenced genes involved in secondary metabolic pathways in other species. These include clones that show high homology to reticuline oxidase, granaticin polyketide synthase, and flavonol-4-sulfotransferase. These genes may encode enzymes for homologous reactions in known *Arabidopsis* pathways, or they may identify currently unknown pathways of secondary metabolism in *Arabidopsis* whose genes are expressed at such low levels that their products have not yet been observed. In either case, the rapid identification of such genes by massive sequencing efforts will identify many genes of plant secondary metabolism for manipulation both in *Arabidopsis* and in other plant species.

SECONDARY METABOLISM AND SECONDARY METABOLISM MUTANTS

Genetic engineering has come to mean the introduction of novel genes into plants, the overexpression or ectopic expression of native genes, and the sense or antisense suppression of the expression of endogenous genes, via *Agrobacterium* or particle bombardment transformation techniques. While all of these technologies have been successfully applied to *Arabidopsis*, mutagenesis of *Arabidopsis* and the identification of mutant plants that are altered in specific biosynthetic pathways is another important, albeit non-directed, form of genetic engineering. It is also a very economical approach when one considers that it may take years to generate antisense plants in which the residual activity of the target pathway may still be 10% of wild type. In contrast, with a suitable screen, it may be possible to identify *Arabidopsis* mutants that carry null mutations in a given pathway in a single afternoon. Studies of this sort have been extremely important in exploring complex physiological processes in *Arabidopsis* that have been intractable to more conventional approaches. In addition, the characterization of these mutants may reveal caveats that must be considered when performing subsequent genetic engineering experiments with economically important species.

The secondary metabolites accumulated by *Arabidopsis* are typical for other members of the Brassicaceae. These include flavonoids, sinapic acid esters, glucosinolates, and indole phytoalexins. For each class of secondary metabolite, mutations at several loci have been identified that alter the spectrum or completely eliminate the accumulation of these metabolites.

Flavonoids

The flavonoids that have been reported in *Arabidopsis* include the flavonols kaempferol and quercetin, and the 3-hydroxyanthocyanidins pelargonidin and cyanidin;[33] however, none of these structural identifications have been performed in a rigorous fashion, nor have any of their glycosidic forms been characterized. There is a great need and opportunity for some careful natural products biochemistry to be performed in this area.

Eleven loci that are required for flavonoid synthesis have been identified through the characterization of mutants of *Arabidopsis* that fail to accumulate flavonoid derivatives. The isolation of these so-called *tt* (*transparent testa*) mutants capitalized upon the fact that the pigmentation of the *Arabidopsis* seed coat is probably due to the presence of condensed tannins (proanthocyanidins), derivatives of the flavonoid biosynthetic pathway. Mutant plants that are defective in the biosynthesis of these tannins have pale yellow or light brown seed coats instead of the dark brown testa seen in wild type.[34] A number of these mutants have been identified as being defective in structural genes involved in flavonoid biosynthesis. For example, the *tt4* mutant, which has the palest seed coat of all of the *tt* mutants, is defective in the gene encoding chalcone synthase. Using a parsley-derived heterologous probe for chalcone synthase, CHS was cloned from *Arabidopsis*,[35] and was mapped to the top of chromosome 5 by RFLP mapping,[14] the position previously established for *TT4* locus. Although this identification has been recently been supported by the sequencing of two independent *tt4* alleles (B. Shirley *et al.* personal communication), *Agrobacterium*-mediated genetic complementation has not been used to confirm this conclusion.

The *TT5* locus was identifed as the structural gene for chalcone isomerase[36] by designing degenerate PCR probes based upon previously published sequence data from other chalcone isomerase sequences. A similar approach was used to characterize the gene that is defective in *tt3* mutants, dihydroflavonol reductase. In these studies, fast-neutron or X-ray induced *tt5* and *tt3* mutants were found to possess significant DNA translocations both locally and at sites up to 38 centimorgans from the deleted locus in the case of the *tt5* mutant. In the *tt3* allele that was examined, there was a deletion of two putative open reading frames, one of which encoded dihydroflavonol 4-reductase. The characterization of the mutants was very informative because these rearrangements might have had unpredictable effects upon attempts to clone these genes by genomic subtraction. Further application of genomic subtraction may therefore require new methods to generate mutants with less severe rearrangements.

The remaining 8 *TT* loci are much less well characterized, and represent excellent opportunities for future research into flavonoid biosynthesis, its regulation, and the roles of these compounds. Thin layer chromatographic (TLC) analysis has suggested that the *tt7* mutant is defective in flavonoid 3'-hydroxylase activity.[36] The *ttg* mutant is very interesting because of its broad phenotypic effects.[37] Mutants defective at the *TTG* locus lack flavonoids in their leaves and seeds due to lowered activity of dihydroflavonol reductase,[34] but also lack leaf trichomes and seed coat mucilage. The *TTG* gene product may therefore be a factor that regulates all of these diverse functions. Little is known about the other six *tt* loci, but the *tt1*, *tt2*, *tt9*, and *tt10* mutants affect only seed color, and not flavonoid levels in vegetative tissues. Clearly much remains to be done in the characterization of these mutants.

The *Arabidopsis* flavonoid biosynthetic mutants may also be relevant to the study of the role of flavonoids in pollen germination in maize and petunia. Studies on petunia plants carrying antisense CHS constructs indicated that inhibition of flavonoid biosynthesis in pollen results in male sterility.[84] Similar analyses of petunia and maize CHS mutants showed that these plants are also male sterile. Exogenously supplied kaempferol rescues the mutant phenotype and permits pollen germination *in vitro*.[85] In contrast, *Arabidopsis* CHS mutants (*tt4*) are completely fertile. A comparison of pollen germination between maize, petunia and *Arabidopsis* may reveal unique signalling processes in maize and petunia, other mechanisms of pollen germination control in *Arabidopsis*, or may indicate that these processes are not universal in the plant kingdom.

Sinapic Acid Esters

The second group of secondary metabolites in *Arabidopsis* that are derivatives of the general phenylpropanoid pathway are the sinapic acid esters.[38]

The metabolism of these compounds has been extensively studied in *Raphanus sativus* (radish) where it was established that their biosynthesis and accumulation is developmentally regulated. In developing seeds, sinapic acid synthesized by the general phenlypropanoid pathway is esterified to form 1-O-sinapoyl glucose by the action of sinapic acid:UDP-glucose glucosyltransferase (SGT).[39] The linkage of sinapic acid to the anomeric carbon provides a high free energy of hydrolysis and enables sinapoyl glucose to donate its sinapoyl moiety to other acceptors.[40] In the developing seed, choline acts as the terminal acceptor in a reaction catalyzed by sinapoyl glucose:choline sinapoyltransferase (SCT),[41] and sinapoyl choline is the end product of the pathway. Upon germination, the activity of a specific esterase, sinapoyl choline esterase (SCE) increases and brings about the hydrolysis of sinapoyl choline.[42] The liberated choline has been shown to be incorporated into lipids in developing *Raphanus* seedlings and thus sinapoyl choline has been suggested to be an important reserve of choline for the biosynthesis of phospholipids during early seedling growth.[43] Within one day of germination, the activity of SGT increases in developing seedlings and the sinapic acid released from sinapoyl choline hydrolysis is re-esterified. This results in a transient increase in the sinapoyl glucose pool. Several days later there is an increase in the activity of another enzyme, sinapoyl glucose:malate sinapoyl transferase (SMT).[44] This enzyme, which has an acidic pH optimum and has been shown to localized in the vacuole,[45] is responsible for synthesizing sinapoyl malate in *Raphanus* and *Arabidopsis*.[46]

Arabidopsis mutants have been identified by TLC screening that have alterations in the accumulation of sinapic acid esters, and these fall into two complementation groups. The first class fails to accumulate sinapoyl esters in all plant tissues.[38] *In vivo* radiotracer experiments indicated that these mutants are blocked in the general phenylpropanoid pathway between ferulate and sinapate. As a result, these mutants cannot synthesize any sinapic acid-derived metabolite, including the syringyl lignin typical of angiosperms. The defective locus in these mutants was initially named *sin1* (*sin*apic acid deficient),[38] but has been renamed *fah1* (*f*erulic *a*cid *h*ydroxylase) to avoid conflict with a simultaneously published mutation referred to as *sin1* (*s*hort *in*teguments) that affects seed development,[47] and to better describe the defect in the mutant (see below). The second class of mutants identified during the screen for the *fah1* mutant accumulates sinapoyl glucose in its leaves instead of sinapoyl malate (Chapple, unpublished results). This mutant has been tentatively named *sng1* (*s*inapoyl *g*lucose accumulating) and may be defective in SMT activity.

These mutants provide a unique opportunity to investigate the role of sinapic acid esters in *Arabidopsis*. It has been suggested that sinapoyl choline serves as a choline reserve for the developing seedling; however, the germination efficiency and seedling growth rate of the *fah1* mutant is indistinguishable from wild type. This may reflect the fact that the *fah1* mutant accumulates free choline instead of sinapoyl choline in its seeds.[38] These results may be of interest with respect to the genetic engineering of agronomically important relatives of *Arabidopsis* because the presence of sinapoyl choline is generally regarded as a negative quality factor in Canola (*Brassica* sp.) production. When Canola meal is used as a feed in poultry production, it imparts a fishy taint to eggs due to its sinapoyl choline content.[48] The fact that *Arabidopsis* mutants that fail to accumulate sinapoyl choline are as viable as wild type suggests that the production of transgenic sinapoyl choline-free *Brassica* through antisense technology might result in an improved, agronomically viable crop.

In contrast to the apparently dispensable nature of sinapoyl choline, the study of the role of sinapoyl malate has revealed an important function for this metabolite. Previous studies in *Raphanus*[49] found that sinapoyl malate is accumulated primarily in the leaf adaxial (upper) epidermis in this species. Given the high UV extinction coefficient of the sinapic acid moiety, such a location is consistent with a role for sinapoyl malate in UV protection. This hypothesis has now been verified experimentally, since *fah1* homozygotes are very UV sensitive (R. Last and C. Chapple, unpublished results). Previous investigations into the role of flavonoids in UV protection in *Arabidopsis* have shown that chalcone synthase mutants (*tt4*) are UV sensitive and *tt5* homozygotes, defective in chalcone-flavone isomerase, are extremely UV sensitive.[50] The acute sensitivity of the *tt5* mutants appears to be due to the fact that these plants, while lacking flavonoids, also have lower amounts of sinapoyl malate in their leaves than does wild type. The reason for this suppression of sinapoyl ester accumulation is unknown. Taken together, these data indicate that there is an important role of both hydroxycinnamic acid esters and flavonoids in the mitigation of UV damage in *Arabidopsis*. These experiments also demonstrate how the study of *Arabidopsis* mutants may be useful in the future genetic engineering of agricultural crops. Attempts to eliminate sinapoyl choline accumulation in Canola must take into account the importance of sinapate esters in UV resistance. In addition, the characterization of UV protection mechanisms in *Arabidopsis* may aid in, and provide new targets for, the engineering of new plant varieties with increased UV tolerance.

While the synthesis and accumulation of lignin is clearly essential for vascular plants, it is often classified as a secondary "metabolite" because it is a derivative of the general propanoid pathway. The *fah1* mutant also presents a novel approach to the study of lignin because it accumulates only guaiacyl (ferulic acid-derived) lignin.[38] Because the mutant lacks syringyl lignin, it will permit the study of the impact of monomer composition on lignin's physical and chemical characteristics. While similar comparisons can be made between angiosperm and gymnosperm lignin, the myriad other differences between their woods make it difficult to draw firm conclusions. In contrast, the study of lignification in wild type and *fah1* mutants will allow comparisons to be made *in an otherwise genetically identical background.*

Recent work has been directed at cloning the *fah1* locus from *Arabidopsis*. The rosette of wild type *Arabidopsis* appears pale blue-green when viewed under long wave UV light due to the *in vivo* fluorescence of sinapoyl malate in the leaf epidermis. In contrast, the lack of sinapoyl malate fluorescence in *fah1* mutants results in these plants appearing dark red under the same conditions.[38] Using the red-fluorescent phenotype of the *fah1* mutant, a T-DNA tagged allele was identified in the DuPont collection of *Arabidopsis* lines.[25] Preliminary results indicate that the T-DNA insert is very tightly linked (within 15 kB) to the *fah1* mutation in the tagged line. Using plasmid rescue to retrieve the genomic sequences that flank the T-DNA insertion in the tagged line,[21] a right border-containing plasmid was obtained that contains DNA sequences that hybridize to wild type *Arabidopsis* DNA in Southern analysis. Sequence analysis of the DNA flanking the right border in the rescued plasmid indicates that this region of genomic DNA encodes a protein with high homology to other cytochrome P-450 dependent monooxygenases. Given that ferulate-5-hydroxylase (F5H) is a member of this class of hydroxylases,[51] these data strongly indicate that the *fah1* mutant is defective in the structural gene for F5H.

Glucosinolates

The glucosinolates, or mustard oil glycosides, are secondary metabolites that provide the flavor that we associate with the vegetable Brassicas, condiment mustards, and horseradish. A complete review of glucosinolate biosynthesis and distribution is outside of the scope of this review, and the reader is referred to previously published work for this information.[29,52,53,54] Over 75 glucosinolates have been characterized and their nomenclature is based upon their sidechain substituents, followed by the suffix glucosinolate. The catabolism of glucosinolates is catalyzed by myrosinase, a thioglucoside glucohydrolase that hydrolyzes glucosinolates to glucose and an unstable aglucone.[55] This enzyme has been found in every glucosinolate-containing plant that has ever been examined, but is sequestered away from its substrates in "myrosin cells"[56] so that it only acts upon glucosinolates following tissue disruption.[57] The glucosinolate aglucones spontaneously rearrange to yield a mixture of thiocyanates, isothiocyanates, substituted isoxazolidines and nitriles. The *Arabidopsis* genome contains a three member myrosinase gene family.[58]

It is the glucosinolate degradation products that have biological activity, and thus may mediate the role of this class of secondary metabolites. Many insects do not feed on crucifers due to their glucosinolate content; however, others such as the cabbage butterfly, *Pieris rapae* have evolved to use the volatile isothiocyanates to locate their hosts.[59] Glucosinolate biosynthesis has also been found to be induced by insect feeding,[60] and this induction is mediated by a signal that is systematically transmitted through the plant.[61] Agronomically, these same glucosinolate degradation products limits the use of *Brassica* species as animal feedstocks[62] because some of them, particularly thiocyanate ion and isoxazolidine-2-thiones, are goitrogenic.[63] Breeding programs have significantly reduced Canola glucosinolate content[64] by using a naturally-occurring glucosinolate mutant that is blocked in the accumulation of alkenyl and hydroxyalkenyl glucosinolates.[65,66] Another potential role for glucosinolate degradation products may be in the synthesis of indole acetic acid from indole acetonitrile produced through the action of myrosinase on indolyl-3-methyl glucosinolate. An *Arabidopsis* nitrilase has been cloned that can catalyze this conversion,[67] and these observations may be related to the clubroot disease phenotype seen in *Brassica* species infected with *Plasmodiophora brassicae*.[68]

The glucosinolates of *Arabidopsis* make up a family of 23 secondary metabolites.[69] Four glucosinolate mutants of *Arabidopsis* have been isolated by screening leaf extracts by HPLC.[70] Mutations at the *GSM1* locus (line TU1) are defective in the chain elongation process devoted to the synthesis of the methionine-derived amino acid precursors of alkyl glucosinolates. In these reactions, a homologous series of non-protein amino acids are produced by a cycle of reactions analogous to those involved in the synthesis of leucine from valine. In *Arabidopsis*, this cycle can be iterated up to six times, yielding glucosinolate precursors that range from homomethionine (2-amino-5-methylthiopentanoic acid) to 2-amino-10-methylthiodecanoic acid. In *gsm1* homozygotes, the glucosinolates derived from the amino acids corresponding to 2, 3, and 4 rounds of chain extension are missing, while their counterparts derived from 5 and 6 cycles are still accumulated. This phenotype could be complemented biochemically by adminstration of the missing amino acid intermediates via the transpiration stream. These data led Haughn *et al.*[70] to speculate that these mutants are defective in a chain length-specific aminotransferase that channels the intermediate length keto acids into glucosinolate biosynthesis, or in a "releasing factor" that acts specifically on the products of the chain extension pathway after 2 to 4 rounds of elongation. These hypotheses are not mutually exclusive and the identification of another non-allelic mutant (line TU5) with a similar glucosinolate phenotype may indicate that both aminotransferase and "releasing factor" mutants exist. The chain-length specificity of glucosinolate biosynthetic enzymes is further demonstrated by lines TU3 and TU6 which fail to accumulate only the methylthioalkyl- and methylsulfinylalkyl glucosinolates that are produced from the highest methionine homologue, 2-amino-10-methylthiodecanoic acid. These mutants may also be aminotransferase or "releasing factor" mutants.[70] These mutant analyses indicate that there may be as many as 8 genes of this type involved in this aspect of glucosinolate biosynthesis alone, a conclusion that would have been very difficult to reach by conventional biochemical approaches.

Characterization of the *gsm1-1* mutant also substantiates previous work concerning the site of synthesis of seed glucosinolates. *Brassica napus* zygotic embryos cultured *in vitro* fail to accumulate glucosinolates in the absence of an exogenous source of these comopunds.[71] This suggests that glucosinolate biosynthesis occurs in the female sporophyte, and that these compounds may be actively accumulated by the developing embryo. Similarly, *Arabidopsis* plants homozygous for the *gsm1-1* allele have reduced leaf and seed glucosinolate content, as do F1 seeds from [*gsm1-1/gsm1-1*] x [wild type] crosses, thus sdemonstrating that seed glucosinolate content is determined by the genotype of the female sporophyte, not the genotype of the embryo.[70] These observations led the authors to suggest that, assuming glucosinolate uptake to be an active carrier-mediated process, it should be possible to reduce seed glucosinolate content by genetically engineering the expression of genes required for embryo-specific glucosinolate uptake proteins.

The TU8 line has the most profoundly altered glucosinolate profile of all of the mutants identified. Leaves of TU8 plants contain virtually no glucosinolates; however, seed glucosinolate levels are phenotypically wild type. These plants are also dwarfed, and this morphological phenotype was found to co-segregate with the lack of leaf glucosinolates.[70] This mutant may raise new questions about the roles of glucosinolates or the possible relationship between indole glucosinolates and auxin.

A survey of glucosinolate variation between various ecotypes of *Arabidopsis* has identified more loci that influence glucosinolate composition. By following the inheritance of glucosinolate composition in F2 populations derived from crosses between the Landsberg *erecta* and Columbia ecotypes, Mithen and Toroser[72] identified a locus required for side-chain hydroxylation, *Gsl-ohp-Ar*. The *Gsl-ohp-Ar* locus maps to the same position as *Gsl-alk-ar*, a locus required for the methylthio elimination reaction required for the synthesis of prop-2-enyl and but-3-enyl glucosinolate.

Indole Phytoalexins

Cruciferous plants are known to synthesize a group of indole phytoalexins when challenged by pathogens or when treated with heavy metals.[73] These include camalexin[74,75,76,77] which is accumulated by *Arabidopsis* when inoculated with pathogens.[78,79,80]

The biosynthetic pathway leading to camalexin is unclear. Tsuji *et al.*[81] reported that radiolabelled anthranilate, but not tryptophan, was incorporated into camalexin in *Arabidopsis*. The use of tryptophan biosynthetic mutants also suggested that an intermediate of tryptophan biosynthesis is the precursor to camalexin. Nonetheless, the striking similarity of the indole phytoalexins to indole glucosinolate degradation products, specifically indole isothiocyanate, and the fact that the chemotaxonomic distribution of indole phytoalexin biosynthesis is restricted to glucosinolate-producing genera, suggests the possibility that glucosinolate degradation products are diverted into phytoalexin biosynthesis. More studies will be required before this biosynthetic pathway is completely elucidated.

To date, three loci involved in the production of camalexin have been identified. These phytoalexin-deficient (*pad*) mutants have been named *pad1*, *pad2*, and *pad3*.[82] Virulent bacteria multiplied to a greater extent in *pad1*, and *pad2* mutants that accumulate less camalexin than wild type, while the camalexin-free *pad3* mutant inhibited bacterial growth as efficiently as wild type. It was hypothesized that the *pad3* mutant may accumulate a camalexin biosynthetic intermediate that is as bacteriotoxic as camalexin, but the possibility also exists that camalexin does not have an important role as a phytoalexin in *Arabidopsis*.

DIRECTED MANIPULATION OF SECONDARY METABOLISM IN ARABIDOPSIS

There is presently only one example of the directed manipulation of secondary metabolism in *Arabidopsis*. Lloyd *et al.*[83] placed cDNAs encoding the maize anthocyanin regulatory genes *C1* and *R* under the control of cauliflower mosaic virus 35S promoters and used these constructs to transform *Arabidopsis*. All *C1* transgenic plants were phenotypically wild type. Transformants carrying the R construct accumulated higher than normal levels of anthocyanins, but the tissue-specificity of anthocyanin accumulation was unperturbed. When both transgenes were expressed together in F1 progeny from crosses of the original transformants, anthocyanin accumulation occured in roots, petals and sepals, tissues that are not normally pigmented in *Arabidopsis*. These data suggest that the abundance of an *Arabidopsis R* homologue limits anthocyanin accumulation levels while the ectopic expression of *C1* and *R* is sufficient to induce anthocyanin accumulation in tissues that are not normally pigmented. Interestingly, R-expressing plants also developed two to five times more trichomes on their leaf surfaces than wild type. The relationship between anthocyanin accumulation and trichome development in R-expressing transgenic plants suggests that the protein encoded by the *TTG* locus may represent a *myc* transcriptional regulator that is an *Arabidopsis R* homologue.[83]

FUTURE OPPORTUNITIES

        The application of standard techniques as well as new methods of secondary metabolite analysis may allow the identification of new secondary metabolite mutants. For example, to identify eight mutant lines of Arabidopsis that showed altered glucosinolate levels, Haughn *et al.*[70] analyzed only 1200 M2 plants because the HPLC-based mutant screen was laborious. Considering that most mutant screens are conducted on 4000 to 5000 M2 plants, many new glucosinolate mutants may remain undiscovered. A more extensive HPLC screen could be justified, considering how much information was gained by the analysis of the *gsm1* mutant alone. Alternatively, a GC-based mutant screen that could identify alterations in alkyl glucosinolate levels would be very informative, and significantly less labour-intensive. Recently, new HPLC methods have been described for the identification of secondary metabolism mutants of *Arabidopsis*.[86] New *in vivo* staining procedures have also been developed for the analysis of flavonoids.[87,88] These procedures clearly discriminate between wild type and *tt* mutants of *Arabidopsis*, and in the future could be applied to the identification of new mutations that affect the quality, quantity or tissue specificity of flavonoid biosynthesis. Taken together, it is clear that many opportunities are available for the isolation of new *Arabidopsis* secondary metabolism mutants.

        The use of the interaction trap[89] may also permit the isolation of genes of secondary metabolism that are difficult to retrieve using conventional techniques. In this method, yeast reporter genes are activated by protein-protein interactions that occur between the products of a cloned gene ("the bait") and the products of genes that are randomly cloned into another site in the vector. This method has been tested using flavonoid biosynthetic genes that have been cloned previously from *Arabidopsis* (B. Shirley, personal communication), and could easily be applied to secondary metabolic pathways where one or more structural genes have already been cloned.

Another technique that could be applied to *Arabidopsis* secondary metabolism is the so-called "enhancer trap". This technique employs a Ti plasmid carrying cauliflower mosaic virus transcriptional enhancers adjacent to the T-DNA right border. When the T-DNA of this vector inserts into the plant genome upstream of a gene, the enhancers lead to the overexpression of the downstream genes. Using this procedure, Hayashi et al.[90] identified a gene that, when overexpressed in tissue culture, permits callus growth in the absence of an exogenous source of auxin. This approach could be applied to secondary metabolism in *Arabidopsis*, thanks to the recent development of the large scale transformation procedure described earlier.[7] For example, it might be possible to use the "enhancer trap" to turn on the expression of transcriptional factors that are normally silent in the *Arabidopsis* genome. If so, the insertion of an enhancer T-DNA into the *Arabidopsis* genome at the appropriate position may generate *Arabidopsis* plants that have completely novel secondary metabolism.

CONCLUSIONS

In summary, it is apparent that with the genetic tools available for the study of biochemical and developmental process in *Arabidopsis,* rapid advances are possible. Despite these possibilities, the study of secondary metabolism in *Arabidopsis* has been largely ignored. Part of this is understandable because *Arabidopsis* is not as amenable to biochemical studies as are many other plants. For example, an *Arabidopsis* seed weighs only 20 μg. Assuming 50 mg to be a reasonable sample mass for obtaining a ^{13}C-NMR spectrum, half a million seeds would have to be sacrificed to characterize a secondary metabolite present at an abundance of 1% of the seed fresh weight, assuming 50% overall yield throughout its purification. Nevertheless, it has proven possible to characterize many of the secondary metabolites of *Arabidopsis*.[38,69,79]

Arabidopsis provides an excellent opportunity for the cloning of all of the genes of secondary metabolism from a single organism. The availability of these genes will enable the evaluation of hypotheses concerning how secondary metabolic pathways have evolved,[91,92] and of the possibility that all plant species carry the genes for secondary metabolic pathways that are highly expressed in only certain taxa.[93] The availability of these genes will also assist the directed manipulation of secondary metabolic processes in other plants for the production of novel crops and varieties with improved traits.[94,95]

ACKNOWLEDGEMENTS

The preparation of this manuscript was supported by funds from the Purdue Agricultural Experiment Station and the Purdue Research Foundation. This is journal paper number 14075 of the Purdue University Agricultural Experiment Station.

REFERENCES

1. MEYEROWITZ, E.M. 1987. *Arabidopsis thaliana*. Ann. Rev. Genet. 21: 93-111.
2. MEYEROWITZ, E.M. 1989. *Arabidopsis*, a useful weed. Cell 56: 263-269.
3. SOMERVILLE, C. 1989. *Arabidopsis* blooms. Plant Cell 1: 1131-1135.
4. LEUTWILER, L.S., Hough-Evans, B.R., Meyerowitz, E.M. 1984. The DNA of Arabidopsis thaliana. Mol. Gen. Genet. 194: 15-23.
5. HAUGHN, G., SOMERVILLE, C.R. 1987. Selection for herbicide resistance at the whole-plant level. In: Biotechnology in Agricultural Chemistry, (H.M. LeBaron, R.O. Mumma, R.C. Honeycutt, and J.H. Duesing, eds.) American Chemical Society, Washington, D.C. pp 98-108.
6. VALVEKENS, D., VAN MONTAGU, M., VAN LIJSEBETTENS, M. 1988. *Agrobacterium tumefaciens*-mediated transformation of *Arabidopsis thaliana* root explants by using kanamycin selection. Proc. Natl. Acad. Sci. USA 85: 5536-5540.
7. BECHTOLD, N., ELLIS, J., PELLETIER, G. 1993. In planta Agrobacterium mediated gene transfer by infiltration of adult *Arabidopsis thaliana* plants. C. R. Acad. Sci. Paris 316: 1194-1199.
8. KONCZ, C., N_METH, K., RÉDEI, G.P., SCHELL, J. 1992. T-DNA insertional mutagenesis in *Arabidopsis*. Plant Mol. Biol. 20: 963-976.
9. ARONDEL, V., LEMIEUX, B., HWANG, I., GIBSON, S.,

GOODMAN, H.M., SOMERVILLE, C.R. 1992. Map-based cloning of a gene controlling omega-3 fatty acid desaturation in *Arabidopsis*. Science 258: 1353-1355.

10. GIRAUDAT, J., HAUGE, B.M., VALON, C., SMALLE, J., PARCY, F., GOODMAN, H.M. 1992. Isolation of the *Arabidopsis ABI3* gene by positional cloning. Plant Cell 4: 1251-1261.

11. SUN, T.-P., GOODMAN, H.M., AUSUBEL, F.M. 1992. Cloning the *Arabidopsis thaliana GA1* locus by genomic subtraction. Plant Cell 4: 119-128.

12. REDEI, G.P., KONCZ, C. 1992. Classical mutagenesis. In: Methods in *Arabidopsis* Research (C. Koncz, N.-H. Chua, and J. Schell, eds.), World Scientific Publishing Co. Pte. Ltd., Singapore, pp. 16-82.

13. KOORNNEEF, M., HANHART, C.J., VAN LOENEN MARTINET, E.P., VAN DER VEEN, J.H. 1987. A marker line that allows the detection of linkage on all *Arabidopsis* chromosomes. Arabid. Inf. Serv. 23: 46-50.

14. CHANG, C., BOWMAN, J.L., DEJOHN, A.W., LANDER, E.S., MEYEROWITZ, E.M. 1988. Restriction fragment length polymorphism linkage map for *Arabidopsis thaliana*. Proc. Natl. Acad. Sci. 85: 6856-6860.

15. NAM, H.-G., GIRAUDAT, J., DEN BOER, B., MOONAN, F., LOOS, W.D.B., HAUGE, B.M., GOODMAN, H.M. 1989. Restriction fragment length polymorphism linkage map of *Arabidopsis thaliana*. Plant Cell 1: 699-705.

16. REITER, R.S., WILLIAMS, J.G.K., FELDMANN, K.A., RAFALSKI, J.A., TINGEY, S.V., SCOLNIK, P.A. 1992. Global and local genome mapping in *Arabidopsis thaliana* by using recombinant inbred lines and random amplified polymorphic DNAs. Proc. Natl. Acad. Sci. USA 89: 1477-1481.

17. MANDEL, M.A., GUSTAFSON-BROWN, C., SAVIDGE, B., YANOFSKY, M.F. 1992. Molecular characterization of the *Arabidopsis* floral homeotic gene *APETALA1*. Nature 360: 273-277.

18. STRAUS, D., AUSUBEL, F.M. 1990. Genomic subtraction for cloning DNA corresponding to deletion mutations. Proc. Natl. Acad. Sci. USA 87: 1889-1893.

19. MARKS, M.D., FELDMANN, K.A. 1989. Trichome development in *Arabidopsis thaliana*. I. T-DNA tagging of the *GLABROUS 1* gene. Plant Cell 1: 1043-1050.

20. MARKS, M.D., FELDMANN, K.A. 1989. Trichome development in

Arabidopsis thaliana. II. Isolation and complementation of the *GLABROUS 1* gene. Plant Cell 1: 1051-1055.

21. BEHRINGER, F.J., MEDFORD, J.I. 1992. A plasmid rescue technique for the recovery of plant DNA disrupted by T-DNA insertion. Plant Mol. Biol. Reporter 10: 190-198.

22. KONCZ, C., MAYERHOGER, R., KONCZ-KALMAN, Z., NAWRATH, C., REISS, B., REDEI, G.P., SCHELL, J. 1990. Isolation of a gene encoding a novel chloroplast protein by T-DNA tagging in *Arabidopsis thaliana.* EMBO J. 9: 1337-1346.

23. YANOFSKY, M.F., MA, H., BOWMAN, J.L., DREWS, G.N., FELDMANN, K.A., MEYEROWITZ, E.M. 1990. The protein encoded by the *Arabidopsis* homeotic gene agamous resembles transcription factors. Nature 346: 35-39.

24. DENG, X.-W., MATSUI, M., WEI, N., WAGNER, D., CHU, A.M., FELDMANN, K.A., QUAIL, P.H. 1992. COP1, an *Arabidopsis* regulatory gene, encodes a protein with both a zinc-binding motif and a Gβ homologous domain. Cell 71: 791-801.

25. FELDMANN, K.A., MARKS, M.D. 1987. *Agrobacterium*-mediated transformation of germinating seeds of *Arabidopsis thaliana:* a non-tissue culture approach. Mol. Gen. Genet. 208: 1-9.

26. BOUCHEZ, D., CAMILLERI, C., CABOCHE, M. 1993. A binary vector based on Basta resistance for in planta transformation of *Arabidopsis thaliana.* C. R. Acad. Sci. Paris 316: 1188-1193.

27. ANDERSON, J.A., HUPRIKAR, S.S., KOCHIAN, L.V., LUCAS, W.J., GABER, R.F. 1992. Functional expression of a probable *Arabidopsis thaliana* potassium channel in *Saccharaomyces cerevisiae.* Proc. Natl. Acad. Sci. USA 89: 3736-3740.

28. SENTENAC, H., BONNEAUD, N., MINET, M., LACROUTE, F., SALMON, J.-M., GAYMARD, F., GRIGNON, C. 1992. Cloning and expression in yeast of a plant potassium ion transport system. Science 256: 663-665.

29. LARSEN, P.O. 1981. Glucosinolates. In: The Biochemistry of Plants, vol. 7, (P.K. Stumpf and E.E. Conn, eds.), Academic Press, New York, pp. 501-525.

30. ADAMS, M.D., DUBNICK, M., KERLAVAGE, A.R., MORENO, R., KELLEY, J.M., UTTERBACK, T.R., NAGLE, J.W., FIELDS, C., VENTER, J.C. 1992. Sequence identification of 2,375 human brain genes. Nature 355: 632-634.

31. KEITH, C.S., HOANG, D.O., BARRETT, B.M., FEIGELMAN, B.,

NELSON, M.C., THAI, H., BAYSDORFER, C. 1993. Partial sequence analysis of 130 randomly selected maize cDNA clones. Plant Physiol. 101: 329-332.

32. UCHIMIYA, H., KIDOU, S., SHIMIZAKI, T., AOTSUKA, S., TAKAMATSU, S., NISHI, R., HASHIMOTO, H., MATSUBAYASHI, Y., KIDOU, N., UMEDA, M., KATO, A. 1992. Random sequencing of cDNA libraries reveals a variety of expressed genes in cultured cells of rice (*Oryza sativa* L.). Plant J. 2: 1005-1009.

33. KOORNNEEF, M., LUITEN, W., DE VLAMING, P., SCHRAM, A.W. 1982. A gene controlling flavonoid-3'-hydroxylation in *Arabidopsis*. Arabid. Inf. Serv. 19: 113-115.

34. KOORNNEEF, M. 1990. Mutations affecting the testa color in *Arabidopsis*. Arabid. Inf. Serv. 27: 1-4.

35 FEINBAUM, R.L., AUSUBEL, F.M. 1988. Transcriptional regulation of the *Arabidopsis thaliana* chalcone synthase gene. Mol. Cell. Biol. 8: 1985-1992.

36. SHIRLEY, B.W., HANLEY, S., GOODMAN, H.M. 1992. Effects of ionizing radiation on a plant genome: analysis of two *Arabidopsis* transparent testa mutations. Plant Cell 4: 333-347.

37. KOORNNEEF, M. 1981. The complex syndrome of *ttg* mutants. Arabid. Inf. Serv. 18: 45-51.

38. CHAPPLE, C.C.S., VOGT, T., ELLIS, B.E., SOMERVILLE, C.R. 1992. An *Arabidopsis* mutant defective in the general phenylpropanoid pathway. Plant Cell 4: 1413-1424.

39. STRACK, D. 1980. Enzymatic synthesis of 1-sinapoylglucose from free sinapic acid and UDP-glucose by a cell-free system from *Raphanus sativus* seedlings. Z. Naturforsch. 35c: 204-208.

40. MOCK, H.-P., STRACK, D. 1993. Energetics of the uridine 5'-diphosphoglucose: hydroxycinnamic acid acyl-glucosyltransferase reaction. Phytochemistry 32: 575-579.

41. STRACK, D., KNOGGE, W., DAHLBENDER, B. 1983. Enzymatic synthesis of sinapine from 1-*O*-sinapoyl-β-D-glucose and choline by a cell-free system from developing seeds of red radish (*Raphanus sativus* L. var. *sativus*). Z. Naturforsch 38c: 21-27.

42. STRACK, D., NURMANN, G., SACHS, G. 1980. Sinapine esterase II. Specificity and change of sinapine esterase activity during germination of *Raphanus sativus*. Z. Naturforsch 35c: 963-966.

43. STRACK, D. 1981. Sinapine as a supply of choline for the biosynthesis

of phosphatidylcholine in *Raphanus sativus* seedlings. Z. Naturforsch. 36c: 215-221.

44. LINSCHEID, M., WENDISCH, D., STRACK, D. 1980. The structures of sinapic acid esters and their metabolism in cotyledons of *Raphanus sativus*. Z. Naturforsch. 35c: 907-914.

45. SHARMA, V., STRACK, D. 1985. Vacuolar localization of 1-sinapoylglucose: L-malate sinapoyltransferase in protoplasts from cotyledons of *Raphanus sativus*. Planta 163: 563-568.

46. MOCK, H.-P., VOGT, T., STRACK, D. 1992. Sinapoylglucose:malate sinapoyltransferase activity in *Arabidopsis thaliana* and *Brassica rapa*. Z. Naturforsch. 47c: 680-682.

47. ROBINSON-BEERS, K., PRUITT, R.E., GASSER, C.S. 1992. Ovule development in wild-type *Arabidopsis* and two female-sterile mutants. Plant Cell 4: 1237-1249.

48. HOBSON-FROHOCK, A., FENWICK, G.R., HEANEY, R.K., LAND, D.G., CURTIS, R.F. 1977. Rapeseed meal and egg taint: association with sinapine. Br. Poult. Sci. 18: 539-541.

49. STRACK, D., PIEROTH, M., SCHARF, H., SHARMA, V. 1985. Tissue distribution of phenylpropanoid metabolism in cotyledons of *Raphanus sativus* L. Planta 164: 507-511.

50. LI, J., OU-LEE, T.-M., RABA, R., AMUNDSON, R.G., LAST, R.L. 1993. *Arabidopsis* flavonoid mutants are hypersensitive to UV-B irradiation. Plant Cell 5: 171-179.

51. GRAND, C. 1984. Ferulic acid 5-hydroxylase: A new cytochrome P-450-dependent enzyme from higher plant microsomes involved in lignin synthesis. FEBS Lett. 169: 7-11.

52. UNDERHILL, E.W. 1980. Glucosinolates. In: Encyclopedia of Plant Physiology, New Series, Volume 8, Secondary Plant Products, (E.A. Bell and B.V. Charlwood, eds.), Springer-Verlag, Berlin, pp 493-511.

53. MITHEN, R.F., LEWIS, B.G., HEANEY, R.K., FENWICK, G.R. 1987. Glucosinolates of wild and cultivated *Brassica* species. Phytochemistry 26: 1969-1973.

54. DAXENBICHLER, M.E., SPENCER, G.F., CARLSON, D.G., ROSE, G.B., BRINKER, A.M., POWELL, R.G. 1991. Glucosinolate composition of seeds from 297 species of wild plants. Phytochemistry 30: 2623-2638.

55. BJÖRKMAN, R. 1976. Properties and function of plant myrosinases. In: The Biology and Chemistry of the Cruciferae, (J.G. Vaughn, A.J. MacLeod, and B.M.G. Jones, eds.), Academic Press, New York, pp. 191-205.

56. HÖGLUND, A.-S., LENMAN, M., FALK, A., RASK, L. 1991. Distribution of myrosinase in rapeseed tissues. Plant Physiol. 95: 213-221.

57. LÜTHY, B., MATILE, P. 1984. The mustard oil bomb: rectified analysis of the subcellular organization of the myrosinase system. Biochem. Physiol. Pflanzen 179: 5-12.

58. XUE, J., LENMAN, M., FALK, A., RASK, L. 1992. The glucosinolate-degrading enzyme myrosinase in Brassicaceae is encoded by a gene family. Plant Mol. Biol. 18: 387-398.

59. BLAU, P.A., FEENY, P., CONTARDO, L., ROBSON, D.S. 1978. Allylglucosinolate and herbivorous caterpillars: a contrast in toxicity and tolerance. Science 200: 1296-1298.

60. BIRCH, A.N.E., GRIFFITS, D.W., SMITH, W.H.M. 1990. Changes in forage and oilseed rape (*Brassica napus*) root glucosinolates in response to attack by turnip root fly (*Delia floralis*). J. Sci. Food Agric. 51: 309-320.

61. BODNARYK, R.P. 1992. Effects of wounding on glucosinolates in the cotyledons of oilseed rape and mustard. Phytochemistry 31: 2671-2677.

62. BELL, J.M. 1984. Nutrients and toxicants in rapeseed meal: a review. J. Anim. Sci. 58: 996-1010.

63. VAN ETTEN, C.H. 1969. Goitrogens. In: Toxic constituents of plant food stuffs (I.E. Liener, ed.), Academic Press, New York, pp 103-142.

64. DOWNEY, R.K., CRAIG, B.M., YOUNGS, D.G. 1969. Breeding rapeseed for oil and meal quality. J. Am. Oil Chem. Soc. 46: 121-123.

65. JOSEFSSON, E. 1971. Studies of the biochemical background to differences in glucosinolate content in *Brassica napus* L. I. Glucosinolate content in relation to general chemical composition. Physiol. Plant. 24: 150-159.

66. JOSEFSSON, E. 1971. Studies of the biochemical background to differences in glucosinolate content in *Brassica napus* L. II. Administration of some sulphur-35 and carbon-14 compounds and localization of metabolic blocks. Physiol. Plant. 24: 161-175.

67. BARTLING, D., SEEDORF, M., MITH_FER, A., WEILER, E.W. 1992. Cloning and expression of an *Arabidopsis* nitrilase which can convert indole-3-acetonitrile to the plant hormone, indole-3-acetic acid. Eur. J. Biochem. 205: 417-424.

68. SEARLE, L.M., CHAMBERLAIN, K., RAUSCH, T., BUTCHER, D.N. 1982. The conversion of 3-indolylmethylglucosinolate to 3-indolylacetonitrile by myrosinases, and its relevance to the clubroot disease of the Cruciferae. J. Exp. Bot. 33: 935-942.

69. HOGGE, L.R., REED, D.W., UNDERHILL, E.W., HAUGHN, G.W. 1988. HPLC separation of glucosinolates from leaves and seeds of *Arabidopsis thaliana* and their identification using thermospray liquid chromatography / mass spectrometry. J. Chromatogr. Sci. 26: 551-556.

70. HAUGHN, G.W., DAVIN, L., GIBLIN, M., UNDERHILL, E.W. 1991. Biochemical genetics of plant secondary metabolites in *Arabidopsis thaliana*. The glucosinolates. Plant Physiol. 97: 217-226.

71. GIJZEN, M., MCGREGOR, I., SÉGUIN-SWARTZ, G. 1989. Glucosinolate uptake by developing rapeseed embryos. Plant Physiol. 89: 260-263.

72. MITHEN, R., TOROSER, D. 1994. Biochemical genetics of aliphatic glucosinolates in *Brassica* and *Arabidopsis*. In: Amino Acids and Their Derivatives in Higher Plants. (R. Wallsgrove, ed.) Cambridge University Press, Cambridge, UK, in press.

73. ROUXEL, T., SARNIGUET, A., KOLLMAN, A., BOUSQUET, J.-F. 1989. Accumulation of a phytoalexin in *Brassica* spp. in relation to a hypersensitive reaction to *Leptosphaeria maculans*. Physiol. Mol. Plant. Pathol. 34: 507-514.

74. CONN, K.L., TEWARI, J.P., DAHIYA, J.S. 1988. Resistance to *Alternaria brassicae* and phytoalexin-elicitation in rapeseed and other crucifers. Plant Sci. 56: 21-25.

75. BROWNE, L.M., CONN, K.L., AYER, W.A., TEWARI, J.P. 1991. The camalexins: New phytoalexins produced in the leaves of *Camelina sativa* (Cruciferae). Tetrahedron 47: 3909-3914.

76. JEJELOWO, O.A., CONN, K.L., TEWARI, J.T. 1991. Relationship between conidial concentration, germling growth, and phytoalexin production by *Camelina sativa* leaves inoculated with *Alternaria brassicae*. Mycol. Res. 95: 928-934.

77. AYER, W.A., CRAW, P.A., MA, Y.-T., MIAO, S. 1992. Synthesis of camalexin and related phytoalexins. Tetrahedron 48: 2919-2924.

78. AUSUBEL, F.M., GLAZEBROOK, J., GREENBERG, J., MINDRINOS, M., YU, G.-L. 1993. Analysis of the *Arabidopsis* defense response to *Pseudomonas* pathogens. In: Advances in

Molecular Genetics of Plant-Microbe Interactions, Vol. 2 (Nester, E.W., and D.P.S. Verma, eds.), Kluwer Academic Publ., Dordrecht, The Netherlands, pp 393-403.

79. TSUJI, J., JACKSON, E., GAGE, D., HAMMERSCHMIDT, R., SOMERVILLE, S.C. 1992. Phytoalexin accumulation in *Arabidopsis thaliana* during the hypersensitive reaction to *Pseudomonas syringae* pv. *syringae*. Plant Physiol. 98: 1304-1309.

80. HAMMERSCHMIDT, R., TSUJI, J., ZOOK, M., SOMERVILLE, S. 1993. A phytoalexin from *Arabidopsis thaliana* and its relationship to other phytoalexins of Crucifers. In: *Arabidopsis thaliana* as a Model for Studying Plant-Pathogen Interactions (K.R. Davis, and R. Hammerschmidt, eds.), American Phytopathological Society Press, St. Paul, MN, pp. 73-84.

81. TSUJI, J., ZOOK, M., HAMMERSCHMIDT, R., SOMERVILLE, S. 1993. Camalexin biosynthesis in *Arabidopsis thaliana*. Physiol. Mol. Plant Pathol. (in press).

82. GLAZEBROOK, J., AUSUBEL, F.M. 1993. Isolation and characterization of phytoalexin deficient mutants of *Arabidopsis thaliana*. 5th International Conference on *Arabidopsis* Research, Abstracts.pp 21.

83. LLOYD, A.M., WALBOT, V., DAVIS, R.W. 1992. *Arabidopsis* and *Nicotiana* anthocyanin production activated by maize regulators R and C1. Science 258: 1773-1775.

84. VAN DER MEER, I.M., STAM, M.E., VAN TUNEN, A.J., MOL, J.N.M., STUITJE, A.R. 1992. Antisense inhibition of flavonoid biosynthesis in petunia anthers results in male sterility. Plant Cell 4: 253-262.

85. MO, Y., NAGEL, C., TAYLOR, L.P. 1992. Biochemical complementation of chalcone synthase mutants defines a role for flavonols in functional pollen. Proc. Natl. Acad. Sci. USA 89: 7213-7217.

86. GRAHAM, T.L. 1991. A rapid, high resolution high performance liquid chromatography profiling procedure for plant and microbial aromatic secondary metabolites. Plant Physiol. 95: 584-593.

87. SHEAHAN, J.J., RECHNITZ, G.A. 1992. Flavonoid-specific staining of *Arabidopsis thaliana*. BioTechniques, 13, 880-883.

88. SHEAHAN, J.J., RECHNITZ, G.A. 1993. Differential visualization of transparent testa mutants in *Arabidopsis thaliana*. Anal. Chem. 65: 961-963.

89.   ZERVOS, A., GYURIS, J., BRENT, R. 1993. Mxi1, a protein that specifically interacts with Max to bind to Myc-Max recognition sites. Cell 72: 223-232.

90.   HAYASHI, H., CZAJA, I., LUBENOW, H., SCHELL, J., WALDEN, R. 1992. Activation of a plant gene by T-DNA tagging: auxin-independent growth *in vitro*. Science 258: 1350-1353.

91.   WILLIAMS, D.H., STONE, M.J., HAUCK, P.R., RAHMAN, S.K. 1989. Why are secondary metabolites (natural products) biosynthesized? J. Nat. Prod. 52: 1189-1208.

92.   JONES, C.G., FIRN, R.D. 1991. On the evolution of plant secondary chemical diversity. Phil. Trans. R. Soc. Lond. B 333: 273-280.

93.   WINK, M. 1988. Plant breeding: importance of plant secondary metabolites for protection against pathogens and herbivores. Theor. Appl. Genet. 75: 225-233.

94.   KISHORE, G.M., SOMERVILLE, C.R. 1993. Genetic engineering of commercially useful biosynthetic pathways in transgenic plants. Curr. Opin. Biotech. 4:152-158.

95.   SOMERVILLE, C.R. 1993. New opportunities to dissect and manipulate plant processes. Phil. Trans. R. Soc. Lond. B. 339: 199-206.

Chapter Eleven

POLYPHENOL OXIDASE[*]

John C. Steffens[1], Eitan Harel[2], and Michelle D. Hunt[1]

[1]Department of Plant Breeding and Biometry
252 Emerson Hall, Cornell University
Ithaca NY 14853-1901

[2]Institute of Life Sciences, Department of Botany,
The Hebrew University of Jerusalem, Jerusalem, Israel 91904

---

[*]J.C.S. dedicates this chapter to the memory of Frederick Steffens

Genetic Engineering of Plant Secondary Metabolism,
Edited by B.E. Ellis *et al.*, Plenum Press, New York, 1994

INTRODUCTION: POLYPHENOL OXIDASE (PPO)

Polyphenol oxidases (PPO) catalyze the one- and two- electron oxidation of phenols to quinones at the expense of O_2. PPOs are copper metalloproteins catalyzing the o- hydroxylation of a monophenol, followed by its oxidation to the o-diquinone (cresolase activity), or the oxidation of an o-dihydroxyphenol to the o-diquinone (catecholase activity). The quinonoid reaction products formed by PPO are familiar to anyone who has homogenized plant tissues. PPO-generated quinones are highly reactive, electrophilic molecules which covalently modify and crosslink a variety of cellular constituents, including nucleophiles of proteins such as sulfhydryl, amine, amide, indole and imidazole substituents. The formation of quinone adducts (usually brown or black colored) represents the primary detrimental effect of PPO in food processing and postharvest physiology of plant products and is the primary reason for PPO's importance in food technology.[1] Conversely, the potential for quinones to covalently modify and reduce the nutritive value of plant proteins has also generated interest in PPO-based strategies to increase the herbivore resistance of plants.[2] Although the function of PPO in plant metabolism is not well understood, the recent cloning of PPO presents opportunities to explore its function in plants as well as to explore the extent to which its expression can be manipulated. The purpose of this paper is to review current progress toward understanding the structure and function of PPO, and to describe initial results of attempts to modify PPO expression. Other aspects of PPO have recently been reviewed.[3]

REACTION MECHANISM OF PPO

In general, polyphenol oxidases catalyze the formation of quinones from phenols in the presence of molecular oxygen. Specifically, PPOs catalyze two distinct reactions: the hydroxylation of monophenols to o-diphenols (cresolase, tyrosinase, or monophenol oxidase activity [EC 1.14.18.1]) and the dehydrogenation of o-dihydroxyphenols to o-quinones (catecholase or diphenol oxygen oxidoreductase activity [EC 1.10.3.2). PPOs may exhibit either or both of these activities,[4] but the monophenol oxidase activity is often labile or requires priming with reducing agents or trace amounts of o-diphenol. Regardless of the substrate utilized, o-quinones are the final reaction products. Because of their reactivity, they can covalently modify and crosslink a variety of cellular nucleophiles, and also autoxidize, forming black or brown colored melanin polymers.

Lerch[5] has proposed a mechanism of catalysis for *Neurospora crassa* tyrosinase based on information regarding the electronic and geometric structure of the binuclear copper active site. Figure 1 depicts for tyrosinase the two activities of PPO - catecholase (A) and cresolase (B). Catecholase activity (A) involves the oxidation of two *o*-diphenols to form two *o*-quinones with a concomitant 4e- reduction of molecular oxygen to form two water molecules. This activity initiates with the binding of an *o*-diphenol to mettyrosinase. The binuclear copper cluster is reduced, leading to the formation of deoxytyrosinase and the release of *o*-diquinone. With the subsequent binding of molecular oxygen, oxytyrosinase is formed. A second *o*-diphenol then binds and reduces peroxide to water with the formation of another *o*-quinone. Cresolase activity (B) initiates with the binding of a monophenol to one copper of oxytyrosinase. The active site stabilizes the monophenol by rearrangement to a trigonal bipyramidal intermediate. A deprotonated polarised peroxide is bound to the other copper atom. This arrangement facilitates the *o*-hydroxylation of the monophenol to an *o*-diphenol via an electrophilic attack on the aromatic ring. The binuclear copper cluster is reduced, leading to the formation of deoxytyrosinase and the release of *o*-diquinone. The subsequent binding of molecular oxygen regenerates the oxytyrosinase form of the enzyme.

Copper Sites

PPOs from bacteria, fungi, plants and animals all are bicopper metalloenzymes and possess two conserved domains, CuA and CuB, responsible for Cu coordination and the interaction with molecular oxygen and phenolic substrates. Each copper atom is presumed to be coordinated by three histidine residues provided by the CuA and CuB sites. In the X-ray crystal structure of arthropodan hemocyanin from *Panilirus interruptus*, each of the copper atoms of the CuA and CuB sites are coordinated by three histidine residues.[6] Site-directed mutagenesis of histidine residues in the CuA and CuB binding regions of *Streptomyces glaucenscens* tyrosinase[7,8] resulted in synthesis of a protein with no detectable tyrosinase activity and decreased affinity for Cu.

In contrast to the experimental definition of Cu binding sites in the organisms described above, experimental evidence for copper coordination by the conserved histidine residues of CuA and CuB in plant PPOs is not available. Deduced amino acid sequences of tyrosinases cloned from sources ranging from bacteria to humans and plants show very little sequence similarity outside of the putative copper-binding regions. However, similarity within these regions, especially within the CuB site, can exceed 50 percent.[5] A high degree of similarity is also evident between tyrosinase CuB binding site sequence and that

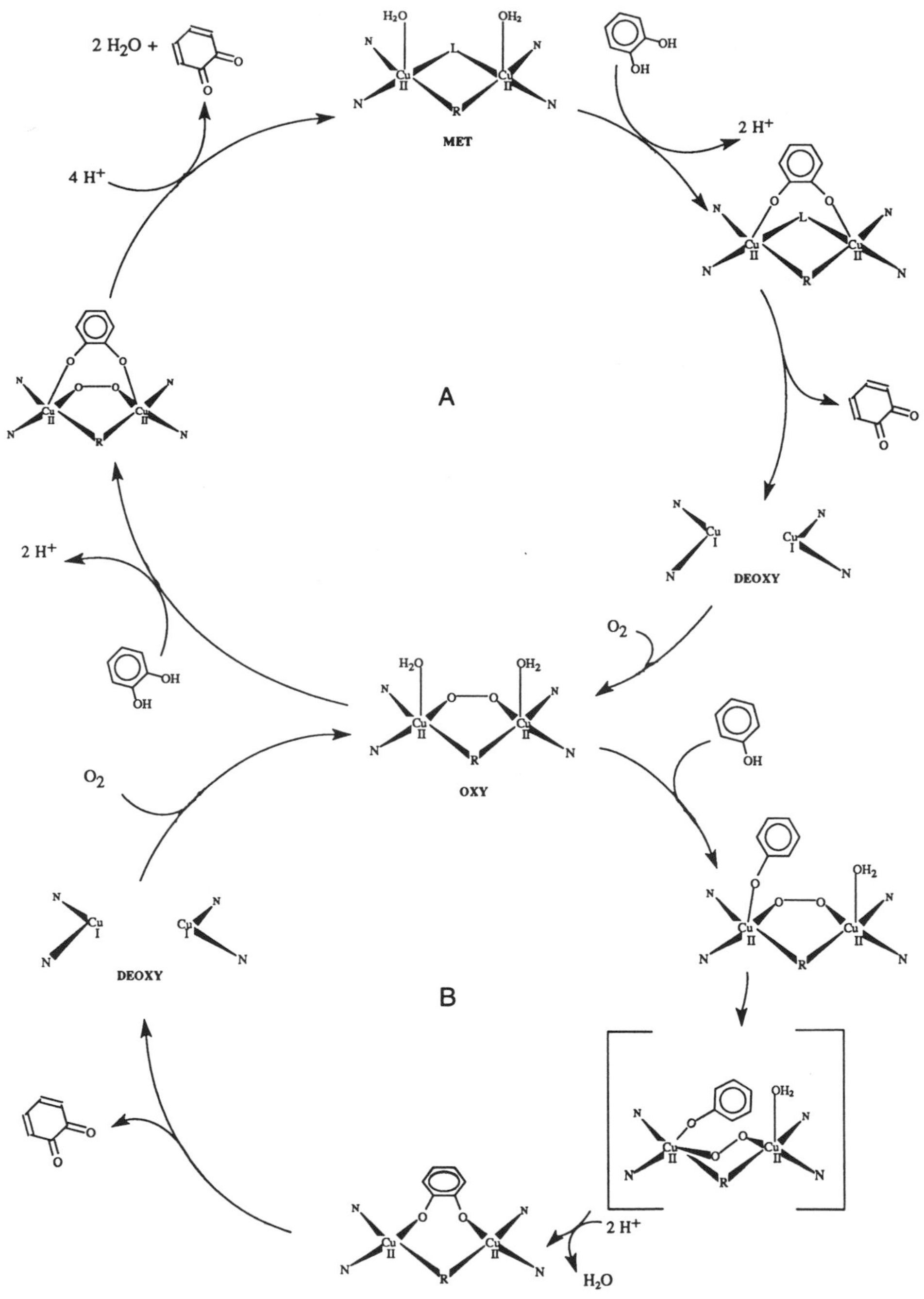
2 H₂O +
4 H⁺
MET
2 H⁺
A
DEOXY
O₂
2 H⁺
OXY
O₂
DEOXY
B
2 H⁺
H₂O

found in the oxygen transport protein, hemocyanin (Fig. 2).[5] Conversely, the CuA-binding site typically shows much less conservation between various tyrosinases and between tyrosinases and hemocyanins (Fig. 3).[9]

Among plant PPOs, however, the CuA site is more highly conserved (90 percent amino acid identity) than the CuB site (70 percent amino acid identity) for the regions compared in Figures 2 and 3. Mammalian tyrosinases do not exhibit such high sequence conservation of the CuA and CuB binding sites (40 and 50 percent, respectively) nor do they exhibit closer CuA sequence similarity relative to the CuB site as evident for plant PPOs (Figs. 2 and 3). It should be noted that the CuA site central histidine residue is not conserved in the mollusc *Helix pomatia*, functional units d and g. Some experimental evidence suggests that CuA of some Mulluscan hemocyanins is comprised of only two histidine ligands and that the conservation of the central histidine may not be crucial[9].

Plant PPOs exhibit other structurally unique features when compared to tyrosinases and hemocyanins. Plant PPOs do not possess the invariant Asn residue evident in other tyrosinases and hemocyanins which putatively functions in an active site hydrogen bonding network bridging the two copper ions.[7] In addition, they do not appear to possess the cysteine residue found in both *N. crassa* tyrosinase and various molluscan hemocyanins thought to be modified posttranslationally and involved in activation.[10] A distinctive feature of plant PPOs is the presence of a third His cluster near the carboxy terminus, which we have designated CuC (Fig. 4). Similarly to the CuA and CuB binding sites, this site possesses three His residues whose positions are highly conserved between *Lycopersicon* and *Solanum*. Although the position of these residues is less conserved when comparing the amino acid sequences of *Vicia faba* and solanaceous PPOs, the 3-His CuC cluster is nonetheless present in all, and perfectly conserved between two of the three *V. faba* cDNAs described by Cary *et al.*.[11] The functional significance of the CuC motif is unknown at present. We are using site-directed mutagenesis to modify the His residues within the putative copper-binding sites of tomato PPO to explore their relationship to enzymic

---

Fig. 1. The reaction mechanism of catecholase and cresolase activity of mushroom tyrosinase.[5] A). Catecholase activity. B). Cresolase activity. R, L = protein ligands that bridge the two tetragonal Ca(II) ions of the active site.

activity and their role in import and processing of the PPO pre-protein into plastids (see below).

```
          * #        *                                                     *
PPOP1:  P |H|T P V|H|I    18   S A G L|D|P|I|F|Y C H|H|A N V|D|R M W D E|W|K
PPOE:   P |H|T P V|H|I    18   S A G L|D|P|I|F|Y C H|H|A N V|D|R M W N E|W|K
PPOA1:  P |H|A P V|H|T    17   S A A R|D|P|I|F|Y S H|H|S N V|D|R L W Y I|W|K

Ode:    A |H|N A I|H|S    14   Y A A Y|D|P|I|F|Y L H|H|S N V|D|R L W V I|W|Q
Hpd:    L |H|N A L|H|S    14   Y T A F|D|P|V F F L H|H|A N T|D|R L W A I|W|Q
Odf:    V |H|N S I|H|Y    14   Y S S F|D|P|I|F|Y V H|H|S M V|D|R L W A I|W|Q
Hpg:    S |H|N A I|H|S    14   Y T A Y|D|P|L F L L H|H|S N V|D|R Q F A I|W|Q
Odg:    G |H|N A I|H|S    14   Y T S Y|D|P|L F Y L H|H|S N T|D|R I W S V|W|Q
Soh:    G |H|N A I|H|S    14   Y T S Y|D|P|L F Y L H|H|S N T|D|R I W A I|W|Q
YSg:    L |H|N R V|H|V     9   M S P N|D|P|L F W L H|H|A Y V|D|R L W A E|W|Q
YNc:    V |H|N E I|H|D     9   V S A F|D|P|L F W L H|H|V N V|D|R L W S I|W|Q
YHs:    M |H|N A L|H|I    10   G S A N|D|P|I|F|L L H|H|A F V|D|S I F E Q|W|L
YM1:    M |H|N A L|H|I    10   G S A N|D|P|I|F|L L H|H|A F V|D|S I F D Q|W|L
YM2:    L |H|N L A|H|L    10   L S P N|D|P|I|F|V L L|H|T F T|D|A V F D E|W|L
```

Fig. 2. Sequence similarity of the CuB binding sites of plant PPOs, tyrosinases and molluscan hemocyanins. Putative copper coordinating histidine residues are marked with an asterisk (*). The position of an invariant asparagine residue thought to function in an active site hydrogen bonding network bridging the two copper ions[7] is marked by a pound sign (#). Boxed sequence denotes complete amino acid identity between the sequences. Code: PPOP1, *Solanum tuberosum* leaf PPO; PPOE, *Lycopersicon esculentum* trichome PPO; PPOA1, *Vicia faba* leaf PPO; Ode, *Octopus dofleini* hemocyanin, chain e; Hpd, *Helix pomatia* hemocyanin, chain d; Odf, *O. dofleini* hemocyanin, chain f; Hpg, *H. pomatia* hemocyanin, chain g; Odg, *O. dofleini* hemocyanin, chain g; Soh, *Sepia officinalis* hemocyanin, chain h; Ysg, *Streptomyces glaucescens* tyrosinase; YNc, *Neurospora crassa* tyrosinase; YHs, human tyrosinase; YM1, mouse tyrosinase 1; and YM2, mouse tyrosinase 2. Taken in part from Lang and van Holde.[9]

```
              *              #  *                      *
PPOP1:  F K Q Q A N I|H|C |14| Q V|H|F S W L|F|F P F|H|R W Y L Y F Y|E|R I L G |9| L|P|Y W N W D
PPOE:   F K Q Q A N I|H|C |14| Q V|H|F S W L|F|F P F|H|R W Y L Y F Y|E|R I L G |9| L|P|Ŷ W N W D
PPOA1:  F T Q Q A N I|H|C |17| Q V|H|G S W L|F|F P F|H|R W Y L Y F Y|E|R I L G |9| L|P|F W N Y D

Ode:    F E A I A S F|H|A |14| C L|H|G M A T|F|P H W|H|R L Y V V Q F|E|Q A L H |7| V|P|Y W D W T
Hpd:    Y E N I A S F|H|G |12| S V S G M P T|F|P S W|H|R L Y V E Q V|E|E A L L |7| V|P|Y F D W I
Odf:    F Q T I A S Y|H|G |12| C L|H|G M P V|F|P H W|H|R V Y L L H F|E|D S M R |7| T|P|Y W D W T
Hpg:    F Q S I A S F|H|G |12| S I G G M A N F P Q W|H|R L Y V K Q W|E|D A L T |7| I|P|Y W D W T
Odg:    Y Q K I A S Y|H|G |13| C Q|H|G M V T|F|P N W|H|R L L T K Q M|E|D A L V |7| I|P|Y W D W T
Soh:    Y Q K I A S Y|H|G |13| C Q|H|G M V T|F|P H W|H|R L Y M K Q M|E|D A M K |7| I|P|Y W D W T
YSg:    Y D E F V T T|H|N |10| T G|H|R S P S|F|L P W|H|R R Y L L E F|E|R A L Q |7| L|P|Y W D W S
YNc:    Y Y Q V A G I|H|G |24| C T|H|S S I L|F|I T W|H|R P Y L A L Y|E|Q A L Y |8| A|P|Ŷ F D W A
YHs:    Y D L F V W M|H|Y |16| F A|H|E A P A|F|L P W|H|R L F L L R W|E|Q E I R |9| I|P|Y W D W R
YM1:    F E N I S V T|H|Y |17| F S|H|E G P A|F|L T W|H|R Y H L L Q L|E|R D M Q |9| L|P|Y W N F A
YM2:    Y D L F V W M|H|Y |16| F A|H|E A G P|F|L P W|H|R L F L L L W|E|Q E I R |9| V|P|Y W D W R
```

Fig. 3. Sequence similarity of the CuA binding sites of plant PPOs, tyrosinases and molluscan hemocyanins. Putative copper coordinating histidine residues are marked with an asterisk (*). The position of a conserved cysteine residue thought to be involved in activation[5] is marked by a pound sign (#). Boxed sequence denotes complete amino acid identitiy between the sequences with the exception of the middle histidine residue (see text). The code is the same as for Fig. 2. Taken in part from Lang and van Holde.[9]

Latency

The enzymology of PPO is plagued by the phenomenon of latent vacuolar substrates. It is only when the cell is disrupted that the two come in contact. In some plant species, the released PPO is immediately active.[12] In others, however, PPO activity remains latent after release from the plastid, and requires activation.[12-16] Depending upon the plant species, *in vitro* activation can be achieved by aging,[17] treatment with ammonium sulfate,[18] anionic and cationic detergents,[14,15,19] proteolytic enzymes,[12] denaturing agents,[18,20] fatty acids,[16] acid and base shock,[19] treatment with cations,[21] and treatment with a 58 kDa lectin isolated from *Daucus carota* cell cultures.[22] Felton *et al.*[23] have shown that passage of PPO through an insect herbivore's gut released latent PPO activity, thereby increasing alkylation of dietary protein by quinones. This is perhaps the first demonstration of an adaptive function for PPO latency and activation phenomena.

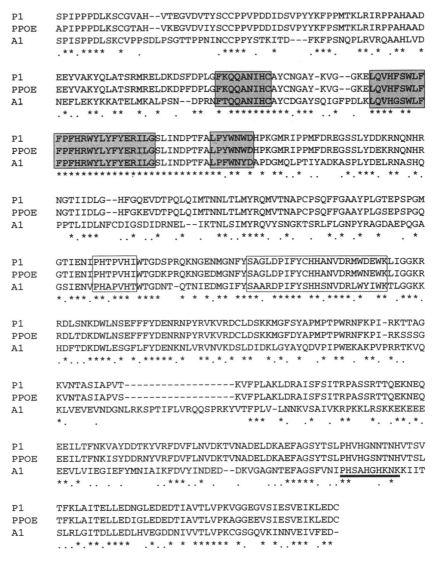

```
P1      SPIPPPDLKSCGVAH--VTEGVDVTYSCCPPVPDDIDSVPYYKFPPMTKLRIRPPAHAAD
PPOE    APIPPPDLKSCGTAH--VKEGVDVIYSCCPPVPDDIDSVPYYKFPSMTKLRIRPPAHAAD
A1      SPISPPDLSKCVPPSDLPSGTTPPNINCCPPYSTKITD---FKFPSNQPLRVRQAAHLVD
        .**.****  *  .              .****  .  *   .***.  **.*.**  *


P1      EEYVAKYQLATSRMRELDKDSFDPLGFKQQANIHCAYCNGAY-KVG--GKELQVHFSWLF
PPOE    EEYVAKYQLATSRMRELDKDPFDPLGFKQQANIHCAYCNGAY-KVG--GKELQVHFSWLF
A1      NEFLEKYKKATELMKALPSN--DPRNFTQQANIHCAYCDGAYSQIGFPDLKLQVHGSWLF
        .*.. **. **  *.*  .  **  * *********.*** ..*  .   **** **


P1      FPFHRWYLYFYERILGSLINDPTFALPYWNWDHPKGMRIPPMFDREGSSLYDDKRNQNHR
PPOE    FPFHRWYLYFYERILGSLINDPTFALPYWNWDHPKGMRIPPMFDREGSSLYDEKRNQNHR
A1      FPFHRWYLYFYERILGSLINDPTFALPFWNYDAPDGMQLPTIYADKASPLYDELRNASHQ
        *********************.****.**.*  **..*. ..    *.***. ** .*.


P1      NGTIIDLG--HFGQEVDTPQLQIMTNNLTLMYRQMVTNAPCPSQFFGAAYPLGTEPSPGM
PPOE    NGTIIDLG--HFGKEVDTPQLQIMTNNLTLMYRQMVTNAPCPSQFFGAAYPLGSEPSPGQ
A1      PPTLIDLNFCDIGSDIDRNEL--IKTNLSIMYRQVYSNGKTSRLFLGNPYRAGDAEPQGA
        *.***  ..*..*  .*    **..****..*.   *.*.*  .*    .  *


P1      GTIENIPHTPVHIWTGDSPRQKNGENMGNFYSAGLDPIFYCHHANVDRMWDEWKLIGGKR
PPOE    GTIENIPHTPVHIWTGDKPRQKNGEDMGNFYSAGLDPIFYCHHANVDRMWNEWKLIGGKR
A1      GSIENVPHAPVHTWTGDNT-QTNIEDMGIFYSAARDPIFYSHHSNVDRLWYIWKTLGGKK
        *.***.**.***.****  * * *.** **** ..***** .****.*  **.***.


P1      RDLSNKDWLNSEFFFYDENRNPYRVKVRDCLDSKKMGFSYAPMPTPWRNFKPI-RKTTAG
PPOE    RDLTDKDWLNSEFFFYDENRNPYRVKVRDCLDSKKMGFDYAPMPTPWRNFKPI-RKSSSG
A1      HDFTDKDWLESGFLFYDENKNLVRVNVKDSLDIDKLGYAYQDVPIPWEKAKPVPRRTKVQ
        .*...****.* * .*****.*.*.*..**..*   * .* .**.*   **. ..*.


P1      KVNTASIAPVT----------------KVFPLAKLDRAISFSITRPASSRTTQEKNEQ
PPOE    KVNTASIAPVS----------------KVFPLAKLDRAISFSITRPASSRTTQEKNEQ
A1      KLVEVEVNDGNLRKSPTIFLVRQQSPRKYVTFPLV-LNNKVSAIVKRPKKLRSKKEKEEE
        *.    .              ***  *. .* . **  *. .**.*.


P1      EEILTFNKVAYDDTKYVRFDVFLNVDKTVNADELDKAEFAGSYTSLPHVHGNNTNHVTSV
PPOE    EEILTFNKISYDDRNYVRFDVFLNVDKTVNADELDKAEFAGSYTSLPHVHGSNTNHVTSL
A1      EEVLVIEGIEFYMNIAIKFDVYINDED--DKVGAGNTEFAGSFVNIPHSAHGHKNKKIIT
        **.*  .   .. ...   ***..*    ***** .** .*.* ..**   *


P1      TFKLAITELLEDNGLEDEDTIAVTLVPKVGGEGVSIESVEIKLEDC
PPOE    TFKLAITELLEDIGLEDEDTIAVTLVPKAGGEEVSIESVEIKLEDC
A1      SLRLGITDLLEDLHVEGDDNIVVTLVPKCGSGQVKINNVEIVFED-
        ...*.**.**** .*  *  * ***** *.  *  .*** .**
```

Fig. 4. Deduced amino acid sequence comparison of mature plant PPOs. P1= *Solanum tuberosum* cDNA PPO-P1;[44] PPOE=*Lycopersicon esculentum* genomic clone;[43] A1=*Vicia faba* cDNA PPO-A1.[11] Shaded boxed sequence denotes the CuA binding site and clear boxed sequence denotes the CuB site.[9] The underlined sequence identifies the "CuC" region. (*)= amino acids conserved in all three sequences; (·)= conservative amino acid substitutions; (-)= gaps introduced to maximize sequence similarity.

Tolbert[12] classified leaf chloroplast PPOs based on their latency characteristics. The first group (spinach, Swiss chard and beet) required light for activation but this requirement diminished upon extract aging. However, activation by trypsinization was equally as effective as light in aged or fresh extracts. Other treatments such as salts and detergents were not effective activators. The second group (pea, wheat, alfalfa, oats and sugarcane) exhibited PPO which required red light for activation, and trypsin was inhibitory to activity. The third group (bean, corn and tomato) exhibited no latency and a light requirement was not evident. The effect of light on PPO activation has also been attributed to photosynthetic oxidation rather than phenolic oxidation.[24]

Much of the work on leaf PPO has been carried out in *V. faba* due to its highly inducible activity and the availability of antibody.[25] The 60 kDa PPO of *V. faba* is extracted in a latent state which can be activated by the addition of sodium dodecyl sulfate.[13,14] Once latency is overcome, the 60 kDa PPO can be cleaved by the protease thermolysin with no loss of activity. This yields an active 42 kDa PPO peptide and inactive peptides of 12 and 18 kDa. This is in contrast to grape PPO, for which the active form is 40 kDa while the inactive form is 60 kDa,[26] suggesting that activation of grape PPO may require proteolytic cleavage. "Bruce's Sport", a grape mutant which exhibits abnormal plastid development in white variegated fruit, provides further evidence that proteolytic conversion is required for activity of grape PPO.[26] In the white sectors of mutant berries, the majority of the PPO is present in an inactive 60 kDa form, with a small fraction at 40 kDa. However, wild-type grape berries primarily possess the active 40 kDa form and a small fraction of the inactive 60 kDa form. The significance of latency, if any, remains unclear and the phenomenon seems likely to continue as a source of bewilderment to those choosing to study the enzymology of PPO. The fact that PPO of many species has no detectable activity *in vivo* without treatments to release latency has led some to suggest that PPO possesses a function unrelated to its phenolic oxidase activity.[24]

STRUCTURE OF PPO

Molecular weight estimates for PPO vary widely.[27,28] Many PPO studies report isoforms of 40-45 kDa, but 59-65, 70-72 kDa isoforms and higher have also been reported.[4,13,29-31] Previous reports of 45 kDa PPO as the primary isoform of this enzyme can in part be attributed to C-terminal susceptibility to proteolysis during isolation and purification.[13] Other PPO isoforms are believed to be artifacts of isolation and purification, generated either by quinone alkylation, partial denaturation, or proteolysis. Artifactual multiplicity of grape PPO was reported by Harel et al.[32] who demonstrated that part of the supposedly native multiplicity could be imitated by incubating the isolated enzyme (ca. 58 kDa) with commercially obtained proteases.

An important source of heterogeneity in PPO M_r estimates has been the failure to completely inhibit quinone formation. In one case, failure to inhibit PPO activity and prevent crosslinking by quinones during isolation led to an M_r estimate in excess of 10^6 for a trichome-localized PPO later shown to possess a M_r of 59,000.[30,33] Similarly, although there is general agreement that PPO is a plastid-localized protein, the presence of covalently attached carbohydrate has been reported in broad bean,[29,34] peach,[35] potato,[36] strawberry,[37] Jerusalem artichoke,[38] and apple.[39] It is not clear whether PPO-associated carbohydrate results artifactually from quinone crosslinking, or whether it represents a *bona fide* component of the enzyme. Ganesa et al.[29] reported that purified *V. faba* PPO contained covalently attached carbohydrates and bound concanavalin A, *Phaseolus vulgaris* erythroagglutinin and *Ricinus communis* agglutinin. However, not all of the isoforms present in *V. faba* PPO preparations were glycosylated. Cary et al.[11] report a substantial M_r discrepancy between the mature *V. faba* M_r deduced from cDNAs (58,000) and the experimental M_r of *V. faba* PPO under fully denaturing conditions in SDS-PAGE (63,000). Robinson and Dry,[13] however, report a mass of 60,000 for the fully-denatured *V. faba* leaf PPO, which is much closer to the predicted M_r of 58,000.

In contrast to the results in *V. faba*, grape PPO does not bind concanavalin A and is apparently not glycosylated.[15] We have not been able to detect glycosylation of potato or tomato PPO (unpublished data). Chloroplast coupling factor 1 (CF_1), which is composed of both nuclear and chloroplast-encoded subunits, is the only other example of a glycosylated plastid-localized enzyme in higher plants[40] of which we are aware. The mechanisms of Golgi-mediated glycosylation and of import of chloroplast precursors are difficult to reconcile with the concept of glycosylated plastid proteins, and confirmation of PPO glycosylation would represent an important milestone.

Another source of variability in PPO molecular weight estimates derives from a technique employing nondenatured or partially denatured PPOs in SDS-PAGE activity gels.[15,41] This analysis is useful for visualizing PPO activity but molecular weight estimations based on SDS-PAGE of such samples can be misleading. If the protein is not fully denatured and saturated with SDS, the relationship between its apparent molecular weight and mobility on a gel is tenuous, as is its relationship to either partially or fully denatured molecular weight standards.[42] Partial denaturation or failure to denature PPO may explain the varied and erratic migration of PPOs reported from many plant species under such electrophoretic conditions. When purified native *V. faba* PPO (which migrates at an M_r of 43) was incubated with varying levels of dithiothreitol, it was converted to the 60 kDa form evident in fully denatured samples.[13] This strongly indicates the presence of disulfide bonding in native *V. faba* PPO and stresses the need for full denaturation when making molecular weight determinations.

An additional point of confusion stems from the fact that several PPOs[13,26] appear to possess protease-sensitive sites at the carboxy-terminal end of the mature peptide. Proteolysis usually yields a 40-45 kDa active peptide and an 18-20 kDa inactive carboxy-terminal peptide. Unfortunately, both intact PPO subjected to the partial denaturation procedure of Angleton and Flurkey,[41] and proteolytically-cleaved fully-denatured PPO, migrate at the same apparent molecular weight in SDS-PAGE, adding to the confusion concerning the structural definition of plant PPOs.

Newman *et al.*[43] recently reported the cloning of the tomato PPO gene family. Slight variations in M_r heterogeneity of the native enzyme could be attributed to expression of different members of a gene family, and could explain the discrete doublet in the range of 57 to 60 kDa which is observed in fully denatured leaf or trichome PPOs from several species.[13,30,44-47]

PPO GENES

In contrast to the wide variability reported for biochemical estimates of PPO size, all PPO genes cloned to date encode 57-62 kDa mature peptides with 8 to 11 kDa putative transit peptides.[11,13,43-45,48] The sequence of putative transit peptides encoded by each cDNA is consistent with localization in the thylakoid lumen (see below). PPO genes have been characterized from tomato,[43,45,48] potato,[44] broad bean[11,13] and grape.[13] Two full-length cDNAs selected by immunoscreening a library constructed from mature *V. faba* leaves encode putative precursor proteins of 68 kDa and mature proteins of 58 kD.[11] Two classes of PPO cDNA were recovered from a tomato epidermis cDNA library screened using polyclonal anti-*Solanum berthaultii* trichome PPO.[48] Internal amino acid sequences of the purified *S. berthaultii* 59 kDa PPO, and N-terminal amino acid sequence of the mature 59 kDa tomato PPOs confirmed the identity of the cDNA clones (Yu *et al.*, in preparation).[47,48] Both cDNAs encoded precursor proteins of *ca.* 67 kDa. Partial peptide mapping of the 67 kDa translation product translated from the transcribed cDNA and mature N-terminal amino acid sequencing demonstrated that the 67 kDa product is a precursor of the 59 kDa mature PPO.[47,48] Complete sequencing of a cDNA initially presumed to represent a 45 kDa PPO revealed that it represented a slightly truncated PPO gene, and its genomic counterpart showed the similar structure of the transit peptide associated with all other PPOs, giving a mature M_r of *ca.* 57,000 for this gene.[43,48] Shahar *et al.*[45] cloned a PPO cDNA from tomato flowers and a genomic clone that appears to encode different members of the PPO gene family, and possesses structural features similar to those described above. All tomato cDNAs hybridize to bands of *ca.* 2.0 Kb on northern blots of tomato leaf, flower, meristem, and epidermal mRNA. [45,46] Two PPO cDNAs were also isolated from potato (*Solanum tuberosum*) leaf by screening with tomato cDNAs.[44] As in tomato, both PPO cDNAs appear to encode polypeptides that are processed to a mature molecular weight of 57,000, and both cDNAs, when hybridized to total leaf RNA, identified a 2 kb transcript class.

Organization of the Tomato PPO Gene Family

Seven nuclear PPO genes were isolated from a λ Charon 35 tomato genomic library. Restriction fragment length polymorphism mapping of these genes on a highly saturated tomato map demonstrated that all seven genes map to the same locus on chromosome 8. The seven genomic clones (PPOs A, A', B, C, D, E, and F) can be differentiated into three classes by restriction mapping. Phage insert mapping demonstrated PPO E and PPO F (both class III), and PPOs B, D, and A (classes I, II, and I respectively) are grouped within separate 12.4 kb clusters. The remaining two genes were isolated on separate phages. Pulsed field gel electrophoresis indicated that the entire PPO gene cluster may be as small as 165 kb. The complete nucleotide sequence was determined for each gene. Comparison to full-length tomato PPO cDNAs revealed that all seven PPO genes lack introns, in agreement with the observation[45] that PPO E and F appeared to lack introns. A transcript of about 2 kb is expected for each PPO. Each PPO encodes a putative transit peptide characteristic of polypeptides targeted to the thylakoid lumen. Predicted precursor polypeptides range in mass from 66 to 71 kDa and predicted mature polypeptides range from 57 to 62 kDa. All the PPOs encode the CuA and CuB putative copper binding sites characteristic of bacterial, fungal, and mammalian tyrosinases. Five of the seven PPOs possess divergent DNA sequences in their 5' promoter regions. PPO A and A' possess identical sequences but their restriction maps outside the sequence regions differ, leading to the conclusion that they are probably distinct genes.

PPOs C and D possess identical 5' flanking sequences, although their coding and 3' flanking sequences differ. These flanking sequence differences may be responsible for regulating the differential expression of PPO genes.[43]

EXPRESSION OF PPO

PPO is ubiquitous among higher plants and is localized in plastids in both photosynthetic and nonphotosynthetic tissues. It is often abundant in leaves, tubers, storage roots, floral parts and fruits. The abundance of PPO in tubers and fruits at early stages of development along with high levels of phenolic substrates has led to suggestions of a possible role in making the unripe fruit and storage organs inedible to predators.[49] Until recent years, little documentation was available on the temporal and spatial expression of PPO genes in various tissues and organs. Antibodies and activity assays allowed characterization of PPO protein[30,50,51] and *in vitro* translation of plant mRNA permitted detection of functional PPO transcripts,[28,52] but detailed studies of gene expression were not possible. Now that genes have been cloned from many plant species,[11,13,43-45,48] a new phase of PPO research is at hand. Molecular probes are now available to document the temporal and spatial expression of PPOs in plant organs and tissues. The expression patterns determined for PPO genes could provide an insight into the function(s) of PPO in plants and, furthermore, aid in the production of improved fruit and vegetable products.

RNA blot analysis of *V. faba* leaf and grape berry RNA reveals a 2.2 kb transcript class, slightly larger than the 2 kb transcript class size reported in tomato[45,48] and potato.[44] In tomato and potato leaves and flowers, and grape berries, PPO transcripts are present predominantly at early developmental stages.[26,44,45] Hunt *et al.*[44] found that in potato a 2 kb PPO transcript class is present at high levels in young leaves and young whole petiole, and at low levels in mature roots, young petiole epidermis and fully open flowers. No PPO message was detected in mature tuber periderm. In potato, PPO appears to be a very stable protein which is synthesized early in development. Potato leaf PPO mRNA is developmentally regulated and only detectable in young foliage (apical leaf nodes 1-3). In contrast, the protein profile of immunologically detectable PPO remains fairly constant from the apical node through the eleventh leaf node. *V. faba* , however, exhibits a different expression pattern than grape, tomato and potato, with PPO transcripts abundant in mature leaves.[11] This is consistent with an earlier report that translatable *V. faba* PPO mRNA is present throughout leaf development.[28]

As mentioned previously, potato PPO transcript is not detectable below the third apical leaf node. However, PPO protein appears to be quite stable once synthesized (Fig.5, upper set). In preliminary analyses of tomato and tobacco, PPO mRNA is not detectable below the first leaf node, suggesting that the developmental expression of PPO mRNA in these species is more restricted than in potato (Hunt et al. unpublished data). Furthermore, the amount of detectable PPO protein, and PPO activity, diminishes basipetally from the apex (Fig. 5, lower set). The differential transcriptional and posttranscriptional regulation of PPO as well as varied PPO turnover rates among these Solanaceous species could reflect different functions for PPOs in these species.

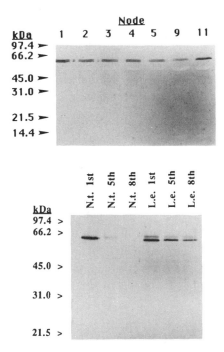

Fig. 5. Nodal analysis of potato (upper), tobacco and tomato (lower) PPO. Total protein was extracted from various nodes beginning at the apex. 10 μg of protein was fractionated by SDS-PAGE. Upper: set: Nodes 1-5, 9 and 11 were analyzed from *Solanum tuberosum* cv. Atlantic. Lower set: N.t.=*Nicotiana tabacum* cv. Petit Havana and L.e.=*Lycopersicon esculentum* cv. Moneymaker.

We are exploiting sequence differences among the PPO gene family members to make gene-specific probes capable of distinguishing the temporal and spatial expression of each PPO gene. Although there are seven individual PPO genes in tomato, PPOs A and A' possess identical promoters, as do PPOs C and D. Therefore we expect differential expression of only 5 PPO genes. We are currently using these probes in *in situ* hybridization experiments to provide a comprehensive description of PPO expression (Joel *et al.*, in preparation). A series of chimeric genes have also been constructed, consisting of putative 5' regulatory regions of the tomato PPO genes fused to β-glucuronidase (GUS)[53-55] to determine the temporal and spatial patterns of expression for each tomato PPO gene.

Inducibility of PPO activity by herbivores or pathogens has been demonstrated in several species[56,57] as has the increased activity of PPO during senescence or deterioration of organelles.[57] These studies have not determined whether the observed increases in enzymic activity were due to an increase in the amount of PPO or were due instead to the loss of PPO latency.[57] We have shown that PPO activity is inducible severalfold by biotic or abiotic wounding. Immunoblotting and ELISA show that inducible activity is accompanied by comparable increases in PPO protein, and northern blots show that PPO-specific steady-state mRNA level increases as well. In potato, when leaflets at node 8-9 are wounded, PPO induction is only detectable in apical leaf nodes 1-4. Interestingly, the apical leaf nodes are the only leaf tissues in which PPO mRNA is detectable in unwounded potato (Tantasawat *et al.*, in preparation).[44] Although PPO exhibits systemic induction (induction of PPO accumulation in tissues remote from the site of wounding), the developmental pattern of PPO expression remains dominant, and increased PPO accumulation only occurs in those tissues which are already competent to express PPO as judged by their possession of steady-state PPO mRNA.

IMPORT AND PROCESSING OF PPO BY PLASTIDS

PPO has been detected in root plastids, potato amyloplasts, leucoplasts, etioplasts and chromoplasts, as well as in plastid-like particles isolated from etiolated sugar beet leaves.[58] Most studies indicate that PPO is membrane-bound in plastids of non-senescing tissues.[3,57,58] Vaughn and Duke,[24] Vaughn *et al.*[59] and Henry *et al.*[60] suggested that the enzyme is located exclusively in plastids and is only released to the cytosol upon wounding, senescence or deterioration of the organelle. So far, no PPO genes encoding a non-plastidic enzyme have been isolated. Although the cytosolic localization of some PPOs may be authentic,[4,57] other reports probably represent lumenal PPOs released by grinding the tissue, or due to disintegration of plastids during senescence or fruit ripening.[3,4,49,59] Although PPO is encoded by nuclear genes,[61,62] relatively little is known about its targeting and import into organelles. Ultrastructural observations have suggested a vesicular mechanism for transport of PPO into plastids,[59] but at present there is little additional evidence to support this suggestion. Deduced amino acid sequences of PPO transit peptides suggest that PPO import follows pathways used by other thylakoid-localized proteins.

The subcellular localization of *Solanum berthaultii* trichome PPO was revealed in ultrastructural studies carried out in collaboration with Dr. K.C. Vaughn (USDA-ARS) (in preparation). In addition to being present in glandular trichomes, 59 kDa PPO can be observed in the outermost layer of leaf epidermal cells. Guard cells of stomata do not express this enzyme. Cytochemical (dihydroxyphenylalanine oxidase activity) and immunogold labelling further demonstrated that 59 kDa PPO is confined to an unusual protein body in leucoplasts of both outer epidermal and trichomal cells. Within the epidermal/trichomal leucoplast, PPO appeared to reside in a soluble form within a modified thylakoid lumen. In mesophyll cells (which possess only low levels of PPO) PPO appeared associated with the thylakoid membrane (Vaughn, Kowalski and Steffens, in preparation).[47] These studies indicate that the products of different PPO genes are localized in plastids in two different ways. Tomato PPOs B and E, which are expressed at high levels accumulate as the dominant constituent of protein bodies within the lumen of chlorenchyma and epidermal plastids lacking thylakoid organization typical of photosynthetic plastids[48]. In contrast, PPOs expressed at lower levels in photosynthetic plastids typically are localized on, or proximal to, the thylakoid membrane.

In contrast to the 59 kDa leucoplast-localized PPOs of trichome and outer epidermal cells, leaf chloroplast PPOs have often been referred to as a different class of proteins. While the trichomal PPO of leucoplasts is freely soluble upon cell disruption,[30] leaf chloroplast PPOs are tightly membrane-associated. These membranes are usually described as thylakoids, although association with plastid envelopes has also been reported.[57,59,69] The frequent observation of 45 kDa thylakoid PPOs, and reports that leaf PPO is translated *in vitro* at the same M_r as the native enzyme added to the belief that leaf chloroplast PPOs are considerably different from those in glandular trichomes, which are translated at 67 kDa and processed to 59 kDa.[47,52,63] The apparent lack of a cleavable transit peptide and observations of PPO-containing vesicles adherent to the inner-envelope membrane in chloroplasts of tentoxin-treated leaves further suggested that PPO is routed into plastids by a unique import mechanism.[59]

However, as described above, all higher plant PPO genes characterized so far encode precursor polypeptides of a similar molecular weight, 66-71 kDa, giving rise to 57-62 kDa mature proteins. Each gene possesses a putative plastidic transit peptide, 83-87 amino acid residues long and typical of lumen proteins.[43] Furthermore, the sequence of the putative transit peptides of *V. faba*, tomato, and potato suggests a common pattern of routing within the chloroplast.[11,13,43-45,48] The characterization of the tomato PPO gene family[43] presented an opportunity to study the pattern of PPO import into plastids. Since tomato appears to possess both soluble and membrane-bound PPOs in plastids, potential differences between these two presumptive classes of PPO may be reflected in their pattern of expression or in the routing of their gene products. We are currently using information derived from the characterization of the tomato PPO gene family to investigate the possible existence of diverse routes for import and processing of these proteins into plastids,[59] as well as the causes of PPO heterogeneity *in vivo*.

The predicted amino terminal portions of the tomato PPO transit peptides are rich in hydroxy amino acid residues, typical of the stromal targeting domain of nuclear-coded plastid protein precursors.[64] The hydrophobic domain that follows is preceded by two Arg residues, typical of the lumen targeting domain of such proteins (Fig. 6A and B).[64] With respect to the mature protein sequence, only three of the seven genes (PPO A, A' and C) have hydrophobic C-terminal α-helical domains which could conceivably confer membrane association or anchoring.[43]

We have subcloned the tomato PPO-B cDNA, Hy-19[43,48] into pGEM-7Z f(+). PPO B was transcribed *in vitro* using SP6 RNA polymerase and the mRNA obtained was translated with reticulocyte lysate in the presence of ^{35}S-methionine. The 67 kDa translation product obtained was used to study the import, targeting and processing of PPO by isolated plastids of tomato, pea, maize and mangold (*Beta vulgaris*) chloroplasts (Sommer, Ne'eman, Steffens, Mayer and Harel, in press). The size and localization of the predominant processed form obtained depended on the plant species, growth conditions and developmental stage of the leaves from which plastids were isolated. In both pea and tomato chloroplasts, PPO was targeted to the lumen, as expected from the deduced sequence of the transit peptide (Fig. 7). Routing took place in two steps. The precursor was first imported by an ATP-dependent process into the stroma, where its N-terminal portion was removed by a stromal peptidase. The resulting *ca.* 62 kDa intermediate continued to the thylakoid lumen in a light-dependent

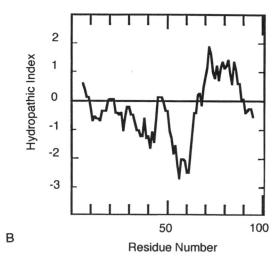

Fig. 6. Amino acid sequence and hydrophobicity plot of the transit peptide of PPO B. (A.) Amino acid sequence of the transit peptide. The arrows indicate the putative first processing site leading to formation of the 62 kD intermediate and the final processing site, deduced from N-terminal sequencing of the native protein.[48] Hydroxy (') and charged amino acid residues and the hydrophobic domain in the lumen targeting part of the transit peptide (underlined) are indicated. (B) Kyte-Doolittle hydrophobicity plot of the first 100 amino acid residues of pPPO B.

reaction. The second processing step, which produces the mature 59 kDa polypeptide, occurred only in the presence of thylakoids.

Tomato chloroplasts processed the PPO precursor completely to its mature 59 kDa form. Pea plastids accumulated both the 62 kDa intermediate and the mature 59 kDa forms (Fig. 7). The relative amounts of the two forms depended on leaf age, illumination regime and mineral nutrition. In maize leaf etiochloroplasts, only the intermediate 62 kDa stromal form was observed. It is

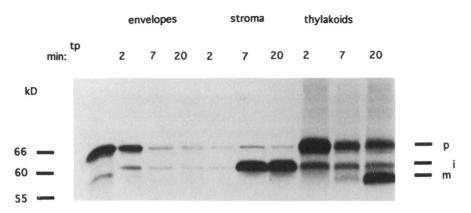

Fig. 7. Import and processing of the precursor of PPO B by isolated chloroplasts. Intact pea chloroplasts were incubated with ^{35}S-labeled precursor of PPO B for 2, 7 or 20 min in light. Intact plastids were re-isolated at the end of each incubation period and fractionated into envelopes, stroma and thylakoids. Some cross-contamination of envelopes and thylakoids is evident. Precursor (p), intermediate (i) and mature (m) forms of PPO are indicated. tp - translation products. The large subunit of Rubisco, GroEL and bovine serum albumin were used as molecular mass markers in the 55 to 66 kD range.

not known whether the precursor or processed forms have enzymatic activity. This question is being investigated using tomato PPOs produced by overexpression in *E. coli*.

When PPO was transiently expressed in *Petunia* protoplasts using a pT7-W-GUS-A50 vector in which the GUS gene was replaced by PPO B, almost all the precursor molecules were processed, with the soluble 62 kDa form predominating (S. Broido, A. Sommer, A. Vainstein, E. Harel and A. Loyter, unpublished observations). In contrast, a *Lemna* LHCPII precursor expressed in the same system was fully processed and bound to the membrane. Taken together, these results suggest that one source of the heterogeneity for which PPO is notorious could, in certain systems, arise from partial processing of the precursor. Partial processing of precursors having a lumen protein transit peptide has been reported previously. For example, biosynthetic threonine deaminase from tomato appears to be a lumen protein judging by the structure of its transit peptide. However, the enzyme isolated from floral tissues had the C-terminal portion of its transit peptide intact, as if only partial processing took place. The protein was soluble and apparently resided in the stroma.[65] Halpin et al.[66] observed that castor bean leucoplasts imported the precursor of plastocyanin but processed it only to the intermediate stromal form. The partial processing was ascribed to lack of a thylakoid protein transport and maturation system.

We have partially purified the stromal processing peptidase that converts the PPO precursor to the 62 kDa intermediate. The stromal peptidase also processed precursors of the Rubisco small subunit and subunit II of PSI to their mature forms, and that of plastocyanin to its intermediate. However, it was unable to process the precursors of intrinsic thylakoid proteins such as the light-harvesting chlorophyll a/b protein I and II. We encountered difficulties in following the second processing step in vitro with isolated thylakoids. Preliminary observations suggest that the enzyme responsible is membrane bound and that processing takes place on the stromal face of the thylakoid and requires unfolding of the 62 KDa intermediate.

Treatment of leaves with tentoxin, a toxin of fungal origin which inhibits CF1, causes loss of plastidic PPO.[67,68] After tentoxin treatment, immunologically recognizable PPO remained associated with the plastid envelope and it was suggested that tentoxin prevents the import and processing of PPO into an active form.[69] We have confirmed this result in in vitro uptake experiments with tomato PPO. Low concentrations of tentoxin (10 μM) prevented import of tomato PPO precursor into isolated plastids. This effect is specific to PPO and the precursor accumulated at the plastid envelope.

Effect of Cu on PPO Import and Processing

The deduced amino acid sequence of all PPO genes isolated so far contains two highly conserved putative copper-binding domains (see above).[11,43-45,48] These are highly homologous to the copper binding domains of fungal, bacterial and mammalian tyrosinases.[5] Outside these domains there is little homology between tyrosinases and plant PPOs. We have determined that Cu^{2+} is not required for the import, targeting or processing of PPO. However, Cu^{2+} (1-5 μM) strongly and specifically inhibits PPO import. Import to the stroma was completely inhibited by 5 μM Cu^{2+}, while import of plastocyanin or LHCP precursors was unaffected. Inhibition was prevented by addition of

EDTA prior to Cu^{2+}. Cyanide, which removes Cu^{2+} from active PPO,[70] relieved Cu^{2+} inhibition of pPPO import. Cu^{2+} has no effect on the import and processing of plastocyanin by isolated chloroplasts[64,71] but was suggested to regulate plastocyanin import in *Chlamydomonas*.[72] Binding of Cu^{2+} to pPPO may prevent unfolding of the precursor which is required for traversing the plastid envelope and possibly also for crossing the thylakoid membrane. We are currently examining this possibility using a series of PPO B mutants in which His residues in the putative Cu binding domains[8] have been replaced (Hunt, Sommer, Steffens and Harel, unpublished data).

Thylakoid-bound PPO

The processed form of Hy-19 (PPO B) does not possess an α-helical domain capable of anchoring the protein in the thylakoid membrane.[43] It is not surprising, therefore, that the mature protein was found in the lumen following import into isolated organelles. PPO A, which possesses hydrophobic α-helical C-terminal domains, also accumulated in the lumen of pea chloroplasts, but a significant proportion remained bound to the thylakoids after rupture by a Yeda press. It appears that association with the thylakoid membrane takes place after conversion of the 62 kDa to the mature 59 kDa form. The mechanism of anchoring to thylakoid membranes is unclear. Post-targeting events could stabilize weak membrane association provided by the C-terminal hydrophobic domain of PPOs A/A' and C. It should be noted, however, that due to the high reactivity of PPO reaction products, binding of the enzyme to thylakoids can also occur artifactually.

Our investigations on tomato PPO import and processing have primarily utilized pea plastids, with the assumption that targeting in a homologous system composed of tomato plastids would proceed in a similar manner. However, in contrast to the results of *in vitro* uptake studies with pea chloroplasts, a significant portion of the tomato 59 kDa PPO remains bound to thylakoids after treatment of tomato leaf thylakoids by sonication, detergents or rupture in a Yeda press. This result suggests the possibility that PPO may be targeted differently in tomato and pea chloroplasts. As we do not currently possess antibodies capable of discriminating among the various PPOs, we are unable to determine whether different PPOs reside in the the lumen or thylakoid fractions or whether plastids originating in different cell types differ in the localization of their PPOs.

FUNCTION OF PPO

Despite intense study, biochemical and physiological studies have provided few answers to the question of PPO function. In addition, the difficulty of obtaining PPO-null plants has thus far minimized the contribution of genetics to understanding the function and expression of these enzymes. A variety of hypotheses concerning the function of PPO have been proposed since the first recognition of its activity in 1895. These proposed functions relate to the oxygen reduction activity of PPO as well as to its ability to oxidize phenolics to quinones. These hypotheses must take into account the fact that PPO ordinarily resides in a subcellular compartment which does not contain phenolic substrates, and that the K_m of PPO for O_2 is far higher than the concentration of O_2 in air-saturated water. PPO has been proposed to be involved in buffering of plastid oxygen levels, biosynthesis of phenolics, wound healing, and anti-nutritive modification of plant proteins to discourage herbivory.[2,4,24,49,60]

Based on its thylakoid membrane association and high K_m for O_2, mesophyll chloroplast PPO has been proposed to function in pseudocyclic photophosphorylation, or the Mehler reaction (ATP production with oxygen as the terminal electron acceptor rather than $NADP^+$), and regulation of plastidic oxygen levels.[12,24,60] There are several problems with such hypotheses. First, there is no evidence for a suitable PPO substrate in this compartment which could allow a PPO-based oxygen reduction cycle to operate. Furthermore, the Mehler reaction occurs on the stromal side of PSI, which is localized primarily in stromal thylakoids. Most evidence places PPO either in the lumen or associated with the lumenal face of the thylakoid membranes (see above) and predominantly associated with PSII, which is present at a higher density in granal thylakoids.[50] Nevertheless, the cofractionation of V. faba PPO with PSII,[50] the observation that albino leaf sectors of a variegated grape mutant ("Bruce's Sport") possess decreased PPO activity,[26] and the fact that all plastid PPOs analyzed so far seem to be routed to the lumenal side of the thylakoid membrane (with at least part of the mature molecules migrating to the grana, Sommer, Ne'eman, Steffens, Mayer and Harel, unpublished observations), the site where oxygen is photosynthetically evolved, suggest that there may be a role for PPO in some aspect of chloroplast metabolism.

On the other hand, the conspicuous enzymic browning generated by PPO in a multitude of plant-pest interactions and its wound inducibility (see above) have led to the view that PPO's primary role is in plant defense. According to this view, sequestration of PPO within the thylakoid prevents its interaction with phenolics until the cell is disrupted by herbivores, pathogens, senescence, or other injury. The quinones thus generated by PPO activity on phenolics cross-link with themselves, proteins, and other nucleophilic cellular constituents. In the case of glandular trichomes, this crosslinking leads to polymerization of trichome exudate. In the case of leaf tissue, covalent modification of protein by quinones may lead to reduced nutritive quality of the plant protein. We will first consider the role of PPO in glandular trichomes.

PPO in Glandular Trichomes

In comparison to many questions of PPO function, the role of PPO in polymerization of trichome exudate and entrapment of small-bodied insects is relatively well accepted.[30,73-76] Type A or Type VI trichomes of *Solanum* or *Lycopersicon* species, respectively, are tetralobulate organs approximately 65 microns in diameter which are found at high densities on the foliar epidermis of many Solanaceous species. These trichomes entrap small-bodied insect pests such as aphids and leafhoppers by oxidative polymerization of trichome exudate. Contact of Type A or Type VI trichomes by an insect causes the membrane-enclosed head of the trichome to rupture, coating the legs and mouthparts of the insect with the trichome contents. Immediately, this coating begins to polymerize, brown and harden. As the insect breaks additional trichomes, the dark polymerized accretions on the tarsi and mouthparts increase dramatically, disrupting insect feeding by restricting movement, occluding the mouthparts, or entrapping the insect on the leaf surface.[73-78] This polymerization process is oxygen-dependent[78] and presumed to be initiated by PPO. PPO is highly abundant in glandular trichomes which undergo oxidative polymerization upon breakage,[30,79,80] constituting 50-70 % of total trichome protein, or *ca.* 1 ng PPO/trichome.

The contribution of peroxidase to oxidative polymerization of trichome exudate remains unclear. Peroxidase is found only at low levels in trichomes,[30] although in the presence of exogenous H_2O_2, peroxidase activity in tomato and potato glandular trichome preparations can be substantial.[30,33,74,75] At present there is no evidence for an endogenous H_2O_2-generating system in glandular trichomes. At the same time, conclusive evidence that PPO is primarily responsible for oxidative polymerization of trichome exudate is lacking. In addition to resistance to small-bodied insects, the PPO- containing type A trichomes of the wild potato, *Solanum berthaultii*, are also implicated in the resistance of this species to Colorado potato beetle (*Leptinotarsa decemlineata*).[73,81-83] However it is not clear whether the contribution of the type A trichome to resistance is based on its content of PPO or on other factors which have yet to be elucidated. Insect resistant, horticulturally-adapted hybrids of *S. berthaultii* and *S. tuberosum* bearing high densities of PPO-containing type A glandular trichome have recently been released.[81]

PPO and Antinutritive Defenses

A second possible role played by PPO in leaves is, upon wounding, to generate quinones which by reacting with the nucleophilic amino acids of protein, act to reduce the digestibility, palatability, and nutritive value of the plant tissue to other organisms.[2,4,23,49] The action of PPO in this role has been most thoroughly explored by Felton *et al.*.[23] Their model suggests that the primary targets of quinones formed by PPO are the nucleophilic amino acids His, Cys, Met, Trp, and Lys, whose low abundance limits insect growth on plant diets.[84,85] Covalent modification of these essential amino acids further decreases their nutritional availability to herbivores and may result in poorer herbivore performance.[23] In support of this hypothesis, Duffey *et al.*[2] showed that relative growth rates of *Spodoptera exigua* (beet armyworm) and *Heliothis zea* (tomato fruitworm) on artificial diet were significantly decreased when the protein source was previously alkylated by incubation in media containing chlorogenic acid and polyphenol oxidase. Similarly, when the protein component of artificial diet was composed of tomato leaf protein from homogenates in which polyphenol oxidase activity was not inhibited, relative growth rates of both *S. exigua* and *H. zea* were significantly decreased relative to insects reared on diets in which tomato leaf protein was isolated in the presence of PPO inhibitors. In addition, 38 to 49% of ^{3}H-chlorogenic acid applied to leaflets prior to consumption by herbivores could not be dissociated from protein components of feces, suggesting that chlorogenoquinone alkylation of protein may be significant *in vivo*.[23]

In contrast to results obtained with *S. exigua* and *H. zea*, Felton *et al.*[56] suggest that a lower amount of quinone cross-linking to dietary protein occurs when PPO-containing diet is consumed by late-instar Colorado potato beetle (*Leptinotarsa decemlineata*). Lower rates of alkylation by quinones is expected due to lower gut pH of colepterans (pH 4.5-6.0) relative to that of lepidopterans (pH 8-9). At lower pH, the equilibrium of amine and sulfhydryl substituents of proteins (with pKa's above 8) lies in the direction of protonation and they are therefore less readily attacked by electrophilic quinones. In the studies of Felton *et al.*,[6] PPO-modified protein was supplied to diet at a single rate, and protein alkylation (covalent binding of ^{3}H-chlorogenic acid to fecal protein) was compared only to the level found in *H. zea*. It remains possible that increased quinone production as a result of increased dietary PPO could lead to significantly increased alkylation of dietary protein in the *L. decemlineata* gut (or alkylation of the cysteine proteinases on which *L. decemlineata* digestion depends)[87] It may then be possible to demonstrate an effect of PPO on growth or developmental rates of *L. decemlineata*.

MODIFICATION OF PPO EXPRESSION

PPO provides an attractive target for modification for several reasons. First, PPO is responsible for significant decreases in post-harvest quality of many crops due to melanin formation associated with injury or senescence. Downregulation of PPO could significantly increase quality attributes of a number of crop commodities. On the other hand, overexpression of PPO may potentially minimize pest damage to crop plants either by antinutritive effects on plant protein or by its ability to initiate polymerization leading to entrapment of insects in trichome exudate. Any attempt to modify PPO expression either to down-regulate or overexpress faces the possibility that PPO possesses vital functions which have heretofore been overlooked. Our current understanding is that PPO resides in an inactive state on the thylakoid membrane or in the thylakoid lumen until the cell is disrupted, leading to oxidation of vacuole-localized phenols by PPO. However, as outlined above, a number of hypotheses suggest that PPO may possess functions unrelated to phenolic oxidation, particularly with regard to photosynthesis and regulation of plastidic oxygen levels. PPO mutants, such as those generated by antisense RNA downregulation or sense overexpression of PPO genes, may exhibit enhanced or inhibited PPO gene expression. In addition to testing the role of PPO in antinutritive resistance to herbivores, transgenic plants exhibiting modified PPO expression will provide a novel means to rigorously test other hypotheses of *in vivo* PPO function as well.[3,4,24,49,60] For example, if chloroplast PPO has a role in O_2 buffering (oxygen reduction during pseudocyclic photophosphorylation),[24,60] under high light intensity regimes the antisense transgenic should exhibit chlorophyll photobleaching, whereas control plants should not express these symptoms. Conversely, PPO overproducers may exhibit increased tolerance to photooxidative conditions. If PPO does possess an important (and heretofore unrecognized) function, it is conceivable that one consequence of altered chloroplast PPO expression could be severe abnormality or lethality. While these would undoubtably prove useful for understanding the function of this enzyme, they would also limit our ability to manipulate PPO expression for particular purposes. Thus, a lack of an obvious function for most PPOs may represent a blessing or a curse for those interested in modifying PPO expression, as a range of unexpected phenotypes could arise if critical but overlooked roles for PPO exist whose disruption could lead to decreased plant fitness or lethality.

Our initial attempts to downregulate PPO expression have faced the question of whether antisense expression of a single PPO could effectively inhibit the entire PPO gene family. We have been able to downregulate PPO expression in tobacco and tomato via antisense RNA expression (Hunt *et al* . in preparation). In both cases, the potato leaf PPO cDNA PPO-P1[44] was used in constructs employing the 35S promoter of CaMV. PPO-P1 was used because it possesses high nucleic acid similarity to all members of the tomato gene family (96 to 70% nucleic acid identity within the family) and was therefore thought to have a greater chance of inhibiting expression of the entire gene family than any single tomato gene. Since the mechanism of antisense inhibition remains unclear, and because the sequences (5' or 3'; coding or non-coding) providing most effective inhibition appear variable and system-dependent,[87-89] the entire cDNA was used to ensure the presence of all important sequences. Foliar PPO levels in transformants ranged from normal to undetectable PPO activity. Immunoblots confirmed the absence of PPO in plants exhibiting undetectable PPO activity. It appears, based on preliminary data, that antisense expression of a single PPO, driven by the CaMV 35S promoter is sufficient to inhibit expression of all members of the PPO gene family in tomato.

Both down-regulated and PPO-null tomato plants grew at rates similar to control plants. They appeared morphologically similar to controls, and flowered and set seed normally. Under greenhouse conditions, the absence of PPO does not appear to affect leaf size or plant vigor. Therefore, it appears that PPO, under these growth conditions, does not play a vital role in plant metabolism, growth, or development (Hunt *et al.*, in preparation).

Attempts to over-express PPO were subject to the same question of whether overproduction could be achieved without disrupting essential plant functions, and whether, given the tight regulation of PPO at the mRNA and protein level (see above), over-expression could be achieved. The full length tomato PPO cDNA HY-19 (PPO B)[48] was ligated to the CaMV 35S promoter in the transformation vector pBI121 in a sense orientation. This vector was then used to transform tobacco and tomato plants. Relative to the antisense experiments, far fewer PPO overexpressing plants were recovered. However several individuals each of tobacco and tomato possessed three- to five-fold higher foliar PPO (based on activity and immunoassays). Analysis of tomato and tobacco R_1 generations has not yet been completed.

Our finding that modification of PPO expression does not visibly affect plant performance allows investigation of additional questions. If PPO can be eliminated from a plant, does feeding behavior and/or survival of insect herbivores change? Conversely, do increased amounts of foliar PPO significantly alter insect herbivore performance? And finally, if absence of PPO does not drastically alter plant performance, can this phenotype be utilized as a strategy of minimizing post-harvest losses due to PPO-induced browning? These questions are considered in the following sections.

Effect of Altered PPO Expression on Insect Herbivores

We view these experiments as a critical first step to assess whether PPO expression in host plants affects insect pest performance. Previous studies seeking to establish a link between PPO and pest tolerance have depended upon correlation between species or non-isogenic hybrids differing in PPO activity (and presumably many other factors), feeding studies with artificial diets, or addition of PPO inhibitors to foliage. These approaches have always remained vulnerable to the limitation that they either do not isolate PPO as a single factor differing in the analysis, or do not represent realistic tests of the hypothesis. Therefore, the preferred test of PPO's putative role in plant-pest interactions is to examine insect herbivore performance on plants differing only in the level of PPO expression.

Colorado Potato Beetle

The proposed role of PPO as an antinutritive defense against herbivores depends upon PPO to generate quinones that oxidatively modify (alkylate) protein or amino acid nucleophiles, thereby decreasing the nutritive quality of plant protein and/or the palatability of the plant.[23] Rates of phenolic oxidation upon homogenization and immunoblotting were used to identify transgenic PPO-null and PPO- overexpressing tomato plants. We examined the influence of these plants on growth and development of a population of tomato-adapted Colorado potato beetle (*Leptinotarsa decemlineata*). Larvae were reared on nontransformed control tomato plants, PPO-null, and transgenic tomato plants possessing threefold overexpression of PPO. Plants were an R_0 generation propagated by stem cuttings. Neonate larvae were placed on excised foliage of greenhouse-grown plants and were confined in feeding chambers throughout development. Foliage was replaced in response to consumption. Developmental rate, weight gain, and larval mortality were recorded for eight days. Leaf area consumption was also calculated to determine whether equal amounts of leaf material were consumed on control plants and plants varying for PPO level. Preliminary results indicate that mortality was lowest and weight gain and developmental rate highest on the PPO-null. Mortality was highest, and developmental rate and weight gain lowest, on the PPO- overexpressing plant. Larval performance on the nontransformed control plants was intermediate to the PPO-overexpressing plants and the PPO-null (Hunt, Castenera, Tingey and Steffens, in preparation).

We are currently analyzing R_1 generation PPO transgenics to provide the genetic evidence that the phenotype of altered herbivore performance is linked to the presence and expression of the PPO-containing construct and is not an artifact due to somaclonal variation arising from the tissue culture process. Further experiments with insect herbivores will employ plants homozygous for gene constructs affecting PPO expression. However, these preliminary results clearly suggest that the presence or absence of PPO in plant tissues affects herbivore performance. The improved performance of insect larvae on the PPO-null clearly demonstrates that PPO is an important factor in plant defense. While the three-fold overexpression of PPO in the high PPO plant is modest, the significant impairment of larval growth and development indicates that higher levels of PPO expression may provide even greater protection against herbivory.

Post-harvest Melanization

As discussed above, the economic impacts of PPO-mediated melanization can be severe for many crop commodities. These post-harvest impacts could be ameliorated by downregulation of PPO expression in the selected tissues and organs whose quality is deleteriously affected by PPO. A primary target for downregulation of PPO is the potato. Potatoes are susceptible to a number of post-harvest syndromes (blackheart, blackspot, internal brown spot, net necrosis, internal heat necrosis) in which PPO is responsible for significant crop losses.[90,91] In particular, blackspot bruising is responsible for significant annual yield losses for both tablestock and processed potato products. Blackspot bruise results from physical impacts on the tuber which occur during mechanical harvesting or in post-harvest handling and storage. Localized discoloration initially appears at the sites of impact some hours after the event, changing from pink to black within 24 h. The discoloration is due to formation of melanins, as PPO released from plastids by tissue damage converts tyrosine and chlorogenic acid to quinones that autoxidize to form various dark-colored products. Although susceptibility to blackspot bruising is cultivar-dependent[92,93] and is also partly dependent upon environmental effects,[93-95] in general higher soluble solids content (which results in greater hardness of the tuber) contributes significantly to bruising susceptibility. In addition, cutting of tubers prior to processing often results in melanization which can only be controlled with sulfite applications or heat treatments to inhibit or inactivate PPO.[93]

The attractiveness of prevention of bruising as a target for PPO modification arises from several factors. First, melanization is apparently due exclusively to PPO activity, with little data available to support a role for either laccase or peroxidase. Therefore, downregulation of PPO could be expected to decrease the rate and extent of melanization in response to physical impacts or cutting. Second, tuber melanization (and blackspot bruising in particular) is frequently not associated with visible tissue damage.[96] Therefore the syndrome can be considered essentially a cosmetic one caused by a relatively simple biochemical process. Successful inhibition of tuber PPO could decrease the frequency and severity of bruising or melanization and result in significantly improved post-harvest tuber quality.

In collaboration with the Keygene company, we have explored the consequences of PPO downregulation by antisense in potato. As discussed above, potatoes with high specific gravity are particularly susceptible to blackspot bruising and many of the potato varieties in production in Europe are severely affected. A series of antisense constructs made by Keygene use either the CaMV 35S, patatin, or granule-bound starch synthase promoters to drive expression of a potato PPO cDNA. Minitubers of over 1000 transformants were assayed for PPO activity in a spectrophotometric assay. Of these, some 100 were found to possess low to undetectable levels of PPO activity. The strongest suppression of PPO activity occurred in constructs utilizing the CaMV 35S or granule-bound starch synthase promoters. Patatin-driven antisense PPO expression was in general less effective in inhibiting PPO. Transgenic mini-tubers exhibiting decreased PPO expression were transplanted to the field in summer of 1992 and cultivated and harvested at maturity following commercial growing practices. After harvest and a period of storage, the tubers were subjected to an assay which involves application of a standard impact sufficient to cause bruising in standard cultivars. The results were in agreement with results derived from spectrophotometric PPO assays: plants with low or null PPO activity failed to develop melanin under conditions favoring development of blackspot bruises (Bachem, Hunt, Steffens and Zabeau, in preparation).

Use of PPO Promoters to Modify Glandular Trichome Biochemistry

The unusual expression patterns of PPO offer unique opportunities for cell-specific modification of secondary metabolism. In current work we are characterizing the promoters of each of the tomato gene family members. Up to 2 kb of DNA 5' to the coding sequences has been used to drive expression of the reporter gene β-glucuronidase (GUS). The sequences responsible for expression of PPO E are of particular interest, since the product of this gene accumulates to very high levels in glandular trichomes and epidermal cells.[30,46] *In situ* hybridization indicates that PPO E transcripts are abundant in these cell types. Thus, the PPO E promoter may prove to be useful for the high-level expression of a number of products in glandular trichome and epidermal cells. As host choice by insects often depends on the chemistry of the epidermis, we hope to use the PPO E promoter to deliver genes capable of altering the profile of volatile or nonvolatile glandular trichome constituents, and by so doing to change the host acceptance behavior of insect herbivores toward host plants.

CONCLUSIONS

Transgenic plants varying only in the level of PPO expression provide an experimental platform that allows direct testing of PPO function, and also permits long-term assessments of the efficacy of PPO-based plant defenses on insect herbivore developmental rates, rates of replacement, and in the long term, the impacts of PPO on non-target organisms and durability of PPO-based antinutritive resistance. A major conclusion derived from our transgenic tomato and tobacco plants analyzed to date is that no gross phenotypic effects are manifest when plants are engineered to possess either a five-fold PPO overexpression or when their PPO activity is undetectable. However, this conclusion is based exclusively on the greenhouse performance of these plants, and abnormal phenotypes may become more apparent under field conditions. The challenge of future studies will be to selectively modify PPO expression in specific organs and tissues to minimize its effects on browning of consumed plant parts, while maximizing its expression in other portions of the plant to minimize herbivory.

ACKNOWLEDGEMENTS

We are grateful for research support from the USDA-NRICGP and from BARD, the United States-Israel Binational Agricultural Research and Development Fund. M.D.H is supported by a fellowship from the NSF/DOE/USDA Plant Science Center.

REFERENCES

1.   MATHEW, A.G., PARPIA, H.A.B. 1971. Food browning as a polyphenol reaction. Adv. Food Res. 19: 75-145.

2.   DUFFY, S., FELTON, G. 1991. Enzymatic antinutritive defenses of the tomato plant against insects. In: Naturally Occurring Pest Bioregulators, P. Hedin, ed., American Chemical Society, Washington, D.C., pp.166-197.

3.   MAYER, A.M., HAREL, E. 1991. Phenoloxidases and their significance in fruit and vegetables. In: Food Enzymology, P.F. Fox, ed., Elsevier, New York, pp. 373-398.

4.   MAYER, A.M., HAREL, E. 1979. Polyphenol oxidases in plants. Phytochemistry 18: 193-215.

5.   LERCH, K. 1987. Molecular and active site structure of tyrosinase. Life Chem. Rep. 5: 221-234.

6.   GAYKEMA, W.P.J, HOL, W.G.J., VEREIJKEN, N.M., SOETER, N.M., BAK, H.J., BEINTEMA, J.J. 1984. 3.2 Å structure of the copper-containing, oxygen-carrying protein *Panulirus interruptus* haemocyanin. Nature 309: 23-29.

7.   JACKMAN, M.P., HAJNAL, A., LERCH, K. 1991. Albino mutants of *Streptomyces glaucescens* tyrosinase. Biochem. J. 274:707-713.

8.   HUBER, M. , LERCH, K. 1988. Identification of two histidines as copper ligands in *Streptomyces glaucescens* tyrosinase. Biochemistry 27: 5610-5615.

9.   LANG, W.H., VAN HOLDE, K.E. 1991. Cloning and sequencing of *Octopus dofleini* hemocyanin cDNA: derived sequences of functional units Ode and Odf. Proc. Natl. Acad. Sci. USA 88:244-248.

10. LERCH, K. 1978. Amino acid sequence of tyrosinase from *Neurospora crassa*. Proc. Natl. Acad. Sci. USA 8:3635-3639.

11. CARY, J.M., LAX, A.R., FLURKEY, W.H. 1992. Cloning and characterization of cDNA coding for *Vicia faba* polyphenol oxidase. Plant Mol. Biol. 20: 245-253.

12. TOLBERT, N.E. 1973. Activation of polyphenol oxidase of chloroplasts. Plant Physiol. 51:234-244.

13. ROBINSON, S.P., DRY, I.B. 1992. Broad bean leaf polyphenol oxidase is a 60-kilodalton protein susceptible to proteolytic cleavage. Plant Physiol. 99: 317-323.

14. MOORE, B.M., FLURKEY, W.H. 1990. Sodium dodecyl sulfate activation of a plant polyphenoloxidase. J. Biol. Chem. 265:4982-4988.

15. SANCHEZ-FERRER, A., BRU, R., GARCIA-CARMONA, F. 1989. Novel procedure for extraction of a latent grape polyphenoloxidase using temperature-induced phase separation in Triton X-114. Plant Physiol. 91:1481-1487.

16. GOLDBECK, J.H., CAMMARATA, K.V. 1981. Spinach thylakoid polyphenol oxidase. Plant Physiol. 67:977-984.

17. MEYER, H.U., BIEHL, B. 1980. Activities and multiplicity of phenolase from spinach chloroplasts during leaf aging. Phytochemistry 19:2267-2272.

18. SWAIN, T., MAPSON, L.W., ROBB, D.A. 1966. Activation of *Vicia faba* tyrosinase as effected by denaturing agents. Phytochemistry 5:469-482.

19. KENTEN, R.H. 1957. Latent phenolase in extracts of broad bean (*Vicia faba* L.) leaves. 2. Activation by anionic wetting agents. Biochem. J. 86:244-251.

20. ROBB, D.A., MAPSON, L.W., SWAIN, T. 1964. Activation of the latent tyrosinase of broad bean. Nature 201:503-504.

21. SODERHALL, K., CARLBERG, I., ERIKSSON, T. 1985. Isolation and partial purification of prophenoloxidase from *Daucus carota* L. cell cultures. Plant Physiol. 78:730-733.

22. SODERHALL, I., BERGENSTRAHLE, A., SODERHALL, K. 1990. Purification and some properties of a *Daucus carota* lectin which enhances the activation of prophenoloxidase by $CaCl_2$. Plant Physiol. 93:657-661.

23. FELTON, G., DONATO, K., DEL VECCHIO, R., DUFFEY, S: 1989 Activation of plant foliar oxidases by insect feeding reduces nutritive quality of foliage for noctuid herbivores. J. Chem. Ecol. 15: 2667-2694.

24. VAUGHN, K.C., DUKE, S.O. 1984. Function of polyphenol oxidase in higher plants. Physiol. Plant. 60: 106-112.

25. HUTCHESON, S.W., BUCHANAN, B.B., MONTALBINI, P. 1980. Polyphenol oxidation by *Vicia faba* chloroplast membranes. Plant Physiol. 66:1150-1154.

26. RATHJEN, A.H., ROBINSON, S.P. 1992. Aberrant processing of polyphenol oxidase in variegated grapevine mutant. Plant Physiol. 99: 1619-1625.

27. FLURKEY, W. 1990. Electrophoretic and molecular weight anomalies associated with broad bean polyphenol oxidase in SDS-PAGE electrophoresis. Phytochemistry 29: 387-391.

28. LANKER, T., KING, T., ARNOLD, S., FLURKEY, W. 1987. Active, inactive and *in vitro* synthesized forms of polyphenoloxidase during leaf development. Physiol. Plant 69: 323-329.

29. GANESA, C., FOX, M.T., FLURKEY, W.H. 1992. Microheterogeneity in purified broad bean polyphenol oxidase. Plant Physiol. 98: 472-479.

30. KOWALSKI, S.P., EANNETTA, N.T., HIRZEL, A.T., STEFFENS, J.C. 1992. Purification and characterization of polyphenol oxidase from glandular trichomes of *Solanum berthaultii*. Plant Physiol. 100: 677-684.

31. HAREL, E., MAYER, A.M. 1968. Interconversion of subunits of catechol oxidase from apple chloroplasts. Phytochemistry 7: 199-204.

32. HAREL, E., MAYER, A.M., LEHMAN, E. 1973. Multiple forms of *Vitis vinifera* catechol oxidase. Phytochemistry 12: 2649-2654.

33. BOUTHYETTE, P.Y., EANNETTA, N., HANNIGAN, K.J., GREGORY P. 1987. *Solanum berthaultii* trichomes contain unique polyphenoloxidases and a peroxidase. Phytochemistry 25: 2949-2954.

34. ROBB, D.A., MAPSON, L.W., SWAIN, T. 1965. On the heterogeneity of the tyrosinase of broad bean (*Vicia faba* L.). Phytochemistry 4: 731-740.

35. FLURKEY,W.H., JEN, J.J. 1980. Purification of peach polyphenol oxidase in the presence of added protease inhibitors. J. Food Biochem. 4:29-41.

36. BALASINGHAM, K., FERDINAND, W. 1970. The purification and properties of a ribonucleoenzyme, *o*-diphenol oxidase from potatoes. Biochem. J. 118: 15-23.

37. WESCHE-EBELING, P., MONTGOMERY, M.W. 1990. Strawberry

polyphenoloxidase: purification and characterization. J. Food Sci. 55: 1315-1319.

38. ZAWISTOWSKI, J., BILIADERIS, C.G., MURRAY E.D. 1988. Purification and characterization of Jerusalem artichoke (*Helianthus tuberosus* L.) polyphenol oxidase. J. Food Biochem. 12: 1-22.

39. STELZIG, D.A., AKHTAR, S., RIBEIRO, S. 1972. Catechol oxidase of Red Delicious apple peel. Phytochemistry 11: 535-539.

40. MAIONE, T.E., JAGENDORF, A.T. 1984. Partial deglycosylation of chloroplast coupling factor 1 (CF1) prevents the reconstitution of photophosphorylation. Proc. Natl. Acad. Sci. USA 81:3733-3736.

41. ANGLETON, E.A., FLURKEY, W.H. 1984. Activation and alteration of plant and fungal polyphenoloxidase enzymes in sodium dodecylsulfate electrophoresis. Phytochemistry 23: 2723-2725.

42. SEE, Y.P., JACKOWSKI, G. 1989. Estimating molecular weights of polypeptides by SDS gel electrophoresis. In: Protein Structure, A Practical Approach, T.E. Creighton, ed., IRL Press, Oxford, United Kingdom, pp 1-21.

43. NEWMAN, S.M., EANNETTA, N.T, YU, H., PRINCE, J.P., DE VICENTE, M.C., TANKSLEY, S.D., STEFFENS, J.C. 1993. Organisation of the tomato polyphenol oxidase gene family. Plant Mol. Biol. 21: 1035-1051.

44. HUNT, M.D., EANNETTA, N.T., YU, H., NEWMAN, S.M., STEFFENS, J.C. 1993. cDNA cloning and expression of potato polyphenol oxidase. Plant Mol. Biol. 21: 59-68.

45. SHAHAR, T., HENNIG, N., GUTFINGER, T., HAREVEN, D., LIFSCHITZ, E. 1992. The tomato 66.3-kD polyphenoloxidase gene: Molecular identification and developmental expression. Plant Cell 4: 135-147.

46. YU, H., KOWALSKI, S.P., STEFFENS, J.C. 1992. Comparison of polyphenoloxidase expression in glandular trichomes of *Solanum* and *Lycopersicon*. Plant Physiol. 100: 1885-1890.

47. STEFFENS, J.C., KOWALSKI, S.P., YU, H. 1990. Characterization of glandular trichome and plastid polyphenol oxidases of potato. In: The Molecular and Cellular Biology of the Potato, M. Vayda and W. D. Park, eds., C.A.B. International, Oxon, pp. 103-112.

48. YU, H. 1992. Cloning, characterization and expression of tomato polyphenol oxidases. Ph. D. thesis. Cornell University, New York, 150 pp.

49. MAYER, E.M., HAREL, E. 1981. Polyphenol oxidases in fruits—

changes during ripening. In: Recent Advances in the Biochemistry of Fruit and Vegetables, J. Friend, M.C. Rhodes, eds., Academic Press, New York, pp 161-180.

50.   LAX, A.R., VAUGHN, K.C. 1991. Co-localization of polyphenol oxidase and photosystem II proteins. Plant Physiol. 96: 26-31.

51.   SHERMAN, T.D., VAUGHN, K.C., DUKE, S.O. 1991. A limited survey of the phylogenetic distribution of polyphenol oxidase. Phytochemistry 30: 2499-2506.

52.   FLURKEY, W.H. 1986. Polyphenoloxidase in higher plants: immunological detection and analysis of *in vitro* translation products. Plant Physiol. 81: 614-618.

53.   JEFFERSON, R.A., WILSON, K.J. 1991. The GUS gene fusion system. In: Plant Molecular Biology Manual, S.B. Gelvin, R.A. Schilperoort, and D.P.S. Verma, eds., Kluwer, Boston, pp. B14/1-B14/33.

54.   JEFFERSON, R.A. 1987. Assaying chimeric genes in plants: the GUS gene fusion system. Plant Mol. Biol. Rep. 5: 387-405.

55.   JEFFERSON, R.A., KAVANAGH, T.A., BEVAN, M.W. 1987. GUS fusions: β-glucuronidase as a sensitive and versatile gene fusion marker. EMBO J. 6:3901-3908.

56.   FELTON, G.W., WORKMAN, J., DUFFEY, S.S. 1992. Avoidance of antinutritive plant defense: role of midgut pH in Colorado potato beetle. J. Chem. Ecol. 18: 571-583.

57.   MAYER, A.M. 1987. Polyphenol oxidases in plants - recent progress. Phytochemistry 26: 11-20.

58.   MAYER, A.M. 1964. Factors controlling activity of phenolase in chloroplasts from sugar beet. Israel J. Bot. 13: 74-81.

59.   VAUGHN, K.C., LAX, A.R., DUKE, S.O. 1988. Polyphenol oxidase: The chloroplast oxidase with no established function. Physiol. Plant. 72: 659-665.

60.   HENRY, E.W., DEPOORE, M.J., O'CONNOR, M.N., DEMORROW, J.M. 1981. Sorbitol-disrupted spinach (*Spinacia oleracea* L.) chloroplasts: cytochemical localization of polyphenol oxidase in discontinuous sucrose density gradient fractions. J. Submicrosc. Cytol. 13: 365-371.

61.   LAX, A.R., VAUGHN, K.C., TEMPLETON, G.E. 1984. Nuclear inheritance of polyphenol oxidase in *Nicotiana*. J. Heredity 75: 285-287.

62.   KOWALSKI, S.P., BAMBERG, J., TINGEY, W.M., STEFFENS, J.C.

1990. Insect resistance in the wild potato *Solanum berthaultii*: inheritance of glandular trichome polyphenol oxidase. J. Heredity 81:475-478.

63. FLURKEY, W.H. 1985. *In vitro* biosynthesis of *Vicia faba* polyphenol oxidase. Plant Physiol. 79: 564-567.

64. DE BOER, A.D., WEISBEEK, P.J. 1991. Chloroplast protein topogenesis. Biochim. Biophys. Acta 1071: 221-253.

65. SAMACH, A., HAREVEN, D., GUTFINGER, T., KEN-DROR, S., LIFSCHITZ, E. 1991. Biosynthetic threonine deaminase gene of tomato: Isolation, structure and upregulation in floral organs. Proc. Natl. Acad. Sci. USA 88: 2678-2682.

66. HALPIN, C., MUSGROVE, J.E., LORD, J.M., ROBINSON, C. 1989. Import and processing of proteins by castor bean leucoplasts. FEBS Letters 258: 32-34.

67. STEELE, J. A., UCHYTIL, T.F., DURBIN, R.D., BHATNAGAR, P., RICH, D.H. 1976. Chloroplast coupling factor 1: A species-specific receptor for tentoxin. Proc. Natl. Acad. Sci. USA 73: 2245-2248.

68. VAUGHN, K.C., DUKE, S.O. 1981. Tentoxin-induced loss of plastidic polyphenol oxidase. Physiol. Plant. 53: 421-428.

69. VAUGHN, K.C., DUKE, S.O. 1984. Tentoxin stops the processing of polyphenol oxidase into an active protein. Physiol. Plant. 60: 257-261.

70. KUBOWITZ, F. 1938. Spaltung und Resynthese der Polyphenoloxidase an des Haemocyanins. Biochem. Z. 299: 299-232.

71. HIBINO, T., DE BOER, A.D., WEISBEEK, P.J., TAKABE, T. 1991. Reconstitution of mature plastocyanin from precursor apo-plastocyanin expressed in *Escherichia coli*. Biochim. Biophys. Acta 1058: 107-112.

72. MERCHANT, M., BOGORAD, L. 1986. Regulation by copper of the expression of plastocyanin and cytochrome c552 in *Chlamydomonas reinhardii*. Mol. Cell Biol. 6:462-469.

73. TINGEY, W.M. 1991. Potato glandular trichomes. In: Naturally Occurring Pest Bioregulators, P.A. Hedin, ed., American Chemical Society, Washington, D.C., pp. 126-135.

74. DUFFEY, S.S. 1986. Plant glandular trichomes: their partial role in defense against insects. In: The Plant Surface and Insects, B. E. Juniper and T. R. E. Southwood, eds., Blackwell, Oxford, pp. 157-178.

75. RYAN, J.D., GREGORY, P., TINGEY, W.M. 1983. Glandular

trichomes: enzymatic browning assays for improved selection of resistance to the green peach aphid. Am. Potato J. 60: 861-868.

76. RYAN, J.D., GREGORY, P., TINGEY, W.M. 1982. Phenolic oxidase activities in glandular trichomes of *Solanum berthaultii*. Phytochemistry 21: 1885-1887.

77. GREGORY, P., TINGEY, W.M., AVE, D.A., BOUTHYETTE, P.Y. 1986. Potato glandular trichomes: A physiochemical defense mechanism against insects. In: Natural Resistance of Plants to Pests. Roles of Allelochemicals, M. B. Green and P. A. Hedin, eds., American Chemical Society, Washington, D.C., pp. 160-167.

78. GIBSON, R.W., TURNER, R.H., 1977. Insect-trapping glandular hairs on potato plants. Proc. Natl. Acad. Sci. USA 23: 272-277.

79. KOWALSKI, S.P., PLAISTED, R.L., STEFFENS, J.C. 1993. Immunodetection of polyphenol oxidase in glandular trichomes of *Solanum berthaultii, S. tuberosum* and their hybrids. Am. Potato J. 70: 185-199.

80. KOWALSKI, S.P. 1989. Insect resistance in potato: purification and characterization of a polyphenol oxidase from type A glandular trichomes of *Solanum berthaultii* Hawkes. Ph.D. thesis. Cornell University. 156 pp.

81. PLAISTED, R.L., TINGEY, W.M., STEFFENS, J.C. 1992. The germplasm release of NYL 235-4, a clone with resistance to the Colorado potato beetle. Am. Potato J. 69: 843-846.

82. NEAL, J.J., STEFFENS, J. C., TINGEY, W.M. 1989. Glandular trichomes of *Solanum berthaultii* and resistance to the Colorado potato beetle. Entomol. exp. appl. 59: 133-140.

83. DIMOCK, M.B., TINGEY, W.M. 1988. Host acceptance behavior of Colorado potato beetle larvae influenced by potato glandular trichomes. Physiol. Entomol. 13: 399-406.

84. BRODERICK, B., STRONG, D. 1987. Amino acid nutrition of herbivorous insects and stress to host plants. In: Insect Outbreaks, P. Barbosa and J.C. Schultz, eds, Academic press, N.Y., pp. 347-365.

85. WALDBAUER, G.P. 1968. The consumption and utilization of food by insects. Adv. Insect Physiol. 5: 229-288.

86. WOLFSON, J.L., MURDOCK, L.L. 1987. Suppression of larval Colorado potato beetle growth and development by digestive proteinase inhibitors. Entomol. Exp. Appl. 44: 235-240.

87. CANNON, M., PLATZ, J., O'LEARY, M., SOOKDEO, C., CANNON, F. 1990. Organ-specific modulation of gene expression in transgenic

plants using antisense RNA. Plant Mol. Biol. 15: 39-47.

88. VAN DER KROL, A.R., MUR, L.A., DE LANGE, P., MOL, J.N.M., STUITJE, A.R. 1990. Inhibition of flower pigmentation by antisense CHS genes: promoter and minimal sequence requirements for the antisense effect. Plant Mol. Biol. 14: 457-466.

89. SANDLER, S.J., STAYTON, M., TOWNSEND, J.A., RALSTON, M.C., BEDBROOK, J.R., DUNSMUIR, P. 1988. Inhibition of gene expression in transformed plants by antisense RNA. Plant Mol. Biol. 11: 301-310.

90. HOOKER, W.J. 1987. Tuber diseases. In: Potato Processing, W.F. Talburt and O. Smith, eds., Van Nostrand Reinhold Company, N.Y., pp 149-181.

91. RICH, A.E. 1983. Potato diseases, Academic Press, N.Y., 238 pp.

92. STARK, J.C., CORSINI, D.L., HURLEY, P.J., DWELLE, R.B. 1985. Biochemical characteristics of potato clones differing in blackspot susceptibility. Am. Potato J. 62:657-666.

93. MAPSON, L.W., SWAIN, T., TOMALIN, A.W. 1963. Influence of variety, cultural conditions and temperature of storage on enzymic browning of potato tubers. J. Sci. Fd Agric. 14:673-684.

94. MATHEIS, G. 1987. Polyphenol oxidase and enzymatic browning of potatoes (*Solanum tuberosum*). II. Enzymatic browning and potato constituents. Chem. Mikrobiol. Technol. Lebensm. 11: 33-41.

95. KUNKEL, R., GARDNER, W.H. 1965. Potato tuber hydration and its effect on blackspot of Russet Burbank potatoes in the Columbia River Basin. Am. Potato J. 42:109-124.

96. REEVE, R.M. 1968. Preliminary histological observation on internal blackspot in potatoes. Am. Pot. J. 45: 157-167.

Chapter Twelve

GENETIC REGULATION OF LIGNIN BIOSYNTHESIS AND THE POTENTIAL MODIFICATION OF WOOD BY GENETIC ENGINEERING IN LOBLOLLY PINE

Ronald Sederoff,[1,2,3]* Malcolm Campbell,[1] David O'Malley,[1] and Ross Whetten[1]

Forest Biotechnology Group, Departments of Forestry[1], Genetics[2], and Biochemistry[3], North Carolina State University, Raleigh, NC 27695

*Correspondence should be addressed to R. Sederoff, Department of Forestry, Box 8008, NCSU, Raleigh NC, 27695, USA.

Genetic Engineering of Plant Secondary Metabolism,
Edited by B.E. Ellis *et al.*, Plenum Press, New York, 1994

INTRODUCTION

Lignin, a highly polymeric product of phenylpropanoid metabolism in plants, is one of the most abundant organic materials on the surface of the earth. The evolution of lignin was a key factor in the appearance and radiation of land plants, and lignin still plays important physiological and developmental roles in terrestrial vascular plants.[1] Hemicelluloses and lignin form the embedding matrix of both primary and secondary plant cell walls, reinforcing the cellulose microfibrils and imparting rigidity to the wall.[2,3] Lignin provides strength, flexibility and impermeability to the cell walls. Thus, lignification is an important step in the terminal differentiation of vascular elements into functional water-conducting tissue.[4,5] Lignin is needed to support the weight of the plant and to prevent the compression of vascular tissue that would collapse the cell walls. Lignification is also induced by environmental effects such as pathogen attack or mechanical stress[6,7] and thus plays an important role in resistance to disease and other adverse environmental stresses.

Lignin constitutes about one fourth of the biomass of woody plants.[2] Much of the world's lignin exists as lignocellulose, where lignin is combined with cellulose and hemicelluloses in wood. Wood components play an important role in the carbon cycle and the global ecology through effects on soils, forest ecology and climate. Some of these effects are imparted by lignin which is degraded slowly and persists as organic matter in solids and water for long periods of time.[3]

Wood is an important raw material and a major fraction of wood harvested today in the industrialized sections of the world is processed for pulp and paper. In the manufacture of pulp and paper, lignin is considered to be an undesirable component of wood; therefore, removal of lignin from wood chips by chemicals is a major step in this process. Lignin also limits the digestibility of forage by herbivores and is degraded slowly by microorganisms.[8,9] Lignin content also affects the utilization of plants for energy from biomass.[10] Biomass from plants is primarily lignocellulose. Lignin is energy rich compared with polysaccharides and other wood components.[10,11] High lignin levels are, therefore, more desirable when wood is used as a fuel because of the higher intrinsic energy content. Low lignin content is more desirable when wood is to be used in conversion processes that utilize the carbohydrate content of wood in fermentation.

Due to the reasons outlined above, modification of lignin content or quality has become a focus of research interest for plant genetic engineering. Even small improvements in the digestibility of forage or in the processing of wood for pulp and paper would be valuable because of the large scale of the industries. Also, the potential to modify lignin content for wood used as fuel could be increasingly important around the world.

Is it feasible to modify lignin or other wood properties by genetic engineering? The physical and chemical properties of wood are derived directly from the composition and morphology of the cell walls that are the substance of wood (mature xylem). The composition and morphology of the wood cell wall depend in turn on the processes of biosynthesis and assembly that are directed by specific genes that control macromolecular synthesis and determine structure. Although the general composition of the cell wall is known, we are at an early stage in understanding the mechanisms and the regulation of cell wall biosynthesis. We have only a rudimentary knowledge of the chemical structure of lignin, and incomplete knowledge of the pathway of lignin biosynthesis. We are just beginning to understand the regulation of some of the individual lignin biosynthetic genes, and the interactions of their gene products in the pathway. We are currently limited by our lack of information at every level of regulation of lignin biosynthesis. Our knowledge of the control of lignification at the level of developmental specificity and the induction of lignification by environmental signals is also in the early stages. Finally, little is known of the genetic factors that determine the quantitative variation of lignin and other related wood properties.

In spite of our current state of knowledge, several lines of evidence argue that lignin content and quality could be modified by directed genetic manipulation. Lignin is known to vary in different organisms over a wide range of content and composition.[2,3] Lignin also varies within populations and varies within individuals in composition during development and in response to environmental cues.[6,12] Lignin is modified during some specific disease states,[13,14] and lignin content can be reduced by chemical inhibitors of lignification [15,16] and by mutations afecting biosynthetic enzymes.[17,18]

One prerequisite to directed genetic modification of lignin is a more thorough understanding of its biosynthesis. Our present understanding of the general pathway for the biosynthesis of lignin is based on chemical analysis, enzymology, isotopic tracer analysis, and cell physiology. The history and progress of the studies that elucidated the general lignin biosynthetic pathway have been extensively reviewed. [2,3,19-23] Although the first proposals that coniferyl alcohol and coniferin act as precursors for lignin biosynthesis were made during the last quarter of the 19th century,[24] the steps that follow the synthesis of monolignol precursors for transport, storage, and polymerization that lead to the final lignin polymer *in plantae* have not yet been established.

This problem is confounded by our limited knowledge of lignin structure at the chemical level and by the heterogeneity of lignin at the subcellular level.[19,25]

In contrast with many biological polymers, lignin does not have a simple ordered structure. Monolignol precursors can form bonds at many locations leading to a highly complex structure.[3] It is difficult to demonstrate, but lignin does depart significantly from the random chemical model.[25-29] Terashima concluded from radiotracer analysis that lignin is a highly heterogeneous polymer formed in a regulated manner.[25,27] Additional evidence for lignin heterogeneity has been obtained using NMR, IR, Raman, and UV spectroscopy.[19,20] The nature of the linkages between lignin and other components of the cell wall is also not understood.[30]

Lignin differs in the extent of methoxylation of the monomeric units involved in biosynthesis and consequently in the linkages that are formed during polymerization (Fig. 1).[2,3,14] In conifers, lignin is typically composed of guaiacyl units and a minor proportion of unmethoxylated *para*-hydroxyphenyl units. In compression wood of gymnosperms, both the lignin content and the proportion of *p*-hydroxyphenyl units are increased, making the wood more difficult to hydrolyse. Hardwood lignins of Angiosperms are typically composed of a mixture of guaiacyl and syringyl units. The presence of methoxylated syringyl units makes hardwood lignin more easily hydrolysed.

In other metabolic or biosynthetic pathways, knowledge of the final chemical structures has provided insight into mechanisms of biosynthesis. However, in lignin biosynthesis, insight into structure may also come from deeper understanding of the enzymology. Considerable effort is currently directed to the identification and characterization of the enzymes involved in lignin biosynthesis. More research is needed to determine the actual pathway of

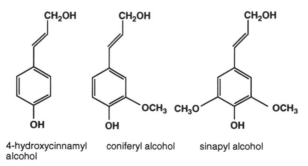

Fig. 1 Structures of the three types of hydroxycinnamyl alcohols (monolignols) that give rise to the three different types of lignin.

biosynthesis from coniferyl alcohol precursors. At several steps in the lignin biosynthetic pathway, multiple genes and enzymes have been identified that could carry out similar reactions, specifically during development or the defense response. Multiple enzymes may also indicate diversity of function or alternative pathways for biosynthesis. The heterogeneity of lignin could be directly related to enzymatic diversity and specificity. In addition, the general pathway for lignin biosynthesis has been inferred from studies of specific steps in several diverse species. There is no example where the entire pathway of lignin biosynthesis has been extensively characterized in a single tissue or a single species.

In this paper, we will discuss our efforts to identify and characterize specific enzymes that carry out key steps in the lignin biosynthetic pathway during the differentiation of xylem (wood formation) in loblolly pine (*Pinus taeda*). We will focus on several proteins and their corresponding genes that we believe to function in the formation of the wood cell wall and compare our results with those obtained for other plant species where lignin biosynthesis is also being studied actively. Several lignin biosynthetic genes are currently targets for specific manipulation of lignin biosynthesis by genetic engineering. In addition, we will discuss the status of gene transfer for conifers. The feasibility of modifying lignin quality or quantity in pine will be discussed in light of data regarding natural variation in lignin, new mutations found in model plant systems, and transgenic plants with modified lignin biosynthetic genes. We will also discuss the assignment of gene function as it relates to lignin biosynthesis.

GENES AND PROTEINS IN DIFFERENTIATING XYLEM OF LOBLOLLY PINE

Differentiating Xylem as a Biochemical Model

Differentiating xylem, early in the growing season, is a rich source of lignin biosynthetic enzymes, as well as other proteins involved in the structure or synthesis of the xylem cell wall. Differentiating xylem has been collected in quantity from large trees for biochemical studies of lignin biosynthetic enzymes.[31,32] We have used the differentiating xylem of loblolly pine to identify enzymes and their corresponding genes involved in lignin biosynthesis during the differentiation of xylem (Fig. 2). The material can be obtained in large quantities from single trees. Although trees are genetically diverse, specific genotypes can be identified. When a particular isozyme is needed, large scale

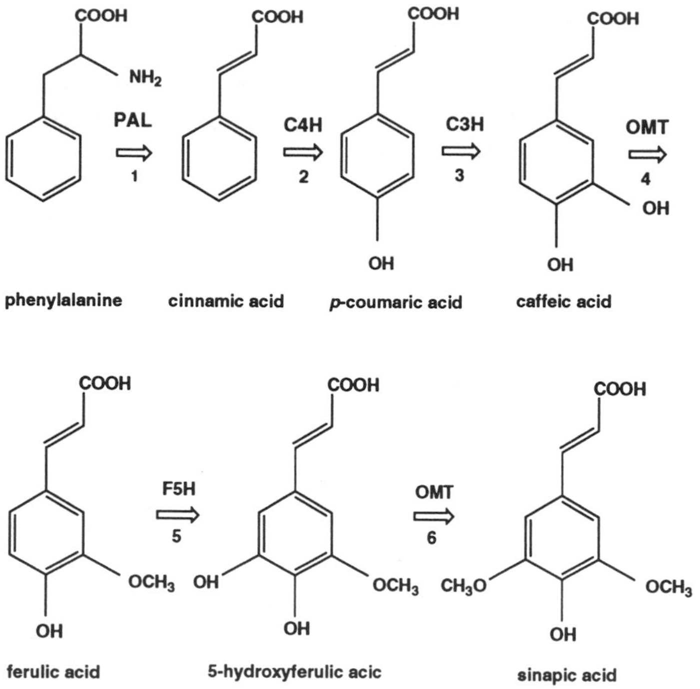

Fig. 2. Biosynthesis of cinnamic and hydroxycinnamic acids. In step 1, phenylalanine is deaminated to cinnamic acid by phenylalanine ammonia-lyase (PAL). In step 2, cinnamic acid is hydroxylated to *p*-coumaric acid by cinnamate-4-hydroxylase (C4H). In step 3, *p*-coumaric acid is further hydroxylated to caffeic acid by *p*-coumarate-3-hydroxylase (C3H), which is subsequently methylated to ferulic acid in step 4, by *O*-methyltransferase (OMT) Ferulic acid can be further hydroxylated (stpe 5) to 5-hydroxyferulic acid by ferulate-5-hydroxylase (F5H), and subsequently methylated (step 6) by a bifunctional *O*-tmethyltransferase (OMT) to produce sinapic acid.

enzyme preparations can be made from a single tree containing the allele coding for a particular electrophoretic form of the enzyme (allozyme) in the homozygous state.

We have concentrated on differentiating xylem because it allows us to purify the proteins that are present during the formation of wood. These proteins are more likely to be the appropriate functional proteins of interest. The sequence information obtained from such proteins allows us to identify the corresponding cDNAs expressed in differentiating xylem. This is done by selection of clones from differentiating xylem cDNA libraries. Subsequently we can identify the appropriate corresponding clones from genomic DNA. These matters are critical to increase confidence that one is studying the appropriate gene.

Phenylalanine Ammonia-Lyase in Pine: The Single PAL Hypothesis

In angiosperms, phenylalanine ammonia-lyase (PAL, EC 4.3.1.5) is typically represented by several isozymes encoded by a small gene family. The best studied systems are those of bean[33,34] and parsley.[35] In bean, for example, different isoforms* before or after a ten fold induction of enzyme activity.[43,44] These results contrast with those of bean, where elicitor treated cultures show multiple have specific patterns of expression in development and in the defense response. PAL2 is apparently involved in lignification of xylem; whereas, PAL3 is implicated in wounding and defense responses.[36] Expression of one PAL or the other is not exclusive. PAL2 and PAL3 have some overlapping responses to developmental and environmental effects. Multigene families for PAL have been reported for bean,[37] parsley,[35] rice,[38] *Arabidopsis*,[39] potato[40] and poplar.[41]

In contrast with these studies, there appears to be only one detectable isoform of PAL in differentiating xylem of loblolly pine. Purified preparations of PAL show no indication of negative cooperativity for substrate binding, a common feature of multiple isoforms. A cDNA clone was identified by anti-PAL antibody and the identity of the specific loblolly pine clone was confirmed by comparison of protein sequence.[42] The PAL cDNA which contains the entire coding region has 60 to 62% DNA sequence identity between pine and angiosperms. Southern blots using the pine cDNA as probe are consistent with a single copy gene. Screening of loblolly pine genomic libraries has thus far yielded only a single clone. The genomic clone differs from the corresponding angiosperm clones because it lacks an intron. Further screening is in progress for confirmation of these results.

---

*The term isoform is used in a general sense, allozymes as allelic forms of the same gene product, isozymes as products of different genes with same functions.

Studies carried out on elicitor treated cell suspension cultures of jack pine (*Pinus banksiana*) by Campbell and Ellis show only one isoform of PAL either before or after a ten fold induction of enzyme activity.[43,44] These results contrast with those of bean, where elicitor treated cultures show multiple isoforms.[33] The elicited jack pine cell cultures produced cell wall bound phenolics with features of true gymnosperm lignin.[45]

These results have led us to propose the "single PAL hypothesis" for pines. If there is only a single gene, an understanding of the regulation of PAL in pines is required . First, it implies that we have identified the gene coding for PAL active during the formation of lignin in xylem, because there is only one PAL. This gene must contain all of the *cis*-acting sequences needed for regulation of expression for lignin biosynthesis during wood formation. The same gene must also carry all of the necessary sequence information for regulation of PAL during the defense response. All of the *cis*-acting regulatory sequence information could be contained within a single genomic clone.

Hydroxycinnamate: CoA Ligase (4CL)

Hydroxycinnamate: CoA ligase (4-coumarate:CoA ligase, 4CL, EC 6.2.1.12) catalyses a reaction that forms an activated CoA thioester intermediate from several cinnamic acids in the phenylpropanoid pathway (Fig. 3).[46-49] In the monolignol biosynthetic pathway, 4CL can activate cinnamate and all of the hydroxycinnamic acids (Fig. 3). Some angiosperm and all gymnosperm 4CL activities tested do not use sinapate as a substrate.[46,48,50] Kutsuki *et al.*[46] have suggested that sinapaldehyde could be formed in angiosperms through an alternative pathway that would form 5-hydroxyconiferaldehyde from 5-hydroxyferulate, because most angiosperms make a guaiacyl-syringyl lignin.

We have recently purified a 4CL from xylem of loblolly pine that has a high affinity for caffeic acid, 4-coumaric acid and ferulic acid, but does not use sinapate as a substrate.[50] Pine xylem 4CL is inhibited noncompetitively by naringenin at low concentrations (13µM) and by coniferin at high concentrations (5mM) suggesting feedback control during synthesis of flavonoids as well as lignification. cDNA clones for the xylem form of 4CL from loblolly pine were identified by antibody screening of an expression library, and by PCR probes made from conserved 4CL sequences. Confirmation of the identity of the clones was made by comparison of protein sequence and DNA sequence.[50]

Isozymes of 4CL with different substrate affinities have been identified in a number of plants, including soybean, parsley, and poplar.[47,49,51,52] Several authors have suggested that differences in the isozyme expression pattern

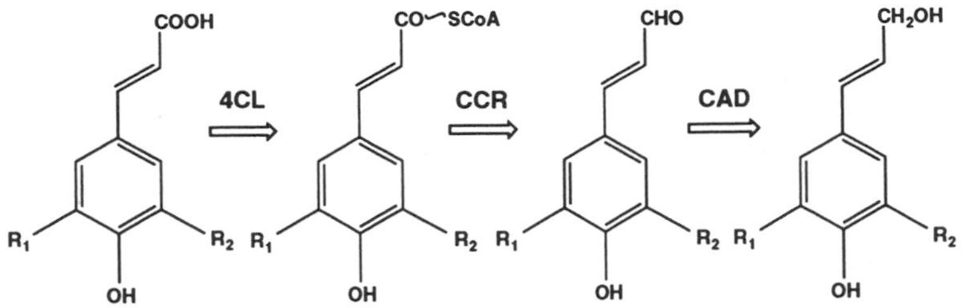

Fig. 3. General model of monolignol biosynthesis from hydroxycinnamic acids. The first step in the reduction of the hydroxycinnamic acids is the activation by 4-coumarate coenzyme A ligase (4CL), followed by the reduction to the aldehyde by hydroxycinnamate coenzyme A reductase (CCR). The reduction of the aldehyde to the alcohol is catalyzed by cinnamyl alcohol dehydrogenase (CAD).

could influence the type of lignin produced in a given tissue. [46,47,51] One example of altered lignin composition in gymnosperms occurs in compression wood. Compression wood has a higher content of p-hydroxyphenyl lignin than does normal wood.[53,54] The activity of lignin biosynthetic enzymes was higher in compression wood than in normal wood.[53] Interestingly, peroxidase activity was the same in both normal and compression wood. An increased level of p-hydroxyphenyl lignin could be due to changes in activity of the same gene products active in normal wood, or to induction of different isoforms of lignin biosynthetic enzymes by mechanical stress. Pine xylem isolated from compression wood contains the same isoform of 4CL found in normal wood.[50] These data argue for a change in the regulation of the lignin biosynthetic pathway, rather than the induction of different isoforms of 4CL, in the formation of compression wood lignin.

Cinnamyl Alcohol Dehydrogenase Encoded by a Single Gene

In loblolly pine, cinnamyl alcohol dehydrogenase (CAD, EC 1.1.1.195) has been purified from differentiating xylem and from developing seeds.[55] Biochemical and genetic characterization of the enzyme from both tissues verified that the same gene is expressed. The protein was partially sequenced, and, from the peptide information, a cDNA clone was identified and sequenced.[55,56] A cDNA clone has been obtained which contains the entire coding region. The sequence of this cDNA shares 67% similarity with a CAD sequence obtained from tobacco.[57] The enzyme shows high affinity for coniferaldehyde compared with sinapaldehyde, and is noncompetitively inhibited by high concentrations of coniferin.

CAD is highly polymorphic in populations of loblolly pine. At least 5 different electrophoretic forms of the enzyme have been detected. All of these variants are allelic (allozymes), i.e., they are variants of the same gene. The pine megagametophyte is a particularly useful tissue for genetic analysis because it is haploid, and each seed contains a megagametophyte that represents a different meiotic product. This tissue has been used extensively for genetic analysis of conifers.[58] Any heterozygous locus in the maternal tree segregates 1:1 in different megagametophytes. Analysis of the different polymorphisms of CAD from several trees indicates that no more than two alleles are present in any diploid, and that all polymorphisms examined segregate as expected for a single functional gene (D.M. O'Malley, unpublished results).

In other species, CAD is found in more than one isozyme.[59] Two forms of CAD were observed in tobacco;[7] *Nicotiana tabacum,* however, is an allotetraploid.[57] In *Eucalyptus gunnii*, a complex but similar set of isoforms of CAD were observed in different lignified tissues.[60,61] One of these CADs has been cloned from *E. gunnii*, and functionally expressed in *E. coli*.[62] Two forms of CAD were also found in soybean suspension cells[63,64] and multiple forms of CAD have been reported also in wheat.[65] Future studies are needed to learn the physiological roles of the different enzymes.

A Blue Copper Oxidase (Laccase) in Xylem

Recently a laccase was purified from differentiating xylem of loblolly pine.[66] The enzyme was solublized from the cell wall fraction using 1M $CaCl_2$. The enzyme was detected as an oxidase that required oxygen, instead of hydrogen peroxide. Histochemical studies using diaminofluorene showed a close association of an oxidase that did not require hydrogen peroxide with lignification in different tissues, particularly in lignifying xylem. The purified enzyme had strong absorbance at 600 to 610 nm, and ESR spectroscopy demonstrated the presence of type I copper. The enzyme is unambiguously a blue copper oxidase (BCO), and has been described as a laccase because of its association with lignification and its substrate specificity.[66] The BCO class of enzymes includes the plant enzymes laccase and ascorbate oxidase as well as the human serum protein ceruloplasmin. The pine enzyme is clearly distinguished from catechol oxidases, which are another class of oxygen-dependent oxidases, by size, type I copper content, substrate specificity and inhibitor specificity. The enzyme is distinguished from the ascorbate oxidases by the use of catechol as a substrate. Ascorbate oxidases are cell wall-associated enzymes of unknown function, found in a wide variety of higher plants [67,68] and thought to function in ascorbate metabolism. Dean and Eriksson[20] have suggested that ascorbate oxidase could have an indirect effect on lignin biosynthesis through alteration of ascorbate levels. Ascorbate oxidase can act on other substrates *in vitro*, including some phenolics. The natural substrate for the enzyme is only inferred; there is no direct evidence that establishes ascorbate as the natural substrate of ascorbate oxidase.

We have identified three different classes of cDNA clones from loblolly pine xylem as BCO clones, based on their sequences (W. Bao, unpublished results). They all share the conserved sequence elements characteristic of copper binding sites and they show high levels of similarity to both ascorbate oxidase and fungal laccase. The level of sequence similarity suggests that all of these enzymes are part of a multigene family. Sequences from more plant laccases are needed to clarify evolutionary relationships and functional roles. It would be very important to characterize a laccase and an ascorbate oxidase from the same plant.

Involvement of laccase in lignification was suggested many years ago by the work of Freudenberg and Richtzenhain.[69] They demonstrated that a fungal extract, later shown to contain a laccase, polymerized coniferyl alcohol in the presence of oxygen.[69] Partially purified laccase from Norway spruce cambial material was more effective at polymerization of coniferyl alcohol than spruce peroxidase.[31,70] Freudenberg[70] postulated that both laccase and peroxidase could be involved in lignin biosynthesis. Siegel *et al.*[71] observed that kidney bean plants contained less lignin when grown under reduced oxygen, and Lesney[72] reported that lignification induced in slash pine cultured cells was inversely correlated with peroxidase activity.

However, within the past twenty years, it had been generally accepted that peroxidases were exclusively involved in the polymerization to produce lignin.[2,14,19,73] This inference was based on the rare occurrence of laccases in plants[74] and on the inability of a laccase, purified from the japanese lacquer tree, to polymerize coniferyl alcohol.[75] Harkin and Obst[76] argued that laccase did not play a role in lignification, based on histochemical localization of peroxidase using syringaldazine as substrate.

In 1983, Bligny and Douce[77] purified a laccase from the spent medium of cell cultures of sycamore maple. These cultures do not make lignin. However, Sterjiades *et al.*[78,79] showed that the purified enzyme preparations can polymerize monolignols into DHP (dehydrierungspolymerizat, also called dehydrogenation product). Driouich *et al.*[80] used antibodies to localize the enzyme in stems, and supported the view that laccase activity in *Acer* is involved in lignification. In addition, Savidge and Udagama-Randeniya[81] reported that a cell wall bound coniferyl alcohol oxidase was associated with lignification in several conifers. Similarly, an insoluble cell wall fraction from *Forsythia* was able to catalyse the formation of pinoresinol from E-coniferyl alcohol with no cofactor other than oxygen added.[82]

At present, several enzymes and enzyme activities have been implicated in the polymerization that produces lignin. It is very difficult to prove that a particular enzyme is necessary and sufficient for polymerization purely by circumstantial evidence, and the ability of enzymes to polymerize monolignols in cell-free systems.It is also quite possible that more than one enzyme is involved, even in a single cell or tissue. Several workers are now arguing for the hypothesis proposed by Freudenberg[70]that both laccase and peroxidase catalyse the biogenesis of the lignin polymer.[83,84] Furthermore, lignification could be regulated differently at various stages of development, and under diverse stress response. Much more work is needed to answer these questions.

Simplicity of Gene Families for Lignin Biosynthesis

It is a matter of some interest that our data to date are consistent with single functional enzymes (PAL, 4CL, CAD) and single functional genes (CAD, PAL) for the small number of enzymes examined. This is not a typical feature of pine, which contains many multigene families and high levels of repeated DNA.[85,86] Two other genes we have studied as abundant xylem specific genes were also single copy genes.[87] The presence of only a single isozyme of PAL in elicited pine cell cultures[44] differs from the results obtained in angiosperms where multiple isozymes coded by different genes are induced. This contrast further supports the hypothesis of genetic simplicity in the organization of the genes coding for the enzymes of the lignin biosynthetic pathway. However, further work is needed to examine other environmental and developmental expression of the lignin associated enzymes to verify this apparent simplicity. If the inference is correct, however, it would raise many questions about the necessity, specificity and regulation of the isoforms of monolignol biosynthetic enzymes in angiosperms. Preliminary work on pine BCO cDNAs supports the conclusion that more than one gene exists (W. Bao, unpublished results), but the function of these BCO cDNAs has not yet been directly established. In addition, no results on the genetic complexity of angiosperm laccase are yet available.

Proteins Immobilized in Cell Walls of Xylem

Proteins immobilized in the wood cell wall represent another target of interest for genetic engineering of wood properties. Although proteins are a minor fraction of the dry weight, (0.2 to 1%), they are likely to represent the major nitrogenous material of wood.[88-90] Structural proteins could play an important role in the differentiation and morphology of the xylem tracheid cell wall. It is also likely that tissue specific cell wall proteins have developmentally regulated effects on cellular differentiation.[30,91,92] Cell wall proteins are therefore potential targets for genetic engineering of the fiber properties of the wood cell wall.

We have isolated and characterized a protein fraction from cell walls of differentiating xylem of loblolly pine.[93] This protein resembles cell wall proteins known as hydroxyproline rich glycoproteins (HRGPs) or extensin-like proteins, which are widely found in other plants.[94,95] The pine protein resembles other HRGPs in its cell wall location, abundance, solubility properties, glycosylation, and amino acid composition.[93] Two extensin-like proteins have also been purified and characterized from Douglas fir cell cultures.[96] Immunolocalization studies indicate that the pine xylem protein is located in wood and is covalently associated with the cell wall. It appears to be deposited continuously during formation of the early wood cell wall. The protein is also found in the cell walls of many other pine cells and tissues. It remains to be determined how a cell wall protein is transported and immobilized in the wood cell wall and what function it might serve in cell wall differentiation. We suggest that genetic modification of wood properties might result if modified proteins could be crosslinked into the wall.

The results observed for the xylem extensin-like HRGP protein may have implications for other proteins in the xylem cell wall. Crosslinking of proteins into the cell wall can be viewed as a pathway with discrete steps. The pine xylem cell wall structural protein is found in the walls of differentiating xylem as an extractable protein and also as an immobilized component of the terminally differentiated wood cell wall.[93] It is likely that it is crosslinked into the wall before lignification begins.This probably occurs through a mechanism that depends on the amino acid sequence of the protein and modifications imposed on the protein before transport to the wall.

Proteins attached to the wall will be trapped irreversibly as lignification proceeds. This idea predicts that many different proteins are immobilized in the lignified wall. Enzymes such as peroxidase can be detected in wood [97] (also W. Bao, unpublished results) suggesting that they have become entrapped during this process. Whitmore[98,99] suggested that a complex composed of lignin, carbohydrate and proteins is formed in the differentiating cell wall. A "crosslinking pathway" has been suggested in part by other studies involving the immobilization of proteins in the cell wall during the defense response.[100,101] The idea of a pathway also suggests a mechanism by which proteins such as the xylem laccase could be immobilized into the differentiating wall during lignification. The glycoprotein form of the laccase we have isolated from xylem cell walls is soluble,[66] whereas, other workers have identified oxidase activities that are immobilized in the conifer xylem cell walls[81] and in extracted cell walls of *Forsythia*.[82] We suggest that a particular enzyme may be found in cell walls in different forms and show altered properties, depending on the extent of modification, the stage of immobilization and the method of extraction, even though the original gene product may be the same.

Tissue-Specific Gene Expression in Differentiating Xylem

Tissue specific gene expression in differentiating xylem is an important aspect of wood formation and may lead to precise regulation of directed genetic modifications in wood. Promoters for two xylem-specific cDNAs have been isolated and characterized.[87] A loblolly pine differentiating xylem cDNA library was screened for tissue-specific cDNAs that were abundantly expressed. Two cDNAs were identified (designated 3H6 and 14A9) as the most abundant tissue-specific cDNAs detectable. One of these clones, 3H6, is the most abundant of the two, and encodes a proline-rich, alanine-rich protein that resembles arabinogalactan proteins.[102] The promoters for both genes (3H6 and 14A9) have been fused to reporter genes and expressed in transgenic tobacco. Both genes are constitutively expressed, suggesting that the sequences included in these constructs are not sufficient for xylem-specific gene expression in a heterologous system.

DNA TRANSFER AND GENETIC ENGINEERING IN CONIFERS

Essential to any effort in genetic engineering is the technology to produce transgenic plants. No methods are currently available for the production of transgenic pine, although significant progress in transient and stable expression of heterologous gene constructions in pine has been made in the past few years.[103-107] Transgenic conifers have been produced in two species, hybrid larch[108,109] and white spruce.[110,111] Stable transformation of spruce cell cultures has been obtained using microprojectile bombardment.[111,112] Ellis *et al.* have extended these results to produce the first transgenic spruce plantlets and have introduced and expressed a gene for insect resistance.[110] In collaboration with Ellis, we have introduced a promoter from the 3H6 genomic clone into white spruce somatic embryos. To date, the 3H6 promoter is expressed in white spruce callus and tobacco plants[87] (also C.A.Loopstra, K. Cheah and D. Ellis, unpublished results).

Natural Variation in Lignin Content or Quality

Natural variation of lignin content provides an indication of the extent to which the properties of wood might be engineered and the kinds of changes that might be tolerated.[18] Lignin content of southern pines is highly heritable, but not highly variable. The average lignin content of loblolly pine reported in different studies has ranged from 28.4% to 29.5%.[113-115] In one study of 24 trees, the lignin content ranged from 26.1% to 30.05% with an average of 28.6 and a standard deviation of 0.9%.[113] Broad sense heritability of lignin content in slash pine was 0.72.[114] Lignin content of wood from individual *Pinus contorta* grown in Germany ranged from 23.8% to 29.4%[116] and from 25.1% to 26.6% for trees grown in North America.[117] Across gymnosperms and angiosperms lignin content ranges from 15% to 36%.[118] Gymnosperms as a group have a more restricted range of average lignin content, from 25% (*Pinus monticola*) to 34% (*Libocedrus decurrens*). Within the genus *Pinus*, the range is even more restricted, from 25% (*Pinus monticola*) to 30% (*Pinus palustris*). The minumum amount of lignin required for normal xylem function and normal plant defense is not known.

New methods in genomic mapping in forest trees [119,120] should make it possible to determine the location of genes presumed to function in the control of lignin biosynthesis. These methods might be extended to investigate the genetic basis of natural variation in lignin content. By correlating the effects of loci where the biochemical product is known, with loci that are known only by their quantitative effects on variation in breeding or in natural populations, one should be able to identify loci which control the natural variation.

Mutations in Lignin Biosynthesis

Mutants that affect lignin biosynthesis have been identified. The most extensively characterized of these mutants are the brown midrib (*bmr*) mutants identified in several grasses, including maize, pearl millet and sorghum.[8,121-123] In comparison to wild-type plants, maize *bmr* mutants have a lower lignin content.[121] In a study of *bmr* in sorghum, no differences in histology or anatomy of lignified tissues were found.[124] Different mutations to the *bmr* phenotype have been attributed to lesions in the genes for OMT[125,126] or CAD.[127] In one case, a single mutation in sorghum resulted in decreased activity of both OMT and CAD.[127] The lignin in the sorghum mutant showed an increase in relative incorporation of hydroxycinnamaldehydes versus hydroxycinnamyl alcohols. It is difficult to interpret the results of these experiments in the absence of more information on the regulation of the pathway. PAL is known to be regulated in many ways, particularly by cinnamic acid,[128,129] by the sense suppression effect (reduction of gene expression by an active sense transgene)[130] and by the regulatory effects of transgenes in cosuppression.[131]

Chapple *et al.*[132] identified a mutation in *Arabidopsis thaliana* that produced only guaiacyl lignin. This is in contrast to the wild-type plants that produce a typical angiosperm guaiacyl-syringyl lignin. The mutant plants are morphologically normal, but have an array of biochemical phenotypes, such as the absence of the main sinapic acid ester in the seed, sinapoyl choline. The mutation was initially designated *sin1* but now awaits a new designation due to conflict with a pre-existing mutation of the same name. Chapple *et al.*[132] identified their mutation by a difference in the fluorescence of sinapic acid esters under UV light. The mutation appears to block the conversion of ferulate to 5-hydroxyferulate in the monolignol biosynthetic pathway, a step catalysed by ferulate 5-hydroxylase (F5H). The mutant is unable to produce sinapyl alcohol and therefore is unable to synthesize syringyl lignin. A T-DNA tagged mutant has been obtained which is allelic to the "*sin1*" mutation. The tagged gene encodes a cytochrome P-450 oxygenase (C. Chapple, personal communication). This result is consistent with the expectation that "*sin1*" is a defect in F5H which is a cytochrome P-450-dependent mixed function oxygenase.[133] Genetic complementation of the mutant with the cloned wild-type F5H gene will verify if the mutation is indeed at this locus. F5H is a step in the pathway with the potential to regulate the amount of ferulate that is utilized to make sinapyl alcohol versus coniferyl alcohol. Overproduction or underproduction of F5H in transgenic plants might be expected to modify lignin composition.

Directed Modification of Lignin Biosynthetic Genes

Many experiments are in progress directed to the genetic engineering of lignin content in plants of agronomic and horticultural importance as well as forest tree species[18,134] Transgenic plants engineered to investigate the mechanism of lignification were first produced by Lagrimini,[135-137] using constructs that produce modified levels of peroxidase in tobacco. The percentage of lignin and lignin-related polymers in cell walls was nearly two-fold greater in pith tissue isolated from peroxidase overproducers compared to control plants. Earlier work[135] showed a wilting effect in peroxidase overproducing transgenic tobacco.

More recently, transgenic tobacco plants have been obtained which have modified expression at other sites in the lignin biosynthetic pathway. Plants with altered levels of expression of PAL[130], CAD[138], and bifunctional caffeate/5-hydroxyferulate O-methyltransferase (OMT, EC 2.1.1.68) have been produced.[139] A construct with a normal but heterologous PAL2 gene from bean (with a 35S enhancer in the promoter) has severe effects when introduced into tobacco.[130] Levels of PAL expression are reduced by sense suppression and morphology of the plants is abnormal.[130] One of the observed abnormalities in these plants is altered lignin content.[130]

A bifunctional OMT implicated in the lignin biosynthetic pathway was cloned from aspen (*Populus tremuloides*)[140] and has been used in an antisense strategy to reduce or modify lignin in tobacco and *Liquidambar styraciflua* (sweetgum).[139] Transgenic tobacco containing a poplar OMT antisense construct had significantly reduced levels of enzyme activity, but showed little change in lignin content. Other OMTs have been cloned from hybrid poplar,[141] alfalfa,[142] maize[143] and tobacco.[144]

Bugos *et al.*[140] have cloned an OMT from poplar as part of a long term effort to introduce syringyl subunits into conifer lignin. In a relatively simple scenario, the addition of a F5H gene and a bifunctional OMT gene might be adequate to produce sinapyl alcohol. At least two of the enzymes (4CL and CAD) needed for the reduction of cinnamic acids to alcohols in conifers show strong preference *in vitro* for ferulic acid and coniferaldehyde relative to sinapic acid and sinapaldehyde.[46,145] It is not clear if these preferences would reflect barriers to the incorporation of the formation of sinapyl alcohol in conifers *in vivo*.

Halpin *et al.*[57,138] have purified CAD from tobacco, isolated cDNA clones, and examined the effects o in transgenic tobacco. Many of the transgenic plants expressing CAD antisense constructs have greatly reduced levels of CAD. CAD can be reduced to 10% of normal levels without affecting the normal development of the plants. However, the phenolic composition of the plants is affected and shares some characteristics with the[127] including a higher incorporation of cinnamyl aldehydes versus alcohols into the polymer. Therefore, CAD apparently can effect the proportion of aldehyde and alcohol monomers incorporated, but may not limit the total incorporation of monomers into lignin. Campbell and Boudet[146] have argued that the enzyme that catalyses the step before CAD, cinnamoyl CoA reductase (CCR; EC 1.2.1.44), could be an important regulatory step in lignin formation. Therefore, CCR is another target for genetic modification of lignin (Fig.3). The many genes and cDNAs encoding enzymes of phenylpropanoid metabolism involved in lignin biosynthesis that have been identified and cloned are summarized in Table 1.

ARE ANY GENES LIGNIN-SPECIFIC?

In considerations of genetic modification of lignin through genetic engineering, the potential specificity of modifications is important. It is therefore of interest to know if any genes involved in the lignin pathway are specific for lignin, and to what extent the expression of such genes might be specific to developmental or environmental signals. Although several genes involved in lignin biosynthesis show some xylem specific effects or differ in response to environmental induction of lignification, there are no genes yet identified that are lignin-specific. Identification of regulatory sequences specific to lignification would be important not only to increase our understanding of lignification, but also to use for directed genetic modification of lignin without other unintended effects. It would also be important to know if the regulation of lignification differs in the environmental induction of lignification in wounding or in the defense response.

Products Other Than Lignin Derived from Monolignols

Monolignols can give rise to molecules other than lignin. For example, condensation of coniferyl alcohol can give rise to dilignols and other lignans such as pinoresinol, secoisolariciresinol, matairesinol and dehydrodiconiferyl alcohol. [2,82,147]. Lignans are structurally diverse, and are widely distributed in vascular plants.[148]. In contrast to lignin, lignans are soluble, optically active and are not necessarily associated with the plant cell wall and therefore may have nonstructural roles. For example, dehydrodiconiferyl alcohol glucosides (DCGs) have been implicated in cytokinin-like activity promoting cell division.[149-150] These glucosides of

Table 1. Cloned genes and cDNAs from the monolignol biosynthetic pathway

1. phenylalanine ammonia-lyase (PAL)

Gene	Species	Common name	Genbank No.	Comments
Pal-1	*Arabidopsis thaliana*		X62747	genomic clone promoter[39]
Pal-1	*Glycine max*	soybean	X52953	genomic clone, full coding sequence[156]
			S46988	partial cDNA[157]
Pal	*Ipomoea batatas*	sweet potato	M29232	full coding cDNA [158]
Pal	*Lycopersicon esculentum*	tomato	M83314	genomic clone, full coding sequence[159]
Pal-5	*Lycopersicon esculentum*	tomato	M90692	genomic clone, full coding sequence[160]
Pal	*Malus* sp.	apple	X68126	partial cDNA, unpublished
Pal	*Medicago sativa*	alfalfa	X58180	full coding cDNA[161]
Pal	*Nicotiana tabacum*	tobacco	X59838	partial cDNA[162]
Pal	*Nicotiana tabacum*	tobacco	D17467	Nagai *et al.*, submitted full coding cDNA
tpa-1	*Nicotiana tabacum*	tobacco	M84466	Fukasawa-Akada, unpublished, genomic clone, full coding cDNA
Pal	*Oryza sativa*	rice	X16099	reference [38]
			Z15085	genomic clone, full coding sequence[163]

(continued)

Table 1. (Continued)

Gene	Species	Common name	Genbank No.	Comments
Pal-1	*Petrosilinum crispum* parsley		X16772	exon 2[35]
			X15473	promoter and exon 1[35]
Pal-4	*Petrosilinum crispum* parsley		X17462	catalytically active cDNA[164]
Pal-5	*Phaseolis vulgaris* bean		M11939	full coding cDNA[165]
Pal	*Pinus taeda* loblolly pine		none	full coding cDNA[42]
Pal	*Pisum sativum* pea		D1001	full coding cDNA[166]
Pal-1	*Pisum sativum* pea		D1002	genomic clone, full coding sequence[167]
Pal-2	*Pisum sativum* pea		D10003	genomic clone, full coding sequence[167]
Pal	*Populus tricocarpa x deltoides* hybrid poplar		L11747	full coding cDNA[41]
Pal-1	*Solanum tuberosum* potato		X63103	genomic clone full coding sequence[40]
Pal-2	*Solanum tuberosum* potato		X63103	genomic clone full coding sequence[40]
wali-4	*Triticum aestivum* wheat		L11883	Snowden and Gardner, unpublished partial cDNA

Table 1. (Continued)

2. cinnamate-4-hydroxylase (C4H)

Gene	Species	Common name	Genbank No.	Comments
C4h	*Helianthus tuberosus* Jerusalem artichoke		Z17369	full coding cDNA[169]
C4h	*Medicago sativa* alfalfa		L11046	full coding cDNA[170]
C4h	*Phaseolus aureus* mung bean		L07634	full coding cDNA[168]

3. *p*-coumarate-3-hydroxylase (C3H) - no clones yet available.

4. *O*-methyl transferase (OMT)

Gene	Species	Common name	Genbank No.	Comments
Omt	*Medicago sativa* alfalfa		M63853	full coding cDNA[142]
Omt	*Populus tremuloides* aspen		X62096, M70523	full coding cDNA[140]
Omt	*Populus deltoides x trichocarpa* clone 064 hybrid poplar	none		full coding cDNA[141]
Omt	*Zea mays* maize		none	genomic clone, full coding sequence, full coding cDNA[143]

5. ferulate-5-hydroxylase (F5H) - no clones yet available

(continued)

Table 1. (Continued)

6. 4-coumarate CoA ligase (4CL)

Gene	Species	Common Name	Genbank No.	Comments
4CL14	*Glycine max*			
		soybean	X69954	partial coding cDNA[171]
4CL16	*Glycine max*			
		soybean	X69955	partial coding cDNA[171]
Pc4CL-2				
	Petroselinum crispum			
		parsley	X05351	5'end genomic clone[173]
			X05353	3' end genomic clone[173]
Pc4CL-1				
	Petroselinum crispum			
		parsley	X05352	3' end genomic clone[173]
			X05350	5' end genomic clone[173]
Pc4CL-1				
	Petroselinum crispum			
		parsley	X13324	full coding cDNA[49]
Pc4CL-2				
	Petroselinum crispum			
		parsley	X13325	full coding cDNA[49]
4CL	*Pinus taeda*			
		loblolly pine	none	partial cDNA[50]
4-CL	*Oryza sativa*			
		rice	X52623	genomic clone, full coding sequence[174]
St4CL-1	*Solanum tuberosum*			
		potato	M62755	genomic clone, full coding sequence[172]

7. hydroxycinnamate CoA reductase (CCR) - no clones yet available.

Table 1. (Continued)

8. cinnamyl alcohol dehydrogenase (CAD)

Gene	Species	Common Name	Genbank	Comments
Cad	*Aralia cordata*		none	full coding cDNA[177]
Cad	*Eucalyptus gunnii*		X65631	full coding cDNA[62]
Cad	*Medicago sativa*	alfalfa	Z19573	Van Doorsselaere *et al*, unpublished full coding cDNA
Cad14	*Nicotiana tabacum*	tobacco	X62343	full coding cDNA[175]
Cad19	*Nicotiana tabacum*	tobacco	X62344	full coding cDNA[175]
Cad	*Picea abies*	Norway spruce	X72675	full coding cDNA[176]
Cad	*Pinus taeda*	loblolly pine	none	full coding cDNA[56]
Cad	*Populus deltoides x trichocarpa* clone 064	hybrid poplar	Z19568	Van Doorsselaere *et al.*, unpublished full coding cDNA

dehydrodiconiferyl alcohol are synthesized by dimerization and subsequent glucosylation of coniferyl alcohol.[151]. Recently, Teutonico *et al.*[152] have suggested that DCGs are a component of signal transduction in cytokinin-mediated cell division.

Clearly, compounds other than lignin can be synthesized from monolignols and the regulation of the monolignol biosynthetic pathway therefore may be complex. Consequently, genetic modification of steps in this pathway may have significant pleiotropic effects. For example, the "*sin1* " mutation in *Arabidopsis* described by Chapple *et al.* [132] has an impact on the biosynthesis and accumulation of soluble sinapic acid esters in addition to lignin. The branch of the monolignol biosynthetic pathway affected by this mutation is often described as being "lignin-specific". The "*sin1*" mutants demonstrate that other phenylpropanoid products share the same biosynthetic origins as monolignols.

Lignin Biosynthetic Enzymes in the Pine Megagametophyte

Lignin biosynthetic enzymes are present and active in the pine megagametophyte, a nonlignifying tissue. This result raises several questions regarding the specificity and roles of lignin biosynthetic enzymes. Why should lignin biosynthetic pathway enzymes be expressed in a nonlignifying tissue? The megagametophyte does not lignify during germination, but contains PAL and 4CL enzyme activity, as well as high levels of CAD. The megagametophyte is rich in peroxidase activity, but shows little if any laccase-like activity by histochemical staining. It is tempting to speculate whether the absence of laccase is responsible for the absence of lignification *per se*. Further work is needed to test this question.

In several experiments, CAD was purified to near homogeneity (a 600-fold purification) from germinating seeds of loblolly pine. CAD has high levels of allelic variation, and the allozymes of the xylem and seeds can be readily compared. It was therefore possible to verify that the same electrophoretic variants (allozymes) are expressed in both xylem and the non-lignifying megagametophyte. As discussed earlier, genetic criteria established that a single functional gene codes for CAD in loblolly pine, therefore, the same alleles of the same enzyne should be present in both tissues. Why? One potential role of monolignols in early development could be based on the postulated function of dehydrodiconiferyl alcohol glucosides in the control of cell division.[150] These results support the view that none of the enzymes of the monolignol precursor pathway is lignin-specific.

Assignment of Specific Gene Function

Within the past few years, new technology for gene transfer, DNA cloning and genetic modification has made possible the molecular genetic analysis of lignin biosynthesis. This technology now allows us to address a fundamental question, that is, what enzymes are necessary and sufficient for the polymerization of monolignols into lignin in plants? Assignment of function for specific enzymes requires both the characterization of the enzyme itself and identification or construction of mutations or variants to establish that the enzyme carries out the postulated functional role. Gene transfer allows for the test of function by the creation of mutations to reduce or eliminate enzyme activity (knockout mutations; antisense suppression; loss of function mutations) or to add a gene to complement the absence of function (gain of function mutation; genetic complementation). Experiments attempting to manipulate lignin in certain plants also test the role of specific enzymes in the overall regulation of the biosynthetic and polymerization pathway by feedback or feedforward mechanisms.

Lewis and Yamamoto[19] recognized the need for clearly defined criteria to associate a specific isozyme with lignification. In a discussion of potential peroxidase isozymes associated with lignification, they argued for four criteria. 1) An association in time and place (development), 2) an association with the lignifying cell wall (subcellular localization), and 3) the ability of the purified enzyme to carry out function *in vitro* (activity), and 4) structural evidence for the identification and classification of the enzyme (sequence). Dean and Eriksson[20] extended the argument to include the ability to remove or inhibit specific enzymes in the plant. All of these criteria may still be difficult to apply where multigene families are involved, and where multiple enzymes can carry out the same function. Tests of specific physiological roles by "loss of function and/or gain of function" experiments are more rigorous than circumstantial associations that attempt to correlate activity with function. No single experimental system is available that presently satisfies all of these requirements.

The difficulty in assigning function to a particular gene is well illustrated by the case of the tomato anionic peroxidase (TAP). TAP is a peroxidase isozyme correlated spatially and temporally with suberization[153] that can be induced to high levels by pathogens, wounding or abscisic acid. In spite of extensive circumstantial evidence implicating the TAP with suberization, antisense constructs that eliminated all detectable TAP protein or TAP enzyme activity, produced transgenic plants that were normal in all respects. Cell wall phenolics and hydroxy and dicarboxylic acids characteristic of suberin were also normal. The authors suggest that TAP is one of several enzymes affecting suberization, and one of these enzymes could carry out the normal function even if another isoform had reduced expression, particularly under laboratory conditions. Phenotypic detection of mutations has long been known to be dependent upon defined environmental conditions. In a recent study, a mutation blocking mannose conversion to asparagine linked glycans[154] was shown to have normal development under laboratory conditions, but showed poor survival in the field when subjected to biotic and abiotic stress. The function of the tomato anionic peroxidase remains to be elucidated.

CONCLUSION

Much work needs to be done to identify the genes and enzymes in the full lignin biosynthetic pathway. Rigorous definition of gene function, through mutation, antisense studies in transgenic plants, for "loss of function" and/or "gain of function" experiments, will be difficult even in model plant systems. Nevertheless, the pace of research is increasing and it is likely that much will be learned in the next few years. It is expected that the next phase of studies will define new boundaries for the larger picture of lignin biosynthesis, with regard to the extent of variation that is biologically allowed, and our ability to make modifications within that framework.

The regulation of lignin biosynthesis is expected to be complex. Regulation of such complex systems depends upon relationships between specific steps in the pathway imposed by feedback or feed forward regulation mechanisms, and also on metabolic elasticity; that is, the extent to which the substrates or products of individual enzymes can change in concentration without affecting the function of the pathway as a whole. Factors that regulate pathways of complex biochemical and developmental specificity are beginning to be characterized in plants. The inhibition of enzymatic activity for both 4CL and CAD by high levels of coniferin suggests the possibility of feedback control for the monolignol biosynthetic pathway. The fact that transgenic tobacco with only 10% of normal CAD enzyme activity have normal growth, appearance, and xylem function suggests a significant degree of metabolic elasticity.

It will be important to extend our knowledge of the regulation of lignin metabolism to transcription factors affecting gene expression and to learn more about the developmental signal transduction pathways in wood formation. In experiments relevant to this issue, interacting regulatory transcriptional activators have been cloned and shown to affect the anthocyanin biosynthetic pathway by a mechanism that is conserved between dicots and monocots.[155]

Molecular genetics provides the technology to modify the cell wall of xylem. Although we are limited by the lack of knowledge of the mechanisms of assembly and biosynthesis, the same technology also provides a means to investigate the mechanisms that regulate the system through directed genetic modifications. Specific mutations could be found or created in genes that regulate key steps or components of the system, and new information obtained from the biochemical, cellular and physiological characterization of mutations. Such information will lead to deeper insight into the process of lignification and cell wall biosynthesis and increase the precision and power of directed modification.

ACKNOWLEDGEMENTS

Many members of the Forest Biotechnology Group at North Carolina State University, both past and present, have contributed to the ideas and the results described in this article. The authors would like to thank, in particular, Wuli Bao, Kui Shin Voo, Carol Loopstra, Wei Wei Liu, John MacKay and Kheng Cheah for their contributions, and for permisson to include unpublished data. Special thanks to Susan McCord and Reenah Schaffer for the roles they played in assisting with the work described herein. We wish to express thanks to Reenah Schaffer for help in the preparation of the manuscript, and to Clint Chapple and David Ellis for permission to include unpublished information. This work has been supported by The USDA Competitive Grants Program; The Department of Energy, Division of Energy Biosciences; The Forest Biotechnology Industrial Associates Program; and The USDA Forest Service.

REFERENCES

1.	CORNER, E.J.H. 1964. The life of plants (Carrington, R., Matthews, L.H., Young, J.Z., eds.) The World Publishing Company, 315 pp.

2.	HIGUCHI, T. 1985. The biosynthesis of lignin. In " Biosynthesis and degradation of wood components". (T. Higuchi. ed.)Academic Press, Orlando FL: pp141-160.

3.	CHEN, C.-L. 1991. Lignin: occurrence in woody tissues, isolation, reactions, and structure. In"Wood Structure and Composition".

(Lewin, M. and Goldstein, I.S., eds.). New York: Marcel Dekker, Inc., pp. 183-261.

4.   WARDROP, A.B. 1981. Lignification and xylogenesis. In "Xylem Cell Development" (J.R. Barnett ed.) Castle House, Turnbridge Wells, Kent, England. pp. 115-152.

5.   SMART, C.C., AMRHEIN, N. 1985. The influence of lignification on the developmet of vascular tissue in *Vigna radiata* L. Protoplasma 124: 87-95.

6.   VANCE, C.P., KIRK, T.K., SHERWOOD, R.T. 1980 Lignification as a mechanism of disease resistance. Ann. Rev. Phytopathol. 18: 259-288.

7.   TIMELL, T.E. 1973. Studies on the opposite wood in conifers. Part I. Chemical composition. Wood Sci. Technol. 7: 1-5

8.   CHERNEY, J.H., AXTELL, J.D., HASSEN, M.M., ANLIKER, K.S. 1988. Forage quality characterization of a chemically induced brown midrib mutant in pearl millet. Crop Sci. 28:783-787.

9.   JUNG, H-J.G., RALPH, J., HATFIELD, R.D. 1991. Degradability of phenolic acid-hemicellulose esters: A model system. J. Sci. Food Agric. 56: 469-478.

10.   BROWN, A. 1985. Review of lignin in biomass. J. Appl. Biochem. 7:371-387.

11.   HARRY, D.E., SEDEROFF, R.R. 1989. Biotechnology in biomass crop production: the relationship of biomass production and plant genetic engineering. Oak Ridge National Laboratory Environmental Science Division Publication 3411:47pp:

12.   ZOBEL, B.J. van BUIJTENEN, J.P. 1989. Wood variation: Its causes and control. Springer Verlag Series in Wood Science, Springer, New York. 363pp.

13.   SONDHEIMER, E., SIMPSON, W.G. 1962. Lignin abnormalities in rubbery apple wood. Can. J. Biochem. Physiol. 40:841-846.

14.   SEDEROFF, R., CHANG, H-M. 1991. Lignin biosynthesis. In "Wood Structure and Composition" edited by M. Lewin & I. Goldstein. Marcel Dekker, Inc. New York. p.263-285.

15.   AMRHEIN, N., FRANK,G., LEMM, G., LUHMANN, H.-B. 1983. Inhibition of lignin formation by L-*alpha*-amino-*beta*-phenylproprionic acid- an inhibitor of phenylalanine ammonia-lyase. Eur. J. Cell Biology 29: 139-144.

16.   GRAND, C., SARNI, F., BOUDET, A.M. 1985. Inhibition of cinnamyl alcohol dehydrogenase activity and lignin synthesis in

poplar (*Populus x euramericana*, Dode) tissues by two organic compounds. Planta 163: 232-237.

17. GRAND, C., BOUDET, A.M., RANJEVA, R. 1982. Natural variation and controlled changes in the lignification process. Holzforschung 36: 217-223.

18. WHETTEN, R., SEDEROFF, R.R. 1991. Genetic engineering of wood. Journal of Forest Ecology and Management 43:301-316.

19. LEWIS, N.G., YAMAMOTO, E. 1990. Lignin: occurrence, biogenesis and biodegradation. Ann. Rev. Plant Physiol. Plant Mol. Biol. 41:455-496.

20. DEAN, J.F.D., ERIKSSON, K.-E.L. 1992. Biotechnological modification of lignin structure and composition in forest trees. Holzforschung 46:135-147.

21. FREUDENBERG, K. 1965. Lignin: Its constitution and formation from *p*-hydroxycinnamyl alcohols. Science 148:595-600.

22. FREUDENBERG, K. 1968. The constitution and biosynthesis of lignin. In "Constitution and Biosynthesis of Lignin" (A.C. Neish and K. Freudenberg, eds.), Springer-Verlag, New York, pp.47-116.

23. FREUDENBERG, K., NEISH, A.C. 1968. "Constitution and Biosynthesis of Lignin." Springer-Verlag, New York.

24. TIEMANN, F., MENDELSOHN, B. 1875. Zur Kenntniss der Bestandteile des Holztheerkreosots. Ber. Dtsch. Chem. Ges. (Chem. Ber.) 8:1136-1139.

25. TERASHIMA, N. 1989. Higher order structure of protolignin in the cell wall of tree xylem. TAPPI Proceedings: International Symposium on Wood and Pulping Chemistry. Raleigh, NC May 22-25, 1989. TAPPI Press, Atlanta. p. 359-364.

26. ATTALA, R.H., AGARWALL, U.P. 1984. Raman microprobe evidence for lignin orientation in the cell walls of native woody tissue. Science 227:636-638.

27. FUKUSHIMA, K., TERASHIMA, N. 1991. Heterogeneity in lignin formation XIV. Formation and structure of lignin in differentiating xylem of *Ginko biloba*. Holzforschung 45:87-94.

28. LEWIS, N.G., YAMAMOTO, E., WOOTEN, J.B., JUST, G., OHASHI, H., TOWERS, G.H.N. 1987. Monitoring biosynthesis of wheat cell-wall phenylpropanoids *in situ*. Science 237:1344-1336.

29. LEWIS, N.G., RAZAL, R.A., YAMAMOTO, E., BOKELMAN, G.H., WOOTEN, J.B. 1989. [13]C specific labelling of lignin in intact plants. In " Plant cell wall polymers: Biogenesis and

biodegradation. ACS Symposium Series. 399, Ch 12, pp.169-181.

30.   CARPITA, N.C., GIBEAUT, D.M. 1993. Structural models of primary cell walls in flowering plants: consistency of molecular structure with the physical properties of the walls during growth. The Plant Journal 3:1-30.

31.   FREUDENBERG, K., HARKIN, J.M., REICHERT, M., FUKUZUMI, T. 1958. Die an der Verholzung beteiligten Enzyme. Die Dehydrierung des Sinapinalkohols. Chemische Berichte 91:581-590.

32.   LUDERITZ, T., GRISEBACH, H. 1981. Enzymatic sysnthesis of lignin precursors-comparison of cinnamoyl-CoA reductase and cinnamyl alcohol: NADP+ dehydrogenase from spruce (*Picea abies* L) and soybean (*Glycine max* L.) Eur. J. Biochem. 119:115-124.

33.   BOLWELL, G.P., BELL, J.N., CRAMER, C.L., SCHUCH, W., LAMB, C.L., DIXON, R. A. 1985. L-phenylalanine ammonia-lyase from *Phaseolus vulgaris*-characterization and differential induction of multiple forms from elicitor-treated suspension cultures. Eur. J. Biochem. 149: 411-419.

34.   LIANG , X., DRON, M., CRAMER, C.L., DIXON, R.A., LAMB, C.J. 1989. Differential regulation of phenylalanine ammonia-lyase genes during plant development and by environmental cues. J. Biol. Chem. 264:14486-14492.

35.   LOIS, R., DIETRICHET, A., HAHLBROCK, K., SCHULZ, W. 1989. A phenylalanine ammonia-lyase gene from parsley: structure, regulation, and identification of elicitor and light responsive *cis* acting elements. EMBO. J. 8:1641-1648.

36.   SHUFFLEBOTTOM, D., EDWARDS, K., SCHUCH, W., BEVAN, M. 1993. Transcription of two members of a gene family encoding phenylalanine ammonia-lyase leads to remarkably different cell specificities and induction patterns. Plant J. 3:835-845.

37.   CRAMER, C.L., EDWARDS, K., DRON, M., LIANG, X., DILDINE, S.L., BOLWELL, G.P., DIXON, R.A., LAMB, C.J., SCHUCH, W. 1989. Phenylalanine ammonia-lyase gene organization and structure. Plant Mol. Biol.12:367-383.

38.   MINAMI, E., OZEKI, Y., MATSUOKA, M., KOIZUKA, N., TANAKA, Y. 1989. Structure and some characterization of the gene for phenylalanine ammonia-lyase from rice plants. Eur. J. Biochem. 185:19-25.

39.   OHL, S., HEDRICK, S.A., CHORY, J., LAMB, C.J. 1990.

Functional properties of a phenylalanine ammonia-lyase promoter from *Arabidopsis*. Plant Cell 2:837-848.

40.   JOOS, H.-J., HAHLBROCK, K. 1992. Phenylalanine ammonia-lyase in potato (*Solanum tuberosum* L.). Eur.J. Biochem. 204:621-629.

41.   SUBRAMANIAM, R., REINHOLD, S., MOLITOR, E.K., DOUGLAS, C.J. 1993. Structure, inheritance, and expression of hybrid poplar (*Populus trichocarpa x Populus deltoides*) phenylalanine ammonia-lyase genes. Plant Physiol. 102: 71-83.

42.   WHETTEN, R.W., SEDEROFF, R.R. 1992. Phenylalanine ammonia-lyase from loblolly pine. Plant Physiol. 98:380-386.

43.   CAMPBELL, M., ELLIS, B. 1992. Fungal elicitor-mediated responses in pine cell cultures. I. Induction of phenylpropanoid metabolism. Planta 186:409-417.

44.   CAMPBELL, M., ELLIS, B. 1992. Fungal elicitor-mediated responses in pine cell cultures. III. Purification and characterization of phenylalanine ammonia-lyase. Plant Physiol. 98:62-70.

45 .   CAMPBELL, M., ELLIS, B. 1992. Fungal elicitor-mediated responses in pine cell cultures. II. Cell wall-bound phenolics. Phytochemistry 31:737-742.

46.   KUTSUKI, H., SHIMADA, M., HIGUCHI, T. 1982. Distribution and roles of p-hydroxycinnamate:CoA ligase in lignin biosynthesis. Phytochemistry 21: 267-271.

47.   KNOBLOCH, K.H., HAHLBROCK, K. 1975. Isozymes of p-coumarate: CoA ligase from cell suspension culture of *Glycine max*. Eur. J. Biochem. 52:311-320.

48.   LUDERITZ, T., SCHATZ, G., GRISEBACH, H. 1982. Enzyme synthesis of lignin precursors. Purification and properties of 4-coumarate:CoA ligase from cambium sap of spruce (*Picea abies* L.). Eur. J. Biochem. 123:583-586.

49.   LOZOYA, E., HOFFMANN, H., DOUGLAS, C., SCHULZ, W., SCHEEL, D., HALBROCK, K. 1988. Primary structures and catalytic properties of isoenzymes encoded by the two 4-coumarate: CoA ligase genes in parsley. Eur. J. Biochem. 176:661-667.

50.   VOO, K.S., WHETTEN, R.W., O'MALLEY, D.M., SEDEROFF, R.R. 1993. 4-Coumarate Co-A ligase from loblolly pine xylem: Isolation, characterization and complementary DNA cloning. Poster 12, Phytochemical Society of North America: Newsletter 33:23.

51.   GRAND, C., BOUDET, A., BOUDET, A.M. 1983. Isozymes of hydroxycinnamate: CoA ligase from poplar stems and tissue distribution. Planta 158:225-229.

52.   ALLINA, S., LEE, D., DOUGLAS, C. 1993. Cloning and characterization of 4-coumarate: coenzyme A ligase (4CL) from poplar and *Arabidopsis*. Poster 27, Phytochemical Society of North America: Newsletter 33:26.

53.   KUTSUKI, H., HIGUCHI, T. 1981. Activities of some enzymes of lignin formation in reaction wood of *Thuja orientalis, Metasequoia glyptostroboides* and *Robinia pseudoacacia*. Planta 152:365-368.

54.   TIMMEL, T.E. 1986. Compression wood in gymnosperms. Springer, Heidelberg, pp.2150.

55.   O'MALLEY, D.M., PORTER, S., SEDEROFF, R.R. 1992. Purification, characterization and cloning of cinnamyl alcohol dehydrogenase in loblolly pine (*Pinus taeda* L.). Plant Physiol. 98:1364-1371.

56.   LIU, W.W. 1993. Cloning and characterization of cinnamyl alcohol dehydrogenase cDNA from loblolly pine. M. Sc. , North Carolina State University. Raleigh, N.C.

57.   HALPIN, C., KNIGHT, M.E., GRIMA-PETTENATI, J., GOFFNER, D., BOUDET, A., SCHUCH, W. 1992. Purification and characterization of cinnamyl alcohol dehydrogenase from tobacco stems. Plant Physiol. 98:12-16.

58.   CONKLE, M.T. 1981. Isozyme variation and linkage in six conifer species, "Proceedings of the symposium on isozymes of North American forest trees and forest insects" M.T. Conkle, ed. Pacific Southwest Forest and Range Experiment Station, Berkeley CA. USDA Forest Service Gen. Tech. Rep. PSW-48.p.11-17.

59.   MANSELL, R.B., BABBEL, G.R., ZENK, M.H. 1976. Multiple forms and specificity of coniferyl alcohol dehydrogenase from cambial regions of higher plants. Phytochemistry 15:1849-1853.

60.   GOFFNER, D., JOFFROY, I., GRIMA-PETTANATI, J., HALPIN, C., KNIGHT, M.E., SCHUCH, W., BOUDET, A.M. 1992. Purification and characterization of isoforms of cinnamyl alcohol dehydrogenase from *Eucalyptus* xylem. Planta 188:48-53.

61.   HAWKINS, S.W., BOUDET, A.M. 1993. Purification and characterization of cinnamyl alcohol dehydrogenase isoforms from the periderm of *Eucalyptus gunnii* Hook. (in press).

62.   GRIMA-PETTENATI, J., FEUILLET, C., GOFFNER, D., BORDERIES, G., BOUDET, A.M. 1993. Molecular cloning and expression of a *Eucalyptus gunni* cDNA clone encoding cinnamyl alcohol dehydrogenase (in press).

63. WYRAMBIK, D., GRISEBACH, H. 1975. Purification and properties of isoenzymes of cinnamyl-alcohol dehydrogenase from soybean cell cultures. Eur. J.Biochem. 59:9-15.

64. WYRAMBIK, D., GRISEBACH, H. 1979. Enzymic synthesis of lignin precursors. Further studies on cinnamyl-alcohol dehydrogenase from soybean suspension cultures. Eur. J. Biochem. 97:503-509.

65. PILLONEL, C., HUNZIKER, P., BINDER, A. 1992. Multiple forms of the constitutive wheat cinnamyl alcohol   dehydrogenase. J. Exp. Bot. 43:299- 305.

66. BAO, W., O'MALLEY, D.M., WHETTEN, R., SEDEROFF, R.R. 1993. A laccase associated with lignification in loblolly pine xylem. Science 260:672-674.

67. BUTT, V.S. 1980. Direct oxidases and related enzymes. In "The Biochemistry of Plants," Vol. 2, (D.D. Davies, ed.), Academic Press, New York, pp.81-123.

68. LIN, L.-S., VARNER, J.E. 1991. Expression of ascorbic acid oxidase in zucchini squash (*Curcurbita pepo* L.). Plant Physiol. 96:159-165.

69. FREUDENBERG, K., RICHTZENHAIN, H. 1943. Enzymatische Versuche zur Enstehung des Lignins. Ber. Dtsch. Chem. Ges. (Chem. Ber.), 76:997-1006.

70. FREUDENBERG, K. 1959. Biosynthesis and constitution of lignin. Nature 183:1152-1155.

71. SIEGEL, S.M., ROSEN, L.A., REWICK, G. 1962. Effects of reduced oxygen tension on vascular plants. Growth and composition of red kidney bean plants in 5% O_2. Physiol. Plant. 15:304-314.

72. LESNEY, M.S. 1990. Effect of 'elicitors' on extracellular peroxidase activity in suspension-cultured slash pine (*Pinus elliottii* Engelm.). Plant Cell Tissue and Organ Culture 20:173-175.

73. HIGUCHI, T. 1990. Lignin biochemistry: Biosynthesis and Biodegradation. Wood Sci. Technol. 24: 23-63.

74. HIGUCHI, T. 1959. Studies on the biosynthesis of lignin. In "Biochemistry of Wood" (Kratzl, K., Billek, G., eds.) New York:Pergamon Press, pp.161-188.

75. NAKAMURA, W. 1967. Studies on the biosynthesis of lignin 1. Disproof against the catalytic activity of laccase in the oxidation of coniferyl alcohol. J. Biochem (Japan) 62:54-60.

76. HARKIN, J.M., OBST, T.R. 1973. Lignification in trees: indication of exclusive peroxidase participation. Science 180:296-297.

77. BLIGNY, R., DOUCE, R. 1983. Excretion of laccase by sycamore (*Acer psuedoplantanus* L.) cells. Biochem. J. 209: 489-496.

78. STERJIADES, R., DEAN, J.F.D., ERIKSSON,. K.-E. L. 1992. Laccase from sycamore maple (*Acer psuedoplantanus*) polymerizes monolignols. Plant Physiol. 99:1162-1168.

79. STERJIADES, R., DEAN, J.F.D., GAMBLE, G., HIMMELSBACH, D.S., ERIKSSON, K.-E. 1993. Extracellular laccases and peroxidases from sycamore maple (*Acer pseudoplantanus*) cell suspension cultures. Reactions with monolignols and lignin model compounds. Planta 190:75-87.

80. DRIOUICH, A., LAINE, A-C., VIAN, B., FAYE, L. 1992. Characterization and localization of laccase forms in stem and cell cultures of sycamore. Plant J. 2:13-24.

81. SAVIDGE, R., UDAGAMA-RANDENIYA, P. 1992. Cell wall-bound coniferyl alcohol oxidase associated with lignification in conifers. Phytochemistry 31:2959-2966.

82. DAVIN, L.B., BEDGAR, D.L., KATAYAMA, T., LEWIS, N.G. 1992. On the stereoselective synthesis of (+)-pinoresinol in *Forsythia suspensa* from its achiral precursor, coniferyl alcohol. Phytochemistry 31: 3869-3874.

83. DAVIN, L.B., PARE, P.W., LEWIS, N.G. 1993. Phenylpropanoid coupling enzymes in lignan/lignin synthesis: non-stereoselective coupling. Phytochemical Society of North America Newsletter 33 (1) Program and Abstracts for Genetic Engineering of Plant Secondary Metabolism, Asilomar, California, June 27-July 31, Oral Paper Abstract 30.

84. DEAN, J.F.D., ERIKSSON, K.-E.L. 1993. Laccase and the evolution of lignin in vascular plants. Holzforschung (in press)

85. GERTULLA, S., KINLAW, C.S. 1993. Complex gene families of pines. Proceedings of the Southern Forest Tree Improvement Conference, Atlanta, Ga. abstracts, p.47.

86. KRIEBLE, H.B. 1993 Molecular structure of forest trees. In "Clonal Forestry I, Genetics and Biotechnology, ed by M.R. Ahuja and W.J. Libby. Springer Verlag, Berlin, Heidelberg: pp.224-240.

87. LOOPSTRA, C.A. 1992. Xylem specific gene expression in loblolly pine. PhD Dissertation. North Carolina State University, Raleigh, NC, 27695 USA.

88. LAIDLAW, R.A., SMITH, G.A. 1965. The proteins of the timber of scots pine (*Pinus sylvestris*). Holzforschung 19:129-134.

89. COWLING, E.B., MERRILL, W. 1966. Nitrogen in wood and its role in wood deterioration. Can. J. Bot. 44:1539-1554.

90. WESTERMARK, U., HARDELL, H.L., IVERSON, T. 1986. Thecontent of protein and pectin in the lignified middle lamella/primary wall from spruce fibers. Holzforschung 40:65-68.

91. VARNER, J.E., LIN, L.-S. 1989. Plant cell wall architecture. Cell 56:231-239.

92. STEBBINS, G.L. 1992. Comparative aspects of plant morphogenesis: a cellular, molecular and evolutionary approach. Am. J. Bot. 79: 589-598.

93. BAO, W., O'MALLEY, D.M., SEDEROFF, R.R. 1992. Wood contains a cell-wall structural protein. Proc. Natl. Acad. Sci. USA 89:6604-6608.

94. TIERNEY, M.L., VARNER, J.E. 1987. The extensins. Plant Physiol. 84:1-2.

95. CASSAB, G.I., VARNER, J.E. 1988. Cell wall proteins. Ann. Rev. Plant. Physiol. Plant Mol. Biol. 39: 321-353.

96. KIELISZEWSKI, M., de ZACKS, R., LEYKAM, J.F., LAMPORT, D.T.A. 1992. A repetitive proline-rich protein from the gymnosperm Douglas fir is a hydroxyproline-rich glycoprotein. Plant Physiol. 98:919-926.

97. STICH, K., EBERMAN, R. 1988. Investigation of the substrate specificity of peroxidase isozymes occuring in wood of different species. Holzforschung 42:221-224.

98. WHITMORE, F.W. 1976. Binding of ferulic acid to cell walls by peroxidases of *Pinus elliottii*. Phytochemistry 15:375-378.

99. WHITMORE, F.W. 1978. Lignin carbohydrate complex formed in isolated cell walls of callus. Phytochemistry 17:421-425.

100. BRADLEY, D.J., KJELLBOM, P., LAMB, C.J. 1992. Elicitor and wound induced oxidative crosslinking of a proline-rich plant cell wall protein: a novel rapid defense response. Cell 70:21-30.

101. KLEIS-SAN FRANSISCO, S.M., TIERNEY, M.L. 1990. Isolation and characterization of a proline-rich cell wall protein from soybean seedlings. Plant Physiol. 94:1897-1902.

102. GLEESON, P.A., McNAMARA, M., WHETTENHALL, R.E.H., STONE, B.A., FINCHER, G.B. 1989. Characterization of the hydroxyproline-rich protein core of an arabinogalactan-protein secreted from suspension-cultured *Lolium multiflorum* (Italian ryegrass) endosperm cells. Biochem. J. 264:857-862.

103.   SEDEROFF, R.R., STOMP, A-M., CHILTON, W.S., MOORE, L.W. 1986. Gene transfer into loblolly pine by *Agrobacterium tumefaciens*. Bio/Technology, 4:647-649.

104.   STOMP, A-M., LOOPSTRA, C., CHILTON, W.S., SEDEROFF, R.R., MOORE, L.W. 1990. Extended host range of *Agrobacterium tumefaciens* in the genus *Pinus*. Plant Physiol. 92:1226-1232.

105.   STOMP, A-M., WEISSINGER, A.K., SEDEROFF, R.R. 1991. Transient expression from microprojectile-mediated DNA transfer in *Pinus taeda*.. Plant Cell Rep.10:187-190.

106.   LOOPSTRA, C.A., STOMP, A-M., SEDEROFF, R.R. 1990. *Agrobacterium* mediated DNA transfer in sugar pine. Plant Mol. Biol. 15:1-9.

107.   LOOPSTRA, C.A., WEISSINGER, A.K., SEDEROFF, R.R. 1992. Transient expression in differentiating wood. Can. J. For.Res. 22:993-996.

108.   HUANG, Y., DINER, A.M., KARNOSKY, D.F. 1991. *Agrobacterium rhizogenes*-mediated genetic transformation and regeneration of a conifer: *Larix decidua*. Vitro Cell. Dev. Biol. 4:201-207.

109.   HUANG, Y., KARNOSKY, D.F. 1991. A system for gymnosperm transformation and plant regeneration: *Agrobacterium rhizogenes* and *Larix decidua*-European larch hairy root culture and propagation. In Vitro 27:153A.

110.   ELLIS, D.D., McCABE, D., McINNIS, S., RAMACHANDRAN, R., McCOWN, B. 1992. Transformation of *Picea glauca*-gene expression and the regeneration of stably transformed embryogenic callus and plants. International Conifer Biotechnology Working Group. abstract.

111.   ELLIS, D.D., McCABE, D.E., McINNIS, S., RAMACHANDRAN, R., RUSSELL, D.R., WALLACE, K.M., MARTINELL, B.J., ROBERTS, D.R., RAFFA, K.F., McCOWN, B.H. 1993. Stable transformation of *Picea glauca* by particle acceleration. Bio/Technology 11:84-89.

112.   ROBERTSON, D., WEISSINGER, A.K., ACKLEY, R., GLOVER,S., SEDEROFF, R.R. 1992. Genetic Transformation of Norway spruce (*Picea abies* (L.) Karst) using somatic embryo explants by microprojectile bombardment. Plant Mol. Biol. 19:925-935.

113. VAN BUIJTENEN, J.P., ZOBEL, B.J., JORANSEN, P.N. 1961. Variation in some wood and pulp properties in an evenly aged loblolly pine stand. TAPPI J. 44:141-144.

114. EINSPAHR, D.W., GODDARD, R.E., GARDNER, H.S. 1964. Slash pine wood and fiber property heritability study, Silvae Genetica 13:103-109.

115. McMILLIN, C.W. 1968. Chemical composition of loblolly pine wood. Wood Sci. Technol. 2:233-240.

116. SCHUTT, P. 1958. Schwankungen in Zellulose- und Ligningehalt bei einigen in Westdeutchland angebauten *Pinus contorta* Herkunften. Silvae Genetica 7:65-69.

117. KIM, W.J., CAMPBELL, A.G., and KOCH, P. 1989. Chemical variation in lodgepole pine with latitude, elevation and diameter class. For. Prod. J. 39:7-12.

118. SARKENEN, K.V., HERGERT, H.L. 1971. Definition and nomenclature. In " K.V. Sarkanen, C.H. Ludwig, eds. Lignins: occurence, formation, structure and reactions. Wiley, Intersciences, New York, pp.43-94.

119. GRATTAPAGLIA, D., CHAPARRO, J., WILCOX, P., McCORD, S., WERNER, D., AMERSON, H., McKEAND, S., BRIDGWATER, F., WHETTEN, R., O'MALLEY, D.M., SEDEROFF, R.R. 1992. Mapping in woody plants with RAPD markers: Application to breeding and horticulture. Procceedings of the symposium; applications of RAPD technology to plant breeding. Minneapolis MN Joint Plant Breeding Symposium: Crop Science, Horticultural Science, American Genetic Association. p.37-40.

120. GRATTAPAGLIA, D., CHAPARRO, J., WILCOX, P., McCORD, S., CRANE, B., AMERSON, H., WERNER, D., LIU, B.-H., O'MALLEY, D.M., WHETTEN,R., McKEAND, S., GOLDFARB, B., GREENWOOD, M., KUHLMAN, G., BRIDGWATER, F., SEDEROFF, R.R. 1993. Application of genetic markers to tree breeding. Proceedings of the Southern Forest Tree Improvement Conference, (SFTIC) Atlanta GA. vol. 26. in press.

121. MULLER, L.D., BARNES, R.F., BAUMAN, L.F., COLENBRANDER, V.F. 1971. Variations in lignin and other structural components of brown midrib mutants of maize. Crop Sci. 11:413-415.

122. WELLER, R.F., PHIPPS, R.H. 1981. The effect of brown midrib

mutation on the *in vivo* digestibility of maize silage. 6th Silage Conference, 1981 pp.73-74.

123.   BITTINGER, T.S., CANTRELL, R.P., AXTELL, J.D. 1981. Allelism tests of the brown midrib mutants of sorghum. J. Heredity 72:147-148.

124.   AKIN, D.E., HANNA, W.W., RIGSBY, L.L. 1986. Normal-12 and brown midrib-12 in sorghum. I. Variations in tissue digestibility. Agronomy J. 78: 827-832.

125.   LAPIERRE, C., TOLLIER, M.T., MONTIES, B. 1988. Occurrence of additional monomeric units in the lignins from internodes of a brown midrib mutant of maize bm3. Compte Rendu de l'Academie des Sciences III 307:723-728.

126.   GRAND, C., PARMENTIER, P., BOUDET, A., BOUDET, A.M. 1985. Comparison of lignin and of enzymes involved in lignification in normal and brown midrib (bm3) mutant corn seedlings., Physiol. Veg. 23:905-911.

127.   PILLONEL, C., MULDER, M.M., BOON, J.J., FORSTER, B., BINDER, A. 1991. Involvement of cinnamyl-alcohol dehydrogenase in the control of lignin formation in *Sorghum bicolor* L. Moench. Planta 185:538-544.

128.   SATO, T., KIUCHI, F., SANKAWA, U. 1982. Inhibition of phenylalanine ammonia-lyase by cinnamic acid derivatives and related compounds. Phytochemistry 21:845-850.

129.   BOLWELL, G.P., CRAMER, C.L., LAMB, C.J., SCHUCH, W., DIXON, R.A. 1986. L-phenylalanine ammonia-lyase from *Phaseolus vulgaris*: Modulation of the levels of active enzyme by trans-cinnamic acid. Planta 169: 97-107.

130.   ELKIND, Y., EDWARDS, R., MAVANDAD, M., HEDRICK, S.A., RIBAK, O., DIXON, R.A., LAMB, C.J. 1990. Abnormal plant development and down regulation of phenylpropanoid biosynthesis in transgenic tobacco containing a heterologous phenylalanine ammonia-lyase gene. Proc. Natl. Acad. Sci. USA 87:9057-9061.

131.   NAPOLI, C., LEMIEUX, C., JORGENSEN, R. 1990. Introduction of a chimeric chalcone synthase gene into petunia results in reversible cosuppression of homologous genes *in trans*. Plant Cell 2:279-289.

132.   CHAPPLE, C.C.S., VOGT, T., ELLIS, B.E., SUMMERVILLE, C. R. 1992. An *Arabidopsis* mutant defective in the general phenylpropanoid pathway. Plant Cell 4:1413-1424.

133.   GRAND, C. 1984. Ferulic acid 5-hydroxylase: a new cytochrome P-

450-dependent enzyme from higher plant microsomes involved in lignin synthesis. FEBS Lett. 169:7-11.

134.   SAVIDGE, R.A. 1985 Prospects for genetic manipulation of wood quality. In "New ways in forest genetics" Proceedings of the 20th meeting of the Canadian Tree Improvement Association:159-165.

135.   LAGRIMINI, L.M., BRADFORD, S., ROTHSTEIN, S. 1990. Peroxidase-induced wilting in transgenic tobacco plants. Plant Cell 2:7-18.

136.   LAGRIMINI, L.M. 1991a. Wound-induced deposition of polyphenols in transgenic plants overexpressing peroxidase. Plant Physiol. 96:577-583.

137.   LAGRIMINI, L.M. 1991b. Altered phenotypes in plants transformed with chimeric tobacco peroxidase genes. In: Molecular and physiological aspects of plant peroxidases, Eds: Penel C, Gaspar T and Lobarowski J, University of Geneva Press, pp. 128-140.

138.   HALPIN, C., KNIGHT, M.E., SCHUCH, W., CAMPBELL, M.M., FOXON, G.A. 1993. Manipulation of lignin biosynthesis in transgenic plants expressing cinnamyl alcohol dehydrogenase. J. Cellular Biochem. Supplement 17A, Abstract A310, p.28.

139.   CHIANG, V., CHEN, Z.-Z., PODILA, G.K., WANG, W.-Y., BUGOS, R.C., CAMPBELL, W.H., DWIVEDI, U.N., YU, J., TSAI, C.-J., TSAY, J.-Y., YANG, J.-C. 1993.   Genetic manipulation of lignification in *Liquidambar stryraciflua* (sweetgum) by introduction of a chimeric sense or antisense O-methyltransferase gene cloned from *Populus tremuloides* (aspen). "Proceedings of the International Conference on Emerging Technologies for Pulp and Paper Industry. Taipei, Taiwan. p.26-29.

140.   BUGOS, R.C., CHIANG, V.L., CAMPBELL, VWH. 1991. cDNA cloning, sequence analysis and seasonal expression of lignin-bispecific caffeic acid/5hydroxyferulic acid O-methyltransferase of aspen. Plant Mol. Biol. 17:1203-1215.

141.   DUMAS, B., vanDOORSSELAERE, J., GIELEN, J., LEGRAND, M., FRITIG, B., van MONTAGU, M., INZE, D. 1992. Nucleotide sequence of a complementary DNA encoding O-methyl transferase from poplar. Plant Physiol. 98:796-797.

142.   GOWRI, G., BUGOS, R.C., CAMPBELL, W.H., MAXWELL, C.A., DIXON, R. 1991. Stress responses in alfalfa (*Medicago sativa* L.) X. Molecular cloning and expression of S-adenosyl-L-methionine:caffeic acid 3-O-methyltransferase, a key enzyme of lignin biosynthesis. Plant Physiol. 97:7-14.

143.   COLLAZO, P., MONTOLIU, L., PUIGDOMENECH, P, RIGAU, J.
       1992. Structure and expression of the lignin O-methyltransferase gene
       from *Zea mays* L. Plant Mol. Biol. 20:857-867.
144.   JAECK, E., DUMAS, B., GEOFFROY, P., FACET, N., INZE, D.,
       van MONTAGU, M., FRITIG, B., LEGRAND, M.   1992.
       Regulation of enzymes involved in lignin biosynthesis: induction of
       O-methyltransferase mRNAs during the hypersensitive reaction of
       tobacco to Tobacco Mosaic Virus. Mol. Plant-Microbe Interact.
       4:294-300.
145.   KUTSUKI, H., SHIMADA, M., HIGUCHI, T. 1982b. Regulatory
       role of cinnamyl alcohol dehydrogenase in the formation of guaiacyl
       and syringyl lignins. Phytochemistry 21:19-23.
146.   CAMPBELL, M. M., BOUDET, A.M. 1933. Hydroxycinnamoyl-
       CoA reductase from *Eucalyptus*: molecular analysis of a key control
       point of lignification. Abstract 305 Keystone Symposium "The
       Extracellular Matrix of Plants: Molecular Cellular and Developmental
       Biology".   The Journal of Cellular Biochemistry Supplement
       17A:p26.
147.   UMEZAWA, T., DAVIN, L.B., LEWIS, N.G. 1991. Formation of
       lignans (-)-secoisolariciresinol and matairesinol with *Forsythia
       intermedia* cell-free extracts. J. Biol. Chem. 266:10210-10217
148.   RAO, C.B.S. 1978. The chemistry of lignans. Audhra University
       Press, Audhra, India.
149.   LYNN, D.G., CHEN, R.H., MANNING, K.S., WOOD, H.N.   1987.
       The structural characterization of endogenous factors from *Vinca rosea*
       crown gall tumors that promote cell division of tobacco cells. Proc.
       Natl. Acad. Science USA 84: 615-619
150.   BINNS, A.N., CHEN, R.H., WOOD, H.N., LYNN, D.G. 1987. Cell
       division promoting activity of naturally occurring dehydroconiferyl
       glucosides: Do cell wall components control cell division? Proc.
       Natl. Acad. Sci. USA 84:980-984.
151.   ORR, J.D., LYNN, D.G. 1992. Biosynthesis of dehydrodiconiferyl
       alcohol glucosides: implications for the control of tobacco cell
       growth. Plant Physiol. 98: 343-352
152.   TEUTONICO, R.A., DUDLEY, M.W., ORR, J.D., LYNN, D.G.,
       BINNS, A.N. 1991. Activity and accumulation of cell division-
       promoting phenolics in tobacco tissue cultures. Plant Physiol. 97:
       288-297

153.   SHERF, B.A., BAJAR, A.M., KOLATTUKUDY, P.E. 1993. Abolition of an inducible highly anionic peroxidase activity in transgenic tomato. Plant Physiol. 101:201-208.

154.   STURM, A., von SCHAEWEN, A., CRISPEELS, M.J. 1993. Identification of a mutant *Arabidopsis thaliana* blocked in the conversion of high mannose to complex asparagine linked glycans. J. Cellular Biochem. Supplement 17A, Abstract A322, p.31.

155.   LLOYD, A.M., WALBOT, V., DAVIS, R.W. 1992. *Arabidopsis* and *Nicotiana* anthocyanin production activated by maize regulators R and C1. Science 258:1773-1775.

156.   FRANK, R.L., VODKIN, L.O. 1991. Sequence and structure of a phenylalanine ammonia-lyase gene from *Glycine max*. DNA Seq. 1: 335-346.

157.   ESTABROOK, E.M., SENGUPTA-GOPALAN, C. 1991. Differential expression of phenylalanine ammonia-lyase and chalcone synthase during soybean nodule development. Plant Cell 3: 299-308

158.   TANAKA, Y., MATSUOKA, M., YAMAMOTO, N., OHASHI, Y., KANO-MURAKAMI, Y. 1989. Structure and characterization of a cDNA clone for phenylalanine ammonia-lyase from cut-injured roots of sweet potato. Plant Physiol. 90: 1403-1407.

159.   BLOKSBERG, L.N. 1991. Studies on the biology of phenylalanine ammonia lyase and plant pathogen interaction. Thesis. Plant Pathology, University of California, Davis.

160.   LEE, S.-W., ROBB, J., NAZAR, R.N. 1992. Truncated phenylalanine ammonia-lyase expression in tomato (*Lycopersicon esculentum*). J. Biol. Chem. 267: 11824-11830.

161.   GOWRI, G., PAIVA, N.L., DIXON, R.A. 1991. Stress responses in alfalfa (*Medicago sativa* L.) XII. Sequence analysis of phenylalanine ammonia lyase (PAL) cDNA clones and appearance of PAL transcripts in elicitor treated cell cultures and developing plants. Plant Mol. Biol. 17: 415-429.

162.   BREDERODE, F., LINTHORST, H.J.M., BOL, J.F. 1991. Differential induction of acquired resistance in tobacco by virus infection, ethephon treatment, UV light and wounding. Plant Mol. Biol. 17: 1117-1125.

163.   MINAMI, E. TANAKA , Y. 1993. Nucleotide sequence of the gene for phenylalanine ammonia-lyase of rice and its deduced amino acid sequence. Biochim. Biophys. Acta 1171: 321-322.

164.   SCHULZ, W., EIBEN, H.G., HAHLBROOK, K. 1989. Expression

in Escherichia coli of catalytically active phenylalanine ammonia lyase from parsley. FEBS Lett. 258: 335-338.

165.   EDWARDS, K., CRAMER, C.L., BOLWELL, G.P., DIXON, R.A., SCHUCH, W., LAMB, C.J. 1985. Rapid transient induction of phenylalanine ammonia-lyase mRNA in elicitor treated bean cells. Proc. Natl. Acad. Sci. USA. 82:6731-6735.

166.   KAWATAMA, S., YAMADA, T., TANAKA, Y., SRIPRASERTSAK, P., KATO, H., ICHINOSE, Y., KATO, H., SHIRAISHI, T., OKU, H. 1992. Molecular cloning of phenylalanine ammonia-lyase from *Pisum sativum*. Plant Mol. Biol. 20: 167-170.

167.   YAMADA, T., TANAKA, Y., SRIPRASERTSAK, P., KATO, H., HASHIMOTO, T., KAWAMATA, S., ICHINOSE, Y., KATO, H., SHIRAISHI, T., OKU, H. 1992. Phenylalanine ammonia-lyase genes from *Pisum sativum*: Structure, organ specific expression and regulation by fungal elicitor and suppressor. Plant Cell Physiol. 33: 715-725.

168.   MIZUTANI, M., WARD, E.R., DiMAIO, RYALS, J., SATO, R. 1993. Molecular cloning and sequencing of a cDNA encoding mung bean cytochrome P450 (P450C4H) possessing cinnamate 4-hydroxylase activity. Biochem. Biophys. Res. Commun. 190: 875-880.

169.   TEUTSCH, H.G., HASSENFRATZ, M., LESOT, A., STOLTZ,C., GARNIER, J., JELTSCH, J., DURST, F., WERCK-REICHHART, D. 1993. Isolation and sequence of a cDNA encoding the Jerusalem artichoke cinnamic acid 4-hydroxylase, a major plant cytochrome P450 involved in the general phenylpropanoid pathway. Proc Natl. Acad. Sci. USA 90: 4102-4106.

170.   FARENDORF, T., DIXON, R.A. 1993. Stress responses in alfalfa (*Medicago sativa* L.). XVIII: Molecular cloning and expression of the elicitor-inducible cinnamic acid 4-hydroxylase cytochrome p450. Arch. Biochem. Biophys. 305: 509-515.

171.   UHLMANN, A., EBEL, J. 1993. Molecular cloning and expression of 4-coumarate: CoA ligase. Plant Physiol. 102: 1147-1156.

172.   HAHLBROCK, K., SCHULZE-LEFERT, P., BECKER-ANDRE, M. 1991. Structural comparison, modes of expression, and putative cis-acting elements of the two 4-coumarate-CoA ligase genes in potato. J. Biol. Chem. 266: 8551-8559.

173.   DOUGLAS, C., HOFFMANN, H., SCHULZ, W., HAHLBROCK, K.

1987. Structure and elicitor or U.V. light-stimulated expression of two 4-coumarate: CoA ligase genes in parsley. EMBO J. 6: 1189-1195.

174. ZHAO, Y., KUNG, S.D., DUBE, S.K. 1990. Nucleotide sequence of rice 4-coumarate:CoA ligase gene, 4-CL.1. Nucleic Acids Res. 18: 6144-6144.

175. KNIGHT, M.E., HALPIN, C., SCHUCH, W. 1992. Identification and characterization of cDNA clones encoding cinnamyl alcohol dehydrogenase from tobacco. Plant Mol. Biol. 19: 793-801.

176. GALLIANO, H., CABANE, M., EKERSKOM, C., LOTTSPEICH, F., SANDERMANN, H., ERNST, D. 1993. Molecular cloning, sequence analysis and elicitor/ozone induced accumulation of cinnamyl alcohol dehydrogenase from Norway spruce (*Picea abies* L.). Plant Mol. Biol. 23: 145-156.

177. HIBINO, T., SHIBATA, D., CHEN,J-Q., HIGUCHI, T. 1993. Cinnamyl alcohol dehydrogenase from *Aralia cordata*: Cloning of the cDNA and expression of the gene in lignified tissues. Plant Cell Physiol. 34: 659-665.

INDEX